Consumer Testing and Evaluation of Personal Care Products

COSMETIC SCIENCE AND TECHNOLOGY

Series Editor

ERIC JUNGERMANN

*Jungermann Associates, Inc.
Phoenix, Arizona*

1. Cosmetic and Drug Preservation: Principles and Practice, *edited by Jon J. Kabara*
2. The Cosmetic Industry: Scientific and Regulatory Foundations, *edited by Norman F. Estrin*
3. Cosmetic Product Testing: A Modern Psychophysical Approach, *Howard R. Moskowitz*
4. Cosmetic Analysis: Selective Methods and Techniques, *edited by P. Boré*
5. Cosmetic Safety: A Primer for Cosmetic Scientists, *edited by James H. Whittam*
6. Oral Hygiene Products and Practice, *Morton Pader*
7. Antiperspirants and Deodorants, *edited by Karl Laden and Carl B. Felger*
8. Clinical Safety and Efficacy Testing of Cosmetics, *edited by William C. Waggoner*
9. Methods for Cutaneous Investigation, *edited by Robert L. Rietschel and Thomas S. Spencer*
10. Sunscreens: Development, Evaluation, and Regulatory Aspects, *edited by Nicholas J. Lowe and Nadim A. Shaath*
11. Glycerine: A Key Cosmetic Ingredient, *edited by Eric Jungermann and Norman O. V. Sonntag*
12. Handbook of Cosmetic Microbiology, *Donald S. Orth*
13. Rheological Properties of Cosmetics and Toiletries, *edited by Dennis Laba*
14. Consumer Testing and Evaluation of Personal Care Products, *Howard R. Moskowitz*

ADDITIONAL VOLUMES IN PREPARATION

Consumer Testing and Evaluation of Personal Care Products

Howard R. Moskowitz
Moskowitz Jacobs Inc.
White Plains, New York

Marcel Dekker, Inc. New York • Basel • Hong Kong

Library of Congress Cataloging-in-Publication Data

Moskowitz, Howard R.
 Consumer testing and evaluation of personal care products / Howard
R. Moskowitz.
   p. cm. — (Cosmetic science and technology ; 14)
 Includes index.
 ISBN 0-8247-9367-6 (hardcover : alk. paper)
   1. Cosmetics—Testing. I. Title. II. Series: Cosmetic science
and technology series ; v. 14.
TP983.M8673 1996
668'.55'0287—dc20   95-31127
                                                                  CIP

The publisher offers discounts on this book when ordered in bulk quantities. For more information, write to Special Sales/Professional Marketing at the address below.

This book is printed on acid-free paper.

Copyright © 1996 by Marcel Dekker, Inc. All Rights Reserved.

Neither this book nor any part may be reproduced or transmitted in any form or by any means, electronic or mechanical, including photocopying, microfilming, and recording, or by any information storage and retrieval system, without permission in writing from the publisher.

Marcel Dekker, Inc.
270 Madison Avenue, New York, New York 10016

Current printing (last digit):
10 9 8 7 6 5 4 3 2 1

PRINTED IN THE UNITED STATES OF AMERICA

This book is dedicated to four wonderful people.

To my dear mother, Leah Moskowitz, who has over the years, inspired me, encouraged me, and when necessary, cajoled me to develop new ideas, write them down, and publish them.

To Arlene, who continues to inspire me, and encourage me to write. It was Arlene's idea to write this book, and present these new ideas to the scientific and business audience.

To my sons, Daniel and David, who as they grow up increasingly appreciate the importance of books, writing, and the joy of intellectual creation.

About the Series

The Cosmetic Science and Technology series was conceived to permit discussion of a broad range of current knowledge and theories of cosmetic science and technology. The series is made up of either books written by a single author or edited volumes with a number of contributors. Authorities from industry, academia, and the government participate in writing these books.

The aim of this series is to cover the many facets of cosmetic science and technology. Topics will be drawn from a wide spectrum of disciplines ranging from chemistry, physics, biochemistry, and analytical and consumer evaluations to safety, efficacy, toxicity, and regulatory questions. Organic, inorganic, physical, and polymer chemistry; emulsion technology; microbiology; dermatology; toxicology; and other fields all play a role in cosmetic science.

There is little commonality in the scientific methods, processes, or formulations required for the wide variety of cosmetics and toiletries manufactured. Products range from hair-care, oral-care, and skin-care preparations to lipsticks, nailpolishes and extenders, deodorants, body powders, and aerosols to over-the-counter products, such as anti-perspirants, dandruff treatments, anti-microbial soaps, and acne and suntan products.

Cosmetics and toiletries represents a highly diversified field with many subsections of science and "art." Indeed, even in these days of high technology, art and intuition continue to play an important part in the development of formulations, their evaluation, and the selection of raw materials. But now there is a strong move toward sophisticated scientific methodologies, particularly in such areas as efficacy evaluations, claim substantiation, safety testing, product evaluation, and analysis.

Emphasis is placed on reporting the current status of cosmetic technology and science in addition to historical reviews. The series includes books on oral-hygiene products, cosmetic product safety, efficacy testing, sunscreen technology, deodorants and anti-perspirants, key cosmetic raw materials (for example, glycerine, clinical and physical testing procedures in the area of skin care and hair care, and the role of preservatives). Contributions range from highly sophisticated and scientific treatises to primers, practical applications, and pragmatic presentations. Authors are encouraged to present their own concepts as well as established theories. They are asked not to shy away from fields that are in a state of transition, nor to hesitate to present detailed discussions of their own work. Altogether, we intend to develop this series as a collection of critical surveys and ideas covering many phases of the cosmetic industry.

Consumer Testing and Evaluation is the fourteenth book published in this series. It is an extension of an earlier book by Dr. Moskowitz that covered a modern psychological approach to cosmetic product testing. In the intervening decade, as competition has increased and the cost of new product introductions has skyrocketed, the need for meaningful, statistically sound consumer testing has become ever more important. The need to understand and cater to consumer perceptions and changing fashions plays an important role in developing and positioning new personal-care products. Aesthetic considerations, such as fragrance, color, or packaging concept, can greatly affect a product's acceptance by the customer.

This book covers and details the consumer testing steps used to define and guide the technical development of a new product from concept development and formula optimization to the final selection and positioning of the new product in the marketplace. Methodologies are described in detail and illustrated with case studies. This reference will help scientists who develop new products to follow marketing and management directions. It will also help marketing people who must communicate their objectives and explain their rationale to laboratory personnel.

I want to thank the author, Dr. Howard R. Moskowitz, for contributing a second book to this series. Special recognition is also due to Sandra Beberman and the editorial staff at Marcel Dekker, Inc., and above all, to my wife Eva, without whose constant support and editorial help I would never have undertaken this project.

Eric Jungermann, Ph.D.

Preface

Those of us who work with personal products at the development or marketing end face, to our dismay, an increasingly competitive industry. More than ever, that fickle criterion, consumer acceptance, drives market success, corporate profits, and job security. The price of failure is high, the rewards of success are temporary. Given this state of affairs, how does one ensure consumer success with commercial products on an ongoing basis? There are no standard rules. Preferences change and products enter and leave the market at a dizzying pace.

In light of the shifting sands of consumer preference, it is far better to begin with fundamental principles that can be applied again and again, to new problems. It is the general approach that is key, with each specific case history a particular manifestation of that general approach. I wrote this book to provide a disciplined approach to development, testing, and evaluation. The underlying message of the book is therefore quite simple and direct. Success will be a much more frequent companion when the developer creates many alternatives (whether in concept or product), tests these among consumers, identifies patterns of characteristics driving acceptance, and then harnesses the knowledge from these patterns to improve the concept, advertising, or product. Once the theme of systematic development and disciplined exploration is recognized everything else becomes commentary.

I would like to thank Diane Chiodo, my literary assistant, for all her ongoing help and support. Without her, this book could not have been written.

Howard R. Moskowitz

Contents

About the Series   *v*
Preface   *vii*

Introduction   1

1   Concept Development, Testing, and Optimization   10
2   Benchmarking a Product Category   80
3   Practical Considerations for Product Testing   158
4   Experimental Design and Product Optimization   219
5   The Role of Psychophysics and Experimental Psychology   288
6   Individual Differences: Conventional Subgroups, Preference Segmentation and the Construction of a Product Line   321
7   Interrelating Data Sets: Physical Measures, Expert Panel and Consumer Ratings   393
8   Streamlining the Product Development Process   424
9   Advice to a Young Researcher   459

Index   *485*

Consumer Testing and Evaluation of Personal Care Products

Introduction

Need for Speed in Product Development

As manufacturers experience increasing pressure to create successful products they are pushed by business realities to look outside their own expertise. Ten, twenty, thirty years ago it was quite common for developers in the health and beauty aids field to rely on instinct, and "shoot from the hip." Those days are gone because failure has become increasingly costly. The old luxury, the "kinder and gentler world" of product development and marketing, has given way to a more rough and tumble, risky reality.

Where does this change leave the product developer? Every sea change in the business world affects the way people do business sooner or later. In the mid-1970's, the notion of systematically developing and testing products in the health and beauty aids business was acceptable but not particularly popular. Some large-scale companies did, of course, fund regular consumer evaluation of their products. Occasionally a visionary company might even "benchmark" its products against the in-market competition in order to assess strengths and weaknesses. This outward thinking was not prevalent in those early days. Today with the increasing risks has come the growing recognition that consumer needs should guide the product, and that product testing is both respectable and essential.

The changing business climate demands new and more practical solutions to old problems. Previously it sufficed to report that one's product was better than, equal to, or worse than the competitor's. Now, this basic information is just the tip of the iceberg. By itself it is only information. Manufacturers, product developers, and marketers must understand their category more profoundly. For instance, what attribute "drives a category?" To what do consumers attend to make a decision? (Knowledge of the consumer's criteria eases the developer's task because it points the developer in the right direction). Manufacturers also need to know whether there exist any potential niches or segments of consumers in the population which transcend typical demographic breakouts. By dividing consumers into homogeneous preference groups, the marketer creates products and thus opportunities that competitors have overlooked.

The increased pressure on manufacturers has produced at least one great benefit for product developers. It has opened up their eyes to the possibility that consumer assessment of products can guide them towards market success, rather than be relegated to an after-the-fact report card which tells them only that they have either succeeded or failed. If nothing else, this change in the use of consumer research provides continuing benefits to manufacturers long after the research has been completed and reported.

The cultural transition from the "shoot from the hip" approach to the empirical, data-based approach is not finished. The health and beauty aids industry relies extensively on the opinion of "experts," whether these be the marketing management or trained panelists who can distinguish fine differences between products. However, as time progresses the use of consumer guidance continues to increase and replace (or at least modify) the judgment of experts.

Risk Reduction Through Product And Concept Testing

The main contribution of consumer testing is to guide development, thereby reducing risk. Markets are becoming increasingly fragmented. Testing provides a mechanism to understand the consumer and assess/enhance the likelihood of success. The better one can test the final product or concept and secure valid, action-oriented consumer feedback, the more likely the research will significantly reduce risk. This book provides a broad view of the development process. It provides both principles and case histories which show how testing significantly increases the chances of product success.

Fragmentation Of In-House Empires Due To New Demands

A cursory glance at the history of product testing in the cosmetic industry will show the trend. In the early days, when testing was a novelty and when methods were crude, most testing was done by informal consensus. The expert panelist was the perfumer, the cosmetic chemist, or (more often than would be admitted) the president of the company. As management recognized the value of testing there was a concerted effort to bring the expertise "in-house." Management learned the business value of testing and embedded that value in a resident "expert panel." The panel took on the role of the resident "expert."

The corporate love affair with trained panelists originated in the 1940's, following the development by the Arthur D. Little Company of "in-house" expert panels for foods. The same approach was then used for health and beauty aids after a lag of several years. The 1950's and 1960's saw this culture of "in-house expertise" grow to dominate the personal product testing field. In the 1970's the stranglehold of in-house testing began to weaken as marketers recognized the need for consumer data in addition to their expert panel data. Competition between manufacturers made it increasingly clear that the expert panel data and consumer data would have to coexist with each other. (A similar recognition was then dawning on researchers in the food industry, who began to understand the criticality of consumer data). It no longer sufficed to appeal to the elitist claim that consumers simply could neither perceive nor report nuances of product attributes.

As the 1970's progressed an increasing number of manufacturers wrestled with the problem of integrating their long standing expert panels with consumer data. At that time, with competition just beginning to heat up, manufacturers were still complacent. Times were still good, at least to all outward appearances. The adoption of consumer testing was relatively slow because no one recognized the impact of competition on profits. Growth was still strong, expert panelists and consultants alike were in demand, and it seemed that the only way to go was up.

In the 1980's competition hit the industry with repeatedly brutal blows. The once profitable fragrance industry began to slow. The competition for a constant dollar grew furious as more and more companies hurled entries into the marketplace at an accelerated pace. As the pace of introductions increased and profits declined, the gentlemanly (or often gentlewomanly) approach to product development ceased. R&D was increasingly pressed to come up with winning products. It was tempting to avoid testing altogether and simply launch, but in many cases that apparently expedient step led to reduced consumer acceptance and a subsequent loss of profits.

We are now in the 1990's, surveying the industry from a 40-year vantage point. Much has changed. The fragrance companies that were so happily gobbled up by multinational drug companies (as well as other large manufacturers) have, in the main, been spun off. The heady optimism so characteristic of the field in the 1960's through the mid-1980's has evaporated. Manufacturers are beset with intense competition and the continuing need to introduce products to an already saturated market. Many of the more successful companies, developed through the efforts and wisdom of their founders, have grown middle-aged. As a consequence they too flounder around, looking for the product(s) to restore their once dominant position in the category.

The Growth Of The Consumer Research Industry

In the 1960's a variety of researchers in different fields came together in various informal meetings to share procedures for consumer research. At that time, consumer research was a collection of procedures, not particularly well-established, designed as needed to measure consumer opinions. Over time those involved in this nascent field of consumer evaluation formalized their discipline. The test procedures that had hitherto appeared in such subject-oriented journals as the *Journal Of Food Science*, or *American Perfumer*, or even the *Journal Of Applied Psychology*, were recognized as contributions from a newly developing field.

The 1960's and the early to mid- 1970's witnessed the introduction of new journals focusing on consumer research. Some of these journals were the *Journal Of Marketing Research* for more marketing-oriented testing, and the scientific journal, *Chemical Senses*, for more basic research devoted to taste and smell. The *Journal Of Food Science* featured an increasing number of articles devoted to procedures for consumer food testing. Over time consumer product testing became increasingly respected and more widely accepted. The American Society For Testing And Materials founded Committee E18 on Sensory Evaluation. The committee at first was comprised of primarily laboratory scientists in the food industry, but soon after opened its doors to researchers in cosmetics, toiletries, etc.

In the late 1970's consumer research came into its own. As a technical field it was recognized in the food industry through the growing popularity of the Sensory Evaluation Division, a recognized and well-populated division in the Institute Of Food Technologists. At the same time, Committee E18 (Sensory Evaluation Division) of the American Society For Testing And Materials grew rapidly, with an influx of researchers from fields other than food. Those interested in the sensory analysis of cosmetics, toiletries, drugs, paper products and the like joined Committee E18.

Scientific papers helped the growth of consumer product evaluation. In the 1970's and 1980's numerous publications began to appear. These papers could trace their origin to papers which had appeared in the 1950's and 1960's. Most of these early papers concerned observations about sensory phenomena of interest to researchers of cosmetics and health and beauty aids, but there were no real methodological papers of which to speak. Those early papers comprised observations, or perhaps some recommended test methods, that today appear to us in hindsight to be at best simplistic, or at worst misleading. The later papers better focused on procedures, using rigorous methods. It was no longer a case of discussing procedures that seemed to work (albeit unsubstantiated by scientific test), nor an issue of reporting basic research in sensory perception with the unarticulated hope that this basic research would somehow inspire studies

Introduction

in sensory analysis of fragrances and cosmetics. Rather, the studies published in the 1970's and 1980's were designed specifically to measure aspects of consumer reactions to products. Studies appeared on the descriptive analysis of sensory perceptions applied to cosmetics, on the evaluation of consumer acceptance of blind versus identified fragrances, and on methods to evaluate attitudes about usage of products.

The late 1980's witnessed another contribution to the field of cosmetic product testing. The increasing complexities of the field brought with it the recognition that the "good old days" of omniscience and "shoot from the hip" were ending. Because the cosmetic and fragrance markets were no longer growing at the dizzying pace of the 1970's and early 1980's, and because the cost to launch new products was rising (both in terms of development, and in terms of the cost of failure in the market) it became obvious that the development process would have to be collaborative. Much as the individual maverick would like to create and market products as a solo act, it was becoming increasingly clear that the rule of the day was cooperation rather than the individualist vision. The risks of failure, the competitive threats, and the capital needed to launch and maintain successful products were so high that they effectively discouraged individual efforts, yet promoted the group consensus. There was still room for the individual vision, but a more tamed vision, offered to the group by a forward-looking team player.

New Agendas As An Impetus For Product Testing

Scientific research and testing procedures in the cosmetic (and food) industry do not necessarily develop because of the researcher's desire to do things "better" (although in some cases that desire may be a strong motivating factor). What typically occurs in an industrial setting is a "paradigm shift." Paradigms are accepted ways of thinking about problems, and implementing approaches to answer those problems. When the business situation changes, often the old paradigms no longer work. It is not that some all-knowing individual at the top of the company suddenly realizes that the procedures no longer work, rather, it is a developing recognition that the old ways simply cannot provide the necessary answers.

Business continually faces changing requirements. Paradigm shifts occur frequently, sometimes barely noticed at all, other times leaving a wrenching effect on the organization. For cosmetic and fragrance science, consumer evaluation has never been a major concern compared to other development and marketing issues. Thus, the paradigm shifts are not caused by shifts from one established way of testing to another, rather, the shift arises from the growing recognition of consumers as the determining factor of product acceptance and success. The new agendas are competitive advantage, consumer perceived quality, and maintenance of consumer acceptance in the face of product changes to sustain profitability. These new agendas require better, more practical ways to test product acceptance. To do so they must incorporate the consumer into the early stage development process.

Given the structural changes in the packaged goods industry that began in the 1980's and have proceeded apace in the 1990's, it is likely that there will be further shifts in paradigms, and new methods to answer problems involving consumer product perception and acceptance. Consumer testing procedures in the industry are far from having achieved their fullest potential. In other industries statistical methods for product development (viz., experimental design and optimization) are gaining acceptance because they shorten the costly and lengthy development cycle. We may expect to see the design approach adopted in the cosmetic and fragrance industries, especially after the manufacturing advantages using these procedures become more visible and the ensuing profits garnered.

Changes In Technology And Expertise Affect Testing

As practitioners in every field become versed in new procedures, they inevitably change the field. Product and concept testing is no different. Scientific, procedural advances exert their influence slowly but inexorably. Since the late 1940's researchers in psychology have concentrated on the measurement of sensory properties of products, and the way that the senses translate physical properties of products to sensory (and hedonic) impressions. This is the field of psychophysics, whose objective is to correlate physical properties with private sensory perceptions.

Psychophysics, the oldest of the psychological sciences, did not exert much impact on product testing during the first two thirds of this century. From time to time one might read an article in a scientific journal about procedures to measure the "differential sensitivity" to a stimulus (e.g., the proportional change in color saturation required for a person to recognize that two products have the same hue, but differ noticeably from each other). These types of questions pertaining to differential sensitivity were certainly understood by the commercial community, but the implications and application of the data were not clear. How could a manufacturer apply the knowledge about a consumer's ability to discriminate between colors? The application of scientific data to practical problems appeared tenuous at best. Despite the lack of connection between basic research in perception and commercial application, most practitioners in the early days of cosmetic product testing were familiar with this literature, perhaps more out of a sense of professional duty rather than from actual utility.

In 1953, S. S. Stevens, a research psychologist at Harvard University, reported the results of a series of studies in sensory perception that would have

Introduction

enormous consequences for psychology in general, and consumer product testing in particular (In those early days, however, it was hard as yet to see the connection). Stevens was seeking a general, quantitative relation between physical stimulus intensity on the one hand and subjective intensity perception on the other. For example, as one increases the concentration of fragrance in equal steps what happens to the perceived intensity of the fragrance? Does *perceived* fragrance intensity parallel physical concentration, so that each unit of increasing concentration produces an equivalent increase? Does the perceived fragrance intensity increase dramatically with small increases in concentration? Or can the researcher substantially increase the concentration of the fragrance and yet the consumer will report virtually no change in fragrance at all? Similar questions can be asked for color, roughness, perceived viscosity, etc. Specifically, what is the relation between what the researcher varies and what the consumer perceives?

The early work coming out of Harvard University dealt with the brightness of lights and the loudness of sounds (Stevens, 1953). Stevens discovered surprising regularities in the data. Perceived sensory intensity (e.g., brightness, loudness) followed physical stimulus intensity (e.g., lumens of light, sound pressure of sound) according to a reliable function known as the power law:

Sensory Intensity = $K(\text{Physical Intensity})^n$

What was important about this relation was that it demonstrated that rules governing perception could be applied to product design and testing. For instance, often the exponent n is lower than 1.0 (it is 0.5 for viscosity, and around 0.3 for odor intensity). For low exponents the product developer and evaluator know that given large physical changes in stimulus level, the consumer will perceive only a small subjective change. The exponent provides the researcher with added insight into perception and enhances the actionability of sensory perception studies.

Psychophysics as a contributor to cosmetic and fragrance science in particular, and product testing in general, came on the applied scene in the 1970's, some 20 years after Stevens' breakthrough research. Since then, psychophysics has grown in popularity. Research psychologists have been drawn to the field of product evaluation as an outlet for their technical expertise. As opportunities in the realm of basic research dried up, these researchers looked to companies where they could apply their knowledge of psychophysics. Corporate R&D management recognized the value of the psychophysics expertise, both as a basic foundation of knowledge that made the researcher intrinsically more valuable, and as a technology whose application could eventually provide a desired technological (and thus business) advantage.

Statistics As An Aid

Consumer testing often invokes the specter of statistics. Indeed in consumer research, statistics has played an extremely important, sometimes overwhelming role. Many practitioners working in the 1950's through 1970's learned statistical procedures almost by rote. They performed tests of statistical significance to determine whether two or more products came from the same sampling distribution, or whether the products could be said to differ. In the early days the statistical analysis of data was often handed to a resident expert or consulting statistician. The actual practitioner usually had experience with test design and execution, but relatively little with statistical analysis.

Product testing in a variety of fields was given a boost by the popularization of the new wave of statistics. Part of this new wave was a renewed appreciation of statistical tests of difference, perhaps made more immediate to the practitioner by the increasing availability of computing. Another movement in statistics, *multidimensional scaling,* also gave researchers a renewed appreciation. Multidimensional scaling places products in a geometrical space with the property that products close to each other in the space are qualitatively similar, whereas products far away from each other are qualitatively different. Unlike statistical testing of differences, multivariate methods deal with the *representation* of products. They present data pictorially in a way that helps the practitioner to understand the products and the consumer. In contrast, statistical difference testing simply attempts to determine whether two products are the same or different.

Multidimensional scaling was only the first in a series of procedures that practitioners wholeheartedly adopted. There have been other procedures, including "free choice profiling" (designed to identify and cancel out individual differences in scaling and attribute choice) and experimental design and optimization (designed to help the developer create an optimum product, using a set of specified formulations and a method for finding the product within the set achieving the highest level of liking; Box *et al.* 1978). These newer methods and others have stimulated the practitioners in the field to adopt statistical methods for exploratory data analysis and guidance, rather than reserving statistics for "pass/fail" grading. In so doing, these methods have demonstrably altered the field of product testing.

Overview—Changing Business Requirements And Available Technologies

The past 15 years have witnessed a major shift in the way that cosmetic and health/beauty aids companies do research with consumers. What traditionally had been "shoot from the hip" has become increasingly data-intensive. It is no longer the case that decisions can be made accurately on the basis of "gut feelings," with the expertise provided by a single individual. Today there are simply too many factors which interact to make a product a success. Sheer vision is not enough. The product must fit consumer needs, must be positioned to break through an ever-increasing amount of clutter, must have "staying power" (in terms of truly satisfying a need, whether that need be real or imagined), and must be price competitive. There has developed extensive expertise to provide the marketer and product developer with a sense of what the consumer wants and will accept. The expertise and procedures truly reduce the risk of failure. What is needed is the combination of inspiration from marketers and developers, and a bank of proven methods which insure that these inspired guesses translate into truly successful products blessed with a long, remunerative life cycle.

References

Box, G.E.P., Hunter, J. and Hunter, S. 1978. Statistics for Experimenters. John Wiley & Sons, New York.

Stevens, S.S. 1953. On the brightness of lights and the loudness of sounds. Science, *118*, 576.

Williams, A.A. and Arnold, G.M. 1985. A comparison of the aromas of six coffees characterized by conventional profiling, free-choice profiling and similar scaling methods. Journal of the Science of Food and Agriculture, *36*, 204–214.

1
Concept Development, Testing, and Optimization

1. Creating Concepts

Concepts lie at the foundation of product development and advertising. For product developers the concept lays out the physical characteristics of the product. Table 1 presents three concepts, one for a shampoo, one for a skin lotion and one for a facial soap. The product concept does not contain a selling message, but rather provides a set of verbal instructions for the product developer. Nor does the product concept instruct the developer on how to formulate the product. The concept simply lays out the properties of the product as the consumer would perceive them to be. It is left to the wisdom, experience and ingenuity of the product developer to translate what is written on a sheet of paper to a physical reality.

Occasionally the product concept is created with additional selling messages attached to the product description. The marketing group in a company, perceiving the existence of a gap in the marketplace, designs the characteristics of the product. However, the marketer's language is neither that of the scientist nor that of the product developer. Although they design the product in the least "selling way" possible, the marketers' descriptions may be strongly tinged with sales or positioning. The product developer must translate the marketing language from consumer terminology to ingredients and processes. The task may range from easy to formidable. Table 2 shows three different concepts for a fragrance. The three concepts describe the same fragrance, but with increasing amounts of positioning and sales.

Table 1 Three Product Concepts

Concept 1. Shampoo

This shampoo contains special ingredients to clean and brighten your hair. The shampoo leaves virtually no residue. The shampoo lathers with rich, foamy, creamy lather.

Concept 2. Hand lotion

This lotion provides you with relief from chapped, dry skin. It contains special rejuvenating ingredients to restore the smoothness of your skin, when applied daily in a regular fashion.

Concept 3. Facial soap

This soap is used to clean and freshen your face in the morning. It contains no harsh perfumes or ingredients. It contains only the best ingredients which clean and refresh your skin.

Where Do Concepts Come From?

There is no magic to concepts, although many interested parties in the industry might believe otherwise. Concepts comprise combinations of words, pictures, and phrases that convey a desired message.

There are no magic steps to creating winning product or advertising concepts, although there is inspiration and talent. What is needed is either a great deal of genius/inspiration or a systematic procedure to identify winning elements and combine these elements into concepts.

Table 2 Example of Three Fragrance Concepts with Increasing Amounts of "Positioning" or Sales Emphasis

Concept 1.   (Least sales oriented)

This fragrance is designed with the young consumer in mind. Young women between the ages of 15 and 21 will find this to be the perfect fragrance of choice to wear both during the day and evening.

Concept 2.   (More sales oriented)

This fragrance has been designed by noted perfumers with the young, but sophisticated consumer in mind. Forward thinking, trendy young women will delight in this fragrance. It has been designed to enhance the wearer, both in the day and in the evening.

Concept 3.   (Most sales oriented)

This fragrance, Allura, comes from master perfumers with decades of experience in enhancing lifestyles with perfume. Allura has been designed specifically for those young women who want a trendy, yet sophisticated fragrance. Allura will enhance the wearer and can be worn both in the daytime for school or work and in the evening for those romantic times.

Promise Testing

Sometimes marketers work at the most basic level with single ideas. These ideas are elements, not concepts (in terms of describing a product or a service). Elements are snippets of ideas, or core ideas around which a product can be built. Elements are simply phrases. In consumer evaluation, the marketer, marketing research or research and development (R&D) product developer lists these phrases. The consumer panelist rates the degree to which he would be interested in purchasing the product. Table 3 shows these elements for a shampoo, and shows both the average score assigned to the item (on a 1–5 point purchase interest scale), and the proportion of consumers who were defined to be very interested (the proportion of consumers who assigned a rating of 4 or 5 to the statement). These two numbers, average interest and proportion of "highly interested consumers," measure the potential of the product. In addition, in promise testing the consumer can rate the degree of "uniqueness" on a scale (e.g., 0 = identical to products currently in the category → 100 = quite different from the products currently in the category). Table 3 shows the uniqueness ratings as well.

Table 3 Promise Testing for Shampoos (Ratings from Consumers, R&D)

Shampoo promise/benefit	Consumer data			R&D data	
	Five point[a]	% Top two box[b]	Unique[c]	R&D cost[d]	R&D time[d]
All natural ingredients	3.9	65	46	78	67
Shampoo + conditioner	3.6	54	51	45	23
Improves combability	3.6	57	67	81	45
Rinses out with no residue	3.6	59	62	69	35
No adverse effects on hair	3.2	48	43	57	46
Rich lathering	3.2	51	68	35	59
Rich, thick shampoo	3.2	49	52	45	21
Unique ingredient blend	2.9	40	73	51	24
You can use it daily	2.9	42	65	57	36
Special ingredients	2.8	32	35	45	21
Measured delivery cap	2.6	31	78	79	47
For all hair types	2.6	34	52	75	68
Easy to grip bottle	2.5	28	68	63	43
Special hair treatment	1.4	10	43	56	34

[a] Five point = 5-point purchase scale (1 = definitely not buy → 5 = definitely buy).
[b] % Top two box = percent of consumers rating the concept as 4 or 5 on the purchase intent scale.
[c] Uniqueness = 0 → 100 (0 = identical to current products; 100 = quite different from other products in the category).
[d] R&D rating = cost in relative units, time in relative units.

Concept Development, Testing, Optimization

R&D can also play a major role in promise testing and evaluation of the specific elements for the concept. Only R&D knows the true technical feasibility of the elements. Some promises which score high from the consumer's point of view may be physically difficult to realize, either in terms of time or money. It may simply take too long to develop the product, no matter how attractive the proposition may be to the consumer. Or, given the current technology, it may be too expensive to create a viable product. A license agreement with another company may have to be negotiated to secure the product. These considerations play a role in the assessment of whether or not the idea is a good one to pursue. Table 3 also shows the R&D estimates of the viability of the items, from both cost and timing standpoints.

However, just because a promise or an idea is currently infeasible because of time or financial constraints does not mean that the idea should be disregarded. Things change. What was once deemed to be infeasible may become very feasible with new technologies or with favorable licensing terms. Furthermore, if an idea is so highly compelling, unique and compatible with corporate goals, then the attractiveness of that idea can often transform the product into a reality. With strong commitment the cost may decrease and the idea may become extremely affordable, especially if it promises significant marketplace success.

Gestalt Concepts (Fully Formed, No Underlying Design)

The simplest and most popular way to create concepts is to take them after they emerge fully-formed from a creative group. This group may comprise members of the marketing team, the advertising agency, the R&D group, and occasionally an outside consultant hired to bring a different perspective to the business. Marketers often use outside resources to provide concepts in a short time, at a low price. Various "idea development services" (so-called creative boutiques) abound in the business bazaar, so that the marketer can easily get dozens or even hundreds of ideas. Some of these ideas will be outstanding, many will be average, and a few will be just plain terrible, but large volumes of ideas are available from these services.

Many of the concepts that emerge from the concept development groups are "rough." They need polishing. Before proceeding to consumer testing, the marketer and his team will typically work on these concepts, sometimes formally, sometimes informally. The team may assess the concepts using other criteria. These are:

a. feasibility (is the idea technically feasible?)
b. uniqueness (does the idea differ sufficiently from the competition so that it is adequately unique?)

c. "fit to the core business" (is the idea compatible with the company's current business or does it represent a "way out idea" having no link with the company's current technical or marketing competency?)

Companies that develop new products on an ongoing basis must have a system to assess the viability of the concepts. Usually viability is measured by the average acceptance of a concept or by the proportion of consumers who show a very high degree of acceptance of a concept. For the most part these two measures highly correlate. That is, if consumers rate a dozen concepts on purchase intent using an anchored 5-point scale (1 = definitely not purchase → 5 = definitely would purchase), then the average rating and the percent of consumers selecting categories 4 (probably would purchase) and 5 (definitely would purchase) correlate highly.

Table 4 shows three concepts for a nail polish remover, the ratings of purchase intent (from consumers), as well as R&D/Marketing assessments.

Using In-Market Concepts for "Norms"

Acceptance scores (e.g., purchase intent) are not sufficient. What does a high acceptance score mean in the "real world" of behavior? Does it signal that a product will essentially "fly off the shelf?" How does a manufacturer know or at least estimate how a product will perform, given its purchase intent score?

The easiest way to determine what a concept score means for market success is to have a "normative" bank of concepts tested prior to the product launch, along with a measure of how well products corresponding to these concepts (or some subset thereof) perform in the marketplace. A normative data set allows the researcher to estimate the likely degree of success to be obtained for a concept with a given score. Many "test market simulators" use a bank of norms to estimate the likely success of a product concept, given the concept scores. The big problem with this approach is that manufacturers are reluctant to spend money launching (or even to continue testing) a product concept that performs poorly on a concept test. Normative data banks are biased in favor of concepts

Table 4 Three Concepts for a Nail Polish Remover

Concept	Synopsis of concept	Consumer interest [a]	R&D estimated cost	R&D estimated time
1	Basic product	45%	28	3 Months
2	Remover with an "easy-on" applicator	51%	42	4 Months
3	Remover with nail strengthener	69%	68	8 Months

[a] % Top two box - percentage of consumers rating the concept as 4 or 5 on the purchase intent scale.

that score well. Often, in fact, these are the only concepts for which manufacturers have additional performance data.

Another way to estimate likely success of a product uses in-market concepts along with market data for reference purposes (Table 5). This approach uses concepts describing products currently in the market, and assessing them

Table 5 Steps Needed to Use In-Market Products as Norms for Concept Testing

A. Steps

Step 1—Identify the specific products currently in-market. Use at least six products in the category

Step 2—Obtain their advertisements or point-of-purchase material

Step 3—Reduce the information obtained in Step 2 to simple concepts (bulletized concepts are best)

Step 4—From market data determine the two relevant numbers for each product tested in Step 3
- Share (or total volume)
- Percent distribution

Step 5—Present these concepts to consumers, and instruct them to evaluate interest in buying the product

Step 6—Create a data matrix with the following
- Share (or volume)
- Consumer interest (acceptance)
- Unit price
- Distribution (optional, if available)

Step 7—Create a descriptive equation:

$$\text{Share (or Volume)} = k(\text{Price})^p(\text{Consumer Acceptance})^n$$

B. Database for Worked Example: Data for Nine Competing Products

Product	Market Relative volume	Market Relative price	Consumer acceptance
1	2	0.23	21
2	3	0.45	16
3	5	0.16	32
4	6	0.23	67
5	6	0.19	33
6	8	0.21	36
7	9	0.12	17
8	10	0.13	42
9	10	0.15	45

Equation: Volume = $(e^{-0.85})(\text{price}^{-0.93})(\text{acceptance}^{0.33})$
Goodness-of-Fit: Multiple $r^2 = 0.53$

in one of two ways: without branding (viz., manufacturer) and with branding. Either two matched groups participate (one rating the unbranded concepts, the other rating the branded concepts), or the same consumer participates in both parts of the evaluation (rating the unbranded concepts first, waiting a while to let memory traces fade, and then rating the branded concepts). Data from the unbranded concept provides information about the strength of the basic selling message. Data from the branded concept provides additional information about the contribution of advertising over and above the contribution provided by the selling message. By developing the database, the manufacturer can assess the likely degree of success associated with a concept score. The method does not, however, account for the fact that market failure may result from the combination of a strong performing concept with a lack of advertising or trade support.

What Happens in Most Companies?

In most companies today, concept assessment is accomplished by testing a battery of different concepts among consumers. One consumer may test all concepts or just a subset of concepts. The concepts may be tested in a mall or through the mail. Rarely are test concepts evaluated along with a full set of competitors. The research provides for a single number which shows the "expected promise level" of the concept. The consumer ratings (e.g., percent top two box purchase intent ratings; definitely or probably would buy on a 5-point scale) are translated to estimates of potential market success by one or another model.

Benefit of the Approach. The approach tests many concepts, increasing the likelihood of success, because the odds are high that one or more of the concepts will successfully achieve a desired level of consumer interest (viz., pass the performance hurdle).

Drawbacks of the Approach. The approach usually does not incorporate current competitor concepts, relying instead on normative data to estimate likely market success. Furthermore, the method does not identify which particular elements in the concept drive ratings. Not all elements of the concept are equally powerful as drivers of purchase intent. Conventional concept testing cannot identify the drivers of purchase intent, except in the most obvious of cases. Occasionally the researcher uses an extensive battery of questions to probe why a concept was liked (or disliked), and what specifically was communicated. These diagnostic questions occasionally provide sufficient information to show that certain messages were successfully communicated, whereas other messages were not. Table 6 shows an example of a concept, diagnostic questions and some marketing conclusions. From Table 6 we see that the researcher or marketer must make several inferences about the link between consumer persuasion measures and communication.

Table 6 Example of a Concept with Diagnostics: Category = Toothpaste (Plaque Remover)

Concept

GreatTooth is a new toothpaste formulated with RK93, a new patented antiplaque chemical recommended by the National Dental Association. GreatTooth is the result of 11 years of research by dentists and oral surgeons, looking for the best preventive product for dental health. GreatTooth comes in three delicious flavors: regular, mint, and exciting cinnamon.

Diagnostics from the concept test

Attribute	% Agree
Would buy the toothpaste	58%
Believe what I read	61%
Different from other products currently on the market	42%
Technologically advanced	61%
Believe that it is a safe product	67%
Would work better than other products	61%
More a medical product than an actual consumer toothpaste	23%
Would use it as part of my daily regimen	63%
Have heard about similar types of products	78%
Understood the basic message	82%
My dentist would recommend it	69%

Marketing conclusions

- Modestly acceptable
- Consumers understand the message, but don't think it's particularly compelling
- Not particularly unique
- Definitely perceived as a toothpaste product, not a medical product
- Overall—modest performer at best, with few unique benefits

Learning More About Concept to Create Winners

Conventional concept testing can *select* winning concepts and, in some cases, identify the reasons for a concept performance. However, most marketers want to *develop* winning concepts on an ongoing basis in order to give their products the best chance for marketplace success. How can the process be improved?

Over the past twenty years, marketers have come to accept the idea that testing fully formed concepts cannot provide the answer if each concept stands alone, unrelated to the other concepts being evaluated. It is easy to determine that one concept performs better than another, but hard to improve the poorer concepts in any systematic, rational way. There must be a better process.

Marketing researchers often use experimentally designed concepts. Experimental design first identifies the components of the concept, and then combines these components in a way that allows the investigator to estimate what each component contributes to concept performance. Performance, in turn, could be either ability to persuade consumers to purchase the product, or ability to communicate specific benefits, whether or not persuasion is changed. The methods used by marketing researchers are subsumed under the research procedure known as "Conjoint Measurement."

Conjoint measurement is a statistically based procedure which estimates the contribution of components of a mixture from ratings assigned to the mixture. The "mixture" may be a concept, in which case conjoint measurement estimates what every phrase (and/or picture) in the concept "brings to the party." The "mixture" may be a product (e.g., a soap), with varying color, texture, or fragrance type, in which case conjoint measurement estimates what each physical feature contributes to the ratings. The ratings are subjective, assigned by a consumer. The features whose contributions are being estimated are under the control of the marketer or the product developer. These features can be manipulated. They may or may not appear in the concept or the product, as dictated by the design.

Like other experimental design procedures, conjoint measurement requires that the investigator create a variety of different stimuli, not just one (the "best" as guessed ahead of time). There is no one concept, but rather, there are many different concepts, the exact number depending upon the number of concept elements to be evaluated. The researcher can determine which elements are most effective by testing these different concepts, discovering which perform well and which perform poorly, and then tracing the performance to the component elements.

2. A Case History—Soap Concepts

To make the conjoint method more concrete, consider the following case history. Baumann Inc. manufactures a line of upscale soaps. Over the past 5 years the category has grown tremendously. The competition has stiffened as different manufacturers jockey for strong positions in this increasingly competitive market. Dollar spending for soaps remained constant, shaking out some of the competitors.

In the midst of this shakeout, Mike Baumann, the CEO, decided to launch a new line of luxurious soaps, aimed at the "middle tier" of soap consumers. Marketing management at Baumann Inc. recognized that they would have a difficult time competing at the very top end because of the advertising and packaging costs, but recognized an opportunity to launch a line of medium-priced soaps with an upper-scale image. The question was "What concept should de-

Concept Development, Testing, Optimization

Table 7 Eight Steps for Concept Development and Optimization

Step 1—Develop elements (snippets of phrases, picture)
Step 2—Polish the grammatical form of the elements, and classify the elements into categories
Step 3—Identify pairwise restrictions (what pairs of elements cannot be in the same concept)
Step 4—Create small test concepts, using experimental design
Step 5—Test concepts among consumers
Step 6—Create a database of concept × attribute × consumer
Step 7—Create additive model showing how many rating points are contributed by each element
Step 8—Create optimized concepts, which deliver specific levels of consumer purchase intent and/or communication

fine the line?" The strategy study that Baumann Inc. had commissioned the year before recommended that concepts emphasize "luxury" feeling, but the word "luxury" was already being played out competitively. How then could Baumann Inc. develop concepts for a line of soaps, and communicate luxury in ways that convinced consumers to buy?

The Stages of Systematic Concept Development

Developing concepts for conjoint measurement and optimizing these concepts follows a specific series of 8 steps, listed in Table 7. If the steps are correctly followed, then the outcome yields a set of concepts which often perform substantially better than the current concepts.

The steps are explained below:

Step 1—Identify the Elements the Concept Can Contain. The objective of Step 1 is to develop a large bank of concept elements. These elements can be words, phrases, etc. For now we will deal only with verbal materials, but subsequent sections in this chapter will go into pictures, voice-overs and music "bites" as concept elements. Table 8 shows the tentative list of elements developed in an "ideation session."

At this very earliest stage of development one need not (and should not) pay much attention to the number nor to the quality of the elements. Before the actual evaluation of concepts the number of elements will be reduced and the quality of the elements will be improved. However, at the element generation stage it is best if all elements are accepted and held for later consideration.

In productive ideation sessions, hundreds of elements emerge. The creative process usually takes several hours, and may be facilitated by "warm up" exercises. Prior to the session the participants can warm up by visiting various

Table 8 Categories and Elements for Soap Concepts

Category = Emotional reward
R1 Feel invigorated
R2 Feel fresh all day
R3 Escape to a more romantic time
R4 Revitalize yourself
R5 Relax your mind and body
R6 Achieve soft and beautiful skin easily
R7 Pamper your whole body
R8 Ease away stress
R9 Luxuriate in soothing lather
R10 Take your skin back to a younger tomorrow

Category = Fragrance
F1 Unscented
F2 A subtle scent
F3 A powdery scent
F4 The scent of baby powder
F5 A flowery scent
F6 A spicy scent
F7 An exotic scent
F8 A fresh scent
F9 A clean scent
F10 A fresh, clean scent
F11 A fresh, clean and natural scent
F12 A soft romantic scent
F13 A sweet scent
F14 The scent of freshly cut roses
F15 A garden-fresh scent

Category = Ingredient
I1 With aloe vera
I2 Does not contain harsh chemicals or heavy perfumes
I3 Totally free of skin irritants
I4 With pure organic ingredients
I5 With only pure and natural ingredients
I6 With essential moisturizing ingredients
I7 With oil absorbers that last all day
I8 With long-lasting deodorant protection

Cateogory = End use
U1 Made for the whole family
U2 Made especially for skin that is dry to slightly dry
U3 Made especially for skin that is oily to slightly oily
U4 Made especially for combination skin types
U5 For face, hands, and body
U6 For complexion care

Table 8 Continued

U7 For body care
U8 Made gentle enough for sensitive skin
U9 Designed for after sports

Category = Benefit
B1 Protects your skin from the effects of aging
B2 Cleanses and conditions your skin
B3 Protects your skin from bacteria
B4 Deeply cleans your pores
B5 Tones and smoothes your skin
B6 Helps control excess oil
B7 Moisturizes while it cleans your skin
B8 Soothes dry, cracked skin

Category = Physical attribute
P1 Produces a luxurious, rich lather
P2 Rinses off completely
P3 Leaves no filmy residue on skin
P4 Dermatologist-tested
P5 Never feels greasy
P6 Biodegradable
P7 Hypo-allergenic
P8 Non-allergenic

Category = Long term benefit
C1 Leaves your skin feeling soft and silky
C2 Leaves your skin feeling clean and fresh
C3 Leaves your skin beautifully clean, soft, and healthy
C4 Firms and tightens skin
C5 Keeps your skin young-looking
C6 Will not irritate even sensitive skin
C7 Doesn't dry like most soaps
C8 For year-round skin care
C9 No dry, flaky patches after use
C10 Minimizes dry skin, fine lines, and wrinkles

Category = Form/Color
L1 Available in an attractive, oversize bar
L2 Available in a small, easy-to-hold bar
L3 Available in an attractive, no-slip bar
L4 Available in an elegant bar
L5 Available as a soft, creamy soap
L6 Available in a solid, hard bar that doesn't melt quickly
L7 Available in pure white
L8 Available in feminine pink

stores to look at products, flipping through magazines to check the advertisements, and talking to their friends and family. During the actual ideation session, the participants should feel free to provide any input that they wish, including words and phrases belonging to the competition. Also, it is often very instructive to test the phrases used by the key competitors.

Step 2—Cull the List to a Reasonable Number of Polished Elements; Classify the Elements into Categories. Typically, the ideation exercise generates many words and phrases which are half-formed and often redundant. As the ideation session progresses, the same idea may appear several different times with similar phrasing. The comparable executions of the same idea can be boiled down to one or two executions. If the same idea appears in many different formats or executions, then several executions of the same idea should appear in the set of elements to be tested. Furthermore, people do not speak in complete ideas or sentences, but rather in fragments. The list of elements should be "cleaned up" by polishing the phrases into coherent wholes rather than using raw, elementary fragments. One need not polish the elements so that they are in perfect English (or whatever language is being used). Rather, the elements merely need to be phrased in reasonably grammatical form so that they convey the message clearly. For the most part, editing requires slight rewriting.

There is no "optimum" number of elements for a concept optimization study. One should attempt to be as general as possible in selecting elements, in order to cover the full gamut of concept possibilities. Usually 50–300 concept elements suffice, depending upon the category. For demonstration purposes we will deal with 76 soap elements here. However, conventional studies in concept optimization often comprise 200–300 elements. The field work and analysis in those larger projects are conducted in the same way as the soap study here.

Categorization of elements is important for purposes of bookkeeping. Categorization sorts the elements into groups which belong together (i.e., fragrance statements, usage occasions). By categorizing, the researcher keeps track of the elements and reduces redundancy. Categorization also helps in the creation of concepts, because in any single concept two elements from the same category cannot appear together. Table 8 shows the elements listed by category. There is no fixed number of categories. The researcher should consider categorization as an aid, not as a constraint.

Step 3—Identify "Restrictions" Among Pairs of Elements. When developing concepts, the writer must be cognizant of "real world" restrictions. Some elements simply do not go together because the combination makes no sense. An example might be a pair of elements, one of which communicates "appropriate for mature skin," whereas the other element communicates "used by teenagers everywhere." In cosmetic and fragrance concepts there are usually relatively few pairs of elements that must be restricted. In contrast, concepts in other categories, such as food, are governed by restrictions that are critical. For instance, it is illogical to combine an element dealing with "full fat" and another element dealing with "dietetic, low fat." Happily those types of combinations are rare in cosmetic science. There may be technological incompatibilities that govern pairs of elements, but not consumer-perceived illogical combinations. Table 9 lists some restrictions among pairs of elements.

Table 9 Restrictions Between Pairs of Elements

Categories
  Emotional reward   R1—R10
  Fragrance description   F1—F15
  Ingredient description   I1—I8
  End uses   U1—U9
  Benefits   B1—B8
  Physical attribute   P1—P8
  Long-term benefit   C1—C10
  Form   L1—L8

Specific pairwise restrictions (partial list)
  If R3 appears, then I8 U9 cannot appear.
  If F2 appears, then I2 P7 P8 cannot appear.
  If F3 appears, then I2 P7 P8 cannot appear.
  If F4 appears, then I2 P7 P8 cannot appear.
  If F5 appears, then I2 P7 P8 cannot appear.
  If F6 appears, then I2 P7 P8 cannot appear.
  If F7 appears, then I2 P7 P8 cannot appear.
  If F8 appears, then I2 P7 P8 cannot appear.
  If F9 appears, then I2 P7 P8 cannot appear.
  If F10 appears, then I2 P7 P8 cannot appear.
  If F11 appears, then I2 P7 P8 cannot appear.
  If F12 appears, then I2 P7 P8 cannot appear.
  If F13 appears, then I2 P7 P8 cannot appear.
  If F14 appears, then I2 P7 P8 cannot appear.
  If F15 appears, then I2 P7 P8 cannot appear.
  If I6 appears, then U3 B6 cannot appear.
  If I7 appears, then U2 C7 cannot appear.
  If I8 appears, then U2 U3 U4 U5 U6 cannot appear.
  If U2 appears, then B6 cannot appear.
  If U3 appears, then B7 cannot appear.
  If U4 appears, then B6 cannot appear.
  If P7 appears, then P8 cannot appear.

Step 4—Develop Test Concepts by Experimental Design. The objective of conjoint analysis is to identify the contribution of each element to consumer reactions. To accomplish this objective requires that the elements be combined by an experimental design into small, easily assimilated combinations or test concepts. There may be between 2 and 6 elements in each design. The designs are created according to the rules listed in Table 10. Each element appears several times in the set of concepts. Every element appears independently of every other element. Furthermore, each element appears against different backgrounds.

The array of concepts provides a means to determine what each element contributes to consumer reactions. Since the elements appear statistically independent of each other in the design, the analytic phase can estimate the individual contribution of each element by regression analysis. The experimental design insures that the estimation of an element's contribution to purchase intent (or communication) will be made in as unbiased a manner as possible. Since every element appears many times against different backgrounds, the odds are low that a single specific concept which scores low will bias the contribution of the component elements in that concept. These same elements appear in many other concepts, so that a single concept itself does not exert much of an effect in determining the performance of its component elements.

Table 10   Rules for Creating an Experimental Design

1. Identify all of the elements in the study. Call the total number N.
2. Since the analysis will be done by regression, insure that there are many more "cases" (concepts) than there are "predictors" (elements).
3. Create at least 2N concepts (to insure adequate degrees of freedom when estimating the contribution of each element).
4. Decide the number of appearances of each element across the full set of 2N concepts. Each element must appear equally often. Call the total number of appearances of each element "T."
5. Decide the average number of elements per concept. Call this number "A."
6. The total number of "slots" available for elements equals the product: 2N × A. (That is, the total number of slots equals the number of concepts, 2N, times the average number of elements per concept, A).
7. Each element appears T times in 2N concepts, and is absent 2N − T times in the 2N concepts.
8. Assign elements to concepts, so that each element appears T times, and never appears with other elements that would violate the restrictions set up for pairs of elements.
9. Select the experimental design of concepts showing the lowest values for pairwise correlations or chi-square values.

The concepts appear in much the same way they do in Table 11, which shows three concepts in the format which is presented to consumers. Although many researchers prefer to use more fully-developed concepts, with finished artwork and dense verbiage (body copy), finished artwork is more appropriate for the later stages in concept development when the final concepts have been selected. For early stage concept development research the effort to create polished final concepts does not pay off. There are simply too many concepts to create, and too many last minute changes to cope with. Furthermore, most consumer panelists have no difficulty understanding the major ideas in the concept, even though the concept is written on plain paper (perhaps with a picture at the top to support the idea).

One of the key additional benefits of an experimental design is that it can be modified to accommodate constraints and restrictions, without (in most cases) sacrificing too much robustness. For the soap study there were numerous restrictions on pairs of elements. A computer program can create literally tens of thousands of concepts, with the property that the elements are statistically independent of each other, and all of the constraints are obeyed so that elements that are designated as never appearing with each other never appear together. As long as there is a reasonable ratio of elements to constraints (which must be determined experimentally), there is no problem. One can usually create an experimental design of concept elements with the property that all constraints are

Table 11 Examples of Three Concepts by Experimental Design (Mixing/Matching)

Not Shown	Shown
	Concept 1
(R1)	Feel invigorated
(F1)	Unscented
(U1)	Made for the whole family
(B6)	Helps control excess oil
	Concept 2
(R6)	Achieve soft and beautiful skin easily
(I3)	Totally free of skin irritants
(B4)	Deeply cleanses your pores
(C8)	For year-round care
(L5)	Available as a soft creamy soap
	Concept 3
(I8)	With long-lasting deodorant protection
(U7)	For body care
(P6)	Biodegradable

satisfied and all elements appear independently of each other. The design becomes unbalanced and the elements lose their statistical independence only when there are an overwhelming number of constraints or unusually burdensome constraints falling on one or two elements or categories.

In addition to the experimentally designed concepts, the researcher should also test a set of reference or benchmark concepts among the same consumers. In their format these concepts should look like the test concepts in terms of the physical layout, number of elements, etc. However, the concepts should describe current in-market products. Half the concepts should present the appropriate brand names (for a measurement of the full impact of the competition), and the other half of the benchmark concepts should be identical, but without the brand name (to measure the impact of the selling message, without the influence of the brand name). Table 12 shows an example of two pairs of benchmark concepts for the soap project.

The Questionnaire. The questionnaire probes for a series of attribute perceptions. For each concept the consumer rates purchase intent as well as other "nonevaluative" attributes that probe communication. These attributes are nonevaluative because either side of the scale is equally acceptable. For example,

Table 12 Example of Two Benchmark Concepts for Soap: Each Concept Presented Two Ways—with Brand and without Brand

Concept 1A
    Jergen's Aloe & Lanolin
    Bursting with extra skin conditioners
    New and improved for softer skin

Concept 1B
    An aloe and lanolin soap
    Bursting with extra skin conditioners
    New and improved for softer skin

Concept 2A
    Unscented Dove
    Wash but don't dry
    Dove contains 1/4 moisturizing cream
    It won't dry your face like soap

Concept 2B
    An unscented soap
    Wash but don't dry
    It contains 1/4 moisturizing cream
    It won't dry your face like soap

Concept Development, Testing, Optimization

Table 13 Questionnaire for Soap Concept Study

Please read the concept in the booklet in front of you.

Now, after having read the concept, please rate the concept on these attributes.

1. How interested are you in buying this soap? Choose the appropriate number.
   - 1 = Definitely would not buy
   - 2 = Probably would not buy
   - 3 = Might or might not buy
   - 4 = Probably would buy
   - 5 = Definitely would buy
2. Who would use this soap?
   - 0 = Young people only → 100 = Mature people only
3. When would you use this soap?
   - 0 = For everyday use → 100 = For special use
4. Where would you use this soap?
   - 0 = For face alone → 100 = For body alone
5. How unique is this soap?
   - 0 = Just like all other soaps → 100 = Very unique soap
6. How elegant is this soap?
   - 0 = Ordinary → 100 = Very elegant

---

if the attribute is "younger" vs. "older," either end of the continuum may represent an equally acceptable concept. Table 13 shows the questionnaire. Purchase intent is rated on a conventional 1–5 point scale (1 = definitely not purchase → 5 = definitely purchase). Communication is rated on anchored 0–100 point scales, with 0 representing one extreme of the attribute and 100 representing the opposite extreme of the same attribute. By anchoring the scales the researcher can minimize any ambiguity.

Step 5—Run the Consumer Evaluation. Conventional concept evaluations require the consumer to assess one to five concepts, and typically entail a relatively short test session (about 10–20 minutes). Concept optimization requires a significantly longer time period because the consumer usually rates from 40 to 100 concepts.

In order to execute the study and maintain consumer cooperation, the researcher should pre-recruit panelists to participate. The typical concept evaluation study is choreographed in a group session lasting from 2 to 4 hours (depending upon the specific number of concepts and the number of attributes used for evaluation). Table 14 lists the sequence of activities in the field, and the rationale (where appropriate).

Step 6—Create a Concept Database. The data is first reduced to a simple matrix, containing two sets of numbers (see Table 15). Each row of the data matrix

Table 14 Sequence of Field Activities for Concept Evaluation

1. Define the appropriate group of consumers (usually base size of 50+).
2. Invite the consumers to participate for an extended session.
   - The session may last 2 to 4 hours
   - Consumers are paid to participate
   - The fieldwork may be done in several markets
3. Orient the consumers in evaluation by means of a short warm-up exercise.
   - This insures that the consumers comprehend the task
   - Consumers who don't understand the task can be coached
4. Present consumers with a test concept, which they evaluate.
   - An interviewer "checks" this concept to insure that the consumer understands the task and the rating scales
5. Each consumer evaluates a pre-defined sequence of concepts.
   - The sequence is randomized by the researcher
   - The randomization reduces order bias
   - The consumer rates each concept on the different attributes using the appropriate scale
6. The consumer data is checked by an attending interviewer.
   - This checking insures that the consumer maintains interest
   - The checking also catches problems with individual consumers
7. After finishing the requisite set of concepts the consumer completes an extensive attitude and usage questionnaire.
8. The consumer is paid and dismissed.

corresponds to one of the concepts (whether this be one of the experimentally designed concepts or a concept from the competitive frame). The first column of data comprises the proportion of consumers who, having read the concept, rate it as either *definitely or probably* would buy (4 or 5 on the 5-point purchase intent scale). This measure is a percentage statistic. The researcher is not interested in the average purchase intent (a 3.7 on the 5-point scale), but rather in the number of consumers who say that they will buy the product. The second column of data corresponds to the number of consumers who said that they will *definitely* purchase the product, based upon reading the concept. Often the "top box" (the percent of consumers choosing the top category on the 5-point scale) is a better indicator of ultimate consumer interest. Many researchers who link market performance to purchase intent data in their "market simulation models" place greater weight on the top box than on the top two boxes (more weight placed on the percent of consumers rating category 5, than on the percent of consumers rating either category 4 or 5). That is because these market researchers believe that only a person who feels really committed to the product will assign it a rating of 5. As a result, the researchers use a composite number, which assigns more weight to the "top box" (5 on the 5-point scale) than to the "fourth box" (4 on the 5-point scale).

The use and analysis of percent data rather than average data for purchase traces its history to the way marketing researchers look at their data. Traditional marketing research looks at the *number* of people who will do something, not at the intensity of a single (or average) person's feeling about doing something. Marketing research traces its intellectual heritage to sociology, where interest focuses on the proportion of the population exhibiting a certain behavior (i.e., interest in buying the soap product, based upon reading the concept). A psychological orientation looks at the average intensity of purchase intent, because the interest focuses on the internal psychological processes involved. Thus, a psychologically based analysis would look at two products, rated 3.0 and 3.5, and conclude that among all the consumers, people want to purchase the second product more than they want to purchase the first product. The emphasis is on the *desire* to purchase the product. The sociologist, however, would be more interested in the fact that a greater number of the consumers say that they would purchase the second product, versus saying that they would purchase the first product. By analyzing percent purchase intent statistics for those two products the sociologist would be satisfied at having "understood" consumer reactions in terms of his world view.

The third and fourth columns of data show the average ratings of each concept on two "tonalities" or "communications." Unlike the purchase intent scores these are averages on the 0–100 scale. They show the intensity or degree of communication. A percentage statistic is inappropriate because we are interested in the *degree* to which a concept communicates a message (and subsequently by analysis the degree to which each element of the concept communicates the message).

The data matrix is fairly large, comprising (for the soap study) a total of 160 rows (one for each experimentally designed soap concept, and another row for each benchmark concept) and seven columns (two for purchase intent, and five for the different communication or tonality attributes).

We can adjoin to that matrix another set of columns, corresponding to the elements in the soap concepts. There are 76 elements. Each element either appears in a concept or does not appear in a concept. If we create an additional 76 columns for the matrix, then we can represent the concepts in terms of a set of 0's and 1's. For a specific row (corresponding to a specific concept) there are 76 columns. Each column corresponds to an element. If the element is absent from the concept, then in that row the column corresponding to that element has a value of 0. If the element is present in that specific concept, then in that

Table 15 Example of a Binary Expansion of the Experimental Design and the Consumer Rating Data

Concept	Elements in binary code (0–1)								Consumer ratings			
	R1	R2	R3...	F1	F2	F3...	U1	U2...	Definitely or probably buy	Definitely buy	Young vs old	Everday vs special
101	0	0	0	1	0	0	0	1	42%	27%	32	52
102	0	1	0	0	0	0	1	0	38%	30%	50	46
103	0	0	0	1	0	0	0	0	51%	40%	28	60
104	0	1	1	0	1	0	1	0	27%	19%	57	35
105	0	0	0	0	0	0	0	0	64%	52%	48	56
106	0	1	0	1	0	0	1	0	41%	36%	61	42

row the column for the element has a value of 1. Thus the large matrix shows the concept architecture (which elements are present or absent, each concept respectively), the interest profile (proportion of interested consumers, either top box or top two boxes), and the communication profile (average rating on the 0–100 scale for each communication attribute). Table 15 shows a small portion of this large matrix.

Step 7—Create the Concept Model by Regression Analysis. The data in Table 15 is set up for regression analysis. Regression analysis estimates how independent variables combine to produce dependent variables. Here the independent variables are the elements, which take on the value 0 (if the element is absent) or 1 (if the element is present in a concept). There are 76 independent variables, one for each element. (In a larger scale concept optimization study involving 300 elements, there would be 300 independent variables in the model. Here there are only 76). Each rating attribute is a separate dependent variable. Thus top box purchase intent, top two boxes purchase intent, etc. are separate variables, as are each of the communication attributes.

The regression model is known as "dummy variable" regression. The "dummy variables" are predictors which take on only one of two values; 0 or 1. (These are the elements in the concept, which by convention have been assigned values of 0 or 1, respectively, for each concept, depending upon whether or not the element appears in the concept.) Dummy variable regression analysis estimates the part-worth contribution of each element to the consumer attribute. Since the elements were combined in a way to be statistically independent of each other, regression analysis can validly estimate their part-worth contributions. If the concepts had been thrown together haphazardly, then we would not have been able to estimate these contributions in as statistically a robust way.

The regression model fits a linear equation of the following form:

Attribute Rating = K_0 + K_1 (Element 1) + K_2 (Element 2) $\cdots$ K_n (Element n)

The regression equation comprises 76 terms (where 76 is the number of elements) and an additive constant K_0. Each coefficient shows the part-worth contribution of the particular element to the attribute rating (whether the rating be of purchase intent or communication). Positive values of the coefficients mean that the element adds to the attribute. For instance, a +8 for % top two box purchase intent means that when the element is present in the concept it adds an additional 8% of the consumers to the group who say that they will probably or definitely buy the soap. A +8 for an attribute (on the face vs. whole body scale, where face = 100 and whole body = 0) means that the attribute moves the concept towards the "facial use" side. The concept may still be overwhelmingly for "whole body," but the element adds a "facial use" tonality to the concept.

The additive constant K_0 is the adjustment factor. It is added to the part-worth contributions of all of the elements in the concept. The total (constant + part-worth contributions of the elements) is the best estimate of what the concept would score, given the empirical data. The additive constant and the part-worth contributions (coefficients in the regression equation) can be thought of as follows:

1. When the constant is high and the coefficients are low, this means that the consumers are positively disposed toward all of the concepts that they have rated. However, there is probably very little discrimination between the concepts, and therefore between the elements. It is the basic idea of the product, rather than the elements, which makes the concept highly acceptable.
2. When the constant is low and the coefficients are low, this means that the consumers are negatively disposed toward all of the concepts that they have rated. Again there is very little discrimination between the elements.
3. When the constant is low but the coefficients vary (they range from low to high), this means that the consumer differentiates between the concepts. The discrimination between the concepts is primarily due to the different elements. (The correct choice of elements can create a winning concept, whereas the incorrect choice of elements, namely, poor performers, will create a losing concept.)

The Additive Model for Soap Concepts—Purchase Intent. Table 16 shows the additive model for concepts, looking at the "% top two box" as the dependent variable. The elements span a wide range, showing that there are elements which are highly impactful and lead to purchase (e.g., a "fresh, clean scent"), and other elements which are negative (e.g., available in feminine pink). In concept work, there are often a substantial number of negative elements, meaning that their presence detracts from acceptability. Table 16 also shows models for two segments of the population (individuals with different points of view) as well as models for the communication attribute.

Statistical Issues—Goodness-of-Fit. Statisticians are vitally concerned with the validity of models, especially because these models are one step removed from the actual data itself. The additive model is no different. Although we begin with full concepts, we end up with part-worth utilities. To what degree does the model truly fit the data? Is the model accurate? Fortunately there are statistical tests to measure the goodness-of-fit of the model to the empirical data. There are two major statistics of interest:

1. The proportion of the variability in the data accounted for by the model: This is called the square of the multiple correlation (viz., r^2). The value of r^2 ranges from a low of 0 (meaning that the model accounts for none of the

Concept Development, Testing, Optimization

Table 16 Models Showing the Additive Contribution of Elements to Ratings

		Attribute (See key)							
		A	B	C	D	E	F	G	H
Additive constant		27	29	25	42	57	40	51	39
1	Feel invigorated	5	10	1	8	-3	6	4	0
2	Feel fresh all day	4	9	0	6	-7	7	-2	2
3	Escape to a more romantic time	-9	-9	-8	-4	8	-1	5	8
4	Revitalize yourself	-4	-2	-6	-2	5	3	5	8
5	Relax your mind and body	-1	-5	5	-6	3	-4	1	5
6	Achieve soft and beautiful skin easily	-8	-10	-6	-4	2	-3	0	5
7	Pamper your whole body	6	2	8	-4	4	-4	1	7
8	Ease away stress	-4	-3	-5	-4	6	-1	3	7
9	Luxuriate in soothing lather	-9	-7	-9	0	4	1	-3	3
10	Take your skin back to a younger tomorrow	-8	-8	-7	-8	1	-6	-1	3
11	Unscented	1	5	-4	-2	-4	0	-3	-8
12	A subtle scent	6	7	4	5	-3	7	5	2
13	A powdery scent	2	2	1	-1	-2	6	5	-4
14	The scent of baby powder	4	4	3	4	-3	7	7	-5
15	A flowery scent	5	8	2	2	1	5	4	3
16	A spicy scent	-1	-5	6	-3	1	4	5	1
17	An exotic scent	-3	-7	4	4	7	7	6	3
18	A fresh scent	9	13	5	5	-6	-1	-1	-6
19	A clean scent	7	10	3	4	-1	6	-1	-7
20	A fresh, clean scent	10	15	5	6	-3	2	-1	-4
21	A fresh, clean, and natural scent	6	9	3	0	-4	3	2	-5
22	A soft romantic scent	7	5	10	3	4	5	4	-1
23	A sweet scent	8	6	11	3	-2	8	5	-2
24	The scent of freshly cut roses	6	1	12	2	1	4	6	2

(*continued*)

Table 16 Continued

					Attribute (See key)				
		A	B	C	D	E	F	G	H
Additive constant		27	29	25	42	57	40	51	39
25	A garden–fresh scent	5	1	10	2	1	6	3	-1
26	With aloe vera	7	6	9	-3	0	-2	5	4
27	Does not contain harsh chemicals or heavy perfumes	10	16	4	1	-2	5	6	4
28	Totally free of skin irritants	10	12	7	-1	1	2	3	2
29	With pure organic ingredients	8	10	6	3	-1	-3	6	3
30	With only pure and natural ingredients	3	7	0	-1	-2	-3	2	1
31	With essential moisturizing ingredients	7	2	12	-1	2	-3	3	2
32	With oil absorbers that last all day	-2	3	-7	4	-1	-1	3	0
33	With long–lasting deodorant protection	2	4	0	-2	-6	8	-8	-6
34	Made especially for the whole family	5	4	5	2	-3	3	1	0
35	Made especially for skin that is dry to slightly dry	9	4	13	-4	-3	-3	5	1
36	Made especially for skin that is oily to slightly oily	-1	4	-5	5	-2	-8	2	-1
37	Made especially for combination skin types	4	4	3	1	-2	-7	3	1
38	For face, hands, and body	4	4	3	-1	-2	-1	2	1
39	For complexion care	3	2	5	2	-6	-8	0	0
40	For body care	7	7	6	-1	0	8	3	3
41	Made gentle enough for sensitive skin	8	7	8	-1	-2	-5	3	3
42	Designed for after sports	3	7	-1	7	5	8	4	-5
43	Protects your skin from the effects of aging	4	0	9	-5	-4	-5	-2	1
44	Cleanses and conditions your skin	7	8	6	-1	-4	-5	0	-1
45	Protects your skin from bacteria	1	3	0	1	-4	-2	2	-2
46	Deeply cleans your pores	12	14	9	1	-2	-8	2	-1
47	Tones and smoothes your skin	12	8	15	-1	-2	-4	2	1
48	Helps control excess oil	0	3	-2	6	-4	-8	3	0
49	Moisturizes while it cleans your skin	9	4	14	-1	-2	-3	3	4

Concept Development, Testing, Optimization

#	Element	A	B	C	D	E	F	G	H
51	Produces a luxurious, rich lather	3	2	5	-5	-2	-5	1	4
52	Rinses off completely	4	4	4	-4	-3	-6	0	-2
53	Leaves no filmy residue on skin	6	5	6	-3	-3	-5	1	2
54	Dermatologist-tested	1	3	0	-2	-5	-8	1	-1
55	Never feels greasy	5	6	4	-5	-1	-2	1	4
56	Biodegradable	4	5	2	-2	-1	-3	2	0
57	Hypo–allergenic	5	7	3	-3	-2	-5	2	3
58	Non-allergenic	4	8	1	-4	-4	-8	0	2
59	Leaves your skin feeling soft and silky	3	1	6	-4	-3	-2	2	5
60	Leaves your skin feeling clean and fresh	2	11	-4	-3	-4	-5	0	1
61	Leaves your skin beautifully clean, soft, and healthy	4	10	0	-4	-5	-7	0	2
62	Firms and tightens skin	9	2	15	-8	-5	-8	4	3
63	Keeps your skin young–looking	10	8	12	-8	-4	-7	1	2
64	Will not irritate even sensitive skin	2	3	1	-5	-2	-4	0	4
65	Doesn't dry like most soaps	9	6	14	-1	-3	-5	3	2
66	For year-round skin care	4	4	2	-2	-4	-8	-2	0
67	No dry, flaky patches after use	11	8	13	-7	-5	-8	3	1
68	Minimizes dry skin, fine lines, and wrinkles	9	2	16	-8	-4	-8	7	7
69	Available in an attractive, oversize bar	1	3	-2	0	-1	3	0	1
70	Available in a small, easy-to-hold bar	4	3	4	0	0	2	2	4
71	Available in an attractive, no-slip bar	6	7	4	1	-2	1	2	3
72	Available in an elegant bar	-1	-3	5	0	3	1	1	6
73	Available as a soft, creamy soap	1	-1	4	-2	0	-5	1	4
74	Available in a solid, hard bar that doesn't melt quickly	5	7	3	-1	-6	6	2	-2
75	Available in pure white	3	5	1	2	-7	1	-8	-3
76	Available in feminine pink	-4	-7	0	-3	3	3	-2	4

Key To Models (Columns):

A, Total panel purchase intent (% Top two box); B, Segment 1 purchase intent (% Top two box) → Fresh, clean seekers; C, Segment 2 purchase intent (% Top two box) → Repair, skin health seekers; D, For older (–) vs. younger (+) users; E, For daily use (–) vs. for special occasions (+); F, For face (–) vs. whole body (+); G, Common (–) vs. unique (+); H, Ordinary (–) vs. elegant (+).

data) to a high of 1.00 (meaning that the model accounts for 100% of the data). To find the proportion of variation in the data for which the model accounts, simply multiply the multiple r^2 value by 100. For this data set the multiple r^2 is 0.72, meaning that 72% of the variability in the values of the % top two box purchase can be accounted for by knowing the elements present in the concept and by assigning them the correct weights. Table 17 lists the multiple r^2 for the various attributes. In general the multiple r^2 values are fairly high, leading us to feel confident that the model accurately describes the data.

2. The standard error of the regression: The regression equation estimates the likely rating that would have been assigned to each concept. The difference between the actual rating (e.g., % top two box) and the estimated rating from the equation generates a distribution across all of the tested concepts. The standard deviation of this distribution of prediction errors is the standard error of the regression. The lower that standard error of regression the lower the prediction error, and therefore the closer the model is to the actual data. For the data here, Table 17 shows the standard errors for the different equations. The standard error varies by attribute.

Building Models for Subgroups. We can also develop an additive model for key subgroups. For soap these subgroups include male and female users, users of various brands, and segments. The same approach applies for subgroups as for the total panel. There is only one exception—the base sizes behind every concept will out of necessity be smaller when the subgroup data is considered. For instance, if each concept was originally rated by 40 consumers, and if the panel is equally divided between female and male consumers, then each concept will be rated by 20 males and 20 females, respectively. Out of necessity the percent top two box statistics will be less robust, because that percent statistic is based upon half the number of observations. The percent top two box for 20 observations will be less stable than the same percent statistic based upon

Table 17 Statistics for the Goodness-of-Fit of the Additive Models

Attribute	Multiple r^2	Standard error of regression
Total Panel (Purchase)	0.72	5.3
Segment 1 (Purchase) → Fresh, Clean	0.68	4.9
Segment 2 (Purchase) → Skin Repair, Skin Health	0.59	5.3
For Older vs. Younger User	0.71	4.1
For Daily Use vs. Special Occasion	0.68	4.8
For Face vs. For Whole Body	0.51	3.2
Common vs. Unique	0.49	4.1
Ordinary vs. Elegant	0.58	3.1

40 observations. Table 16 shows the additive model for two key preference segments of consumers.

Creating Models for Communication (Tonality). Up to now we have dealt only with persuasion. For communication we work with averages on a 0–100 scale, and estimate the part-worth contributions based upon those averages. In general, averages are more robust statistics than are percentages. We will also find the part-worth contributions to be smaller when based upon averages than when based upon percentages. This occurs because percentages show a wider range of variation than do means.

The analysis and statistics are precisely the same for communication as they are for persuasion. The independent variables are the elements, which take on the values of 0 or 1, respectively. The dependent variable is the average rating. Table 16 presents the additive model for communication, as well as for purchase intent. The contributions across the elements show a much narrower range. However, the data appears to be valid because elements that we expect to communicate certain points do so according to the consumers. The key difference is that for the most part the communication values cluster near zero, except in those instances where an element truly communicates what we expect of it. In that case the element shows an absolute level of contribution beyond 5 or 6 (beyond –5 at the lower end and +5 at the upper end). When we see communication values beyond the level of 5, we are dealing with elements which show strong communication.

Table 18 highlights some of these strongly communicating elements. Unlike persuasion, communication is specific. An element may not communicate anything, or it may communicate one characteristic or benefit particularly well. It is rare that a single element strongly communicates more than one, non-evaluative, specific message. Furthermore, it is not necessary for an element to strongly communicate anything in order to be highly persuasive. An element can be persuasive without strongly communicating a single specific message.

On Synergism in Concepts. Synergism refers to the rating of a mixture that exceeds one's expectations, based upon the arithmetic sum of the effects of the components. In concept, research synergism can be established empirically. If two elements are combined into a single element as well as tested alone, then we can compare the arithmetic sum of their part-worth contributions to the contribution of the elements tested together in a single concept element. Synergism exists when the combinations *substantially exceed* the arithmetic sum of the part-worth contributions. For instance, if elements X (part-worth = 10) and Y (part-worth = 10) are synergistic, then we may expect the combination of X and Y in a single element to generate a part-worth far exceeding 20. If the combination is simply additive rather than synergistic, then the combination would

Table 18 Key Elements with High Communication

		D	E	F	G	H
1	Feel invigorated	**8**	-3	6	4	0
10	Take your skin back to a younger tomorrow	**-8**	1	-6	-1	3
3	Escape to a more romantic time	-4	**8**	-1	5	**8**
42	Designed for after sports	7	5	**8**	4	-5
40	For body care	-1	0	**8**	3	3
33	With long-lasting deodorant protection	-2	-6	**8**	**-8**	**-6**
39	For complexion care	2	-6	**-8**	0	0
75	Available in pure white	2	**-7**	1	**-8**	-3
4	Revitalize yourself	-2	5	3	5	**8**
11	Unscented	-2	-4	0	-3	**-8**
23	A sweet scent	3	-2	**8**	5	**-2**
62	Firms and tightens skin	**-8**	-5	**-8**	4	3
67	No dry, flaky patches after use	**-7**	-5	**-8**	3	1
48	Helps control excess oil	6	-4	**-8**	3	0
58	Non-allergenic	-4	-4	**-8**	0	2
68	Minimizes dry skin, fine lines, and wrinkles	**-8**	-4	**-8**	7	7

Bold numbers = Key communicating elements
D = Old (-) vs. Young (+)
E = Daily Use (-) vs. Special Occasions (+)
F = Facial Use (-) vs. Whole Body Use (+)
G = Common (-) vs. Unique (+)
H = Ordinary (-) vs. Elegant (+)

have a part-worth (or coefficient) of 20. If the combination is "suppressive," then the combination would achieve a part-worth far less than 20.

True pairwise synergism is rare, but pairwise suppression is common. It is hard to find two elements which go together so well that their combination scores substantially higher than the sum of the two measured alone. What researchers often think of as "synergism" is really additivity, so that the two elements do very well together, but not any better than would have been predicted from data about each separately. Suppression occurs quite often. Two elements may just not seem to go together particularly well, although each element by itself does well.

Step 8—Create High Scoring Concepts Through the Additive Model. Most researchers use conjoint modeling to identify the "hot buttons" and then create improved concepts. Creating a high scoring concept becomes substantially easier once we have identified the part-worth contributions of each of the elements. If we are interested in purchase intent alone, then we can select the top scor-

Concept Development, Testing, Optimization 39

ing elements in various categories and combine them. We thus create a combination of elements which maximizes overall purchase intent. Furthermore, our concept will not contain logical impossibilities (at least at a pairwise level), because we have a list of restrictions which shows which elements cannot be combined together. Table 19A presents the best concepts, defined as a set of three concepts which maximize overall purchase intent and have no elements in common with each other.

One of the key benefits of concept development and testing in a systematic, experimental manner is the ability to assign to each element its degree of contribution to the overall concept score. Since the concept is built up synthetically by combining elements of known contributions, the researcher has available a table of part-worth contributions for each element. Table 19B shows those contributions for the best concept, enabling the researcher to see which elements truly contribute strongly and which contribute weakly.

Incorporating Tonalities into the Concept. The concepts in Table 19 do not strongly focus on a single direction. Although the concepts are acceptable, we can further refine them by focusing their communication or tonality profile. Recall that the consumers rated the concepts on a series of communication or tonality attributes, and that from the data we created an additive model for each communication attribute (see Table 16). We can carry the optimization one step further in order to develop a combination of concept elements which jointly satisfies several objectives. Each objective is a number. One objective may be to score highly on overall purchase intent (e.g., a goal of 100 on the top two box purchase intent scale). A second objective may be to be perceived as strongly appropriate for "daily use" (e.g., a goal or sum of 1 on the bipolar every day (0) → special occasion (100) scale). A third goal may be perceived as strongly, but not completely, "unique" (e.g., 75 on the uniqueness scale).

To discover the "optimal combination" satisfying multiple objectives, we need to use a computer program which can evaluate hundreds of thousands of element combinations in a systematic (albeit incomplete) fashion. The MJI "Concept Optimizer"™ does this analysis quickly, looking at the various categories, the high scoring elements within the categories, the pairwise restrictions (to avoid combinations that are forbidden), and the objectives to be satisfied.

By using the tonalities or communications as objectives or goals to be satisfied (along with purchase intent as another goal), the researcher creates focused concepts. The concepts are focused because the communication profile or goal sets the stage for the type of concept to be developed (e.g., strongly feminine). One may wish to communicate gender, or perhaps incorporate many different tonalities in the concept. The Concept Optimizer handles the communication objectives in an efficient manner, providing the researcher with new concept combinations for further consideration. If such a computer-based sys-

Table 19A Set of Three Concepts for Soap:
Objective—Maximize Overall Consumer Interest

First of three non-overlapping concepts (purchase intent = 79)
- F10 A fresh, clean scent
- I3   Totally free of skin irritants
- U2   Made especially for skin that is dry to slightly dry
- B5   Tones and smoothes your skin
- C9   No dry, flaky patches after use

Second of three non-overlapping concepts (purchase intent = 73)
- I2   Does not contain harsh chemicals or heavy perfumes
- U8   Made gentle enough for sensitive skin
- B4   Deeply cleans your pores
- C5   Keeps your skin young-looking
- L3   Available in an attractive, no-slip bar

Third of three non-overlapping concepts (purchase intent = 69)
- F8   A fresh scent
- I4   With pure organic ingredients
- U7   For body care
- B7   Moisturizes while it cleans your skin
- C10 Minimizes dry skin, fine lines, and wrinkles

tem seems to lack "soul," keep in mind that the Concept Optimizer may be viewed as merely a computer-based sorting system which identifies some combinations of elements for further work. The main function of the Optimizer is to reduce some of the inevitable drudgery which is part and parcel of the concept development process.

Table 19B Diagnositics for the Optimal Concept: What Every Element Contributes

		Purchase intent total	Old-young	Special	Face-body	Unique	Elegant
Constant		27	42	57	40	51	39
F10	A fresh, clean scent	10	6	-3	2	-1	-4
I3	Totally free of skin irritants	10	-1	1	2	3	2
U2	Made especially for skin that is dry to slightly dry	9	-4	-3	-3	5	1
B5	Tones and smoothes your skin	12	-1	-2	-4	2	1
C9	No dry, flaky patches after use	11	-7	-5	-8	3	1
	Total	79	35	45	29	63	40

Concept Development, Testing, Optimization

To put the Concept Optimizer to its fullest use, let us consider a variety of strategies and operationalize them as objectives to be satisfied. We then use the Concept Optimizer to select the appropriate combination of elements. Table 20 lists these strategies or objectives, and the best combination of elements which satisfies these objectives simultaneously.

Creating Concepts with Different Levels of Importance for Tonalities. Previously we saw how to create concepts with different tonalities (communications), but with all of the goals or objectives having equal importance. Occasionally, some objectives are more important than others. In these cases we may wish to satisfy the same profile of objectives, but with different degrees of importance. Table 21 shows a profile of communication objectives to satisfy, but with varying weights assigned to the communication objectives (goals). By weighting these objectives in different proportions, we see that we can modify the ultimate concept that emerges from the exercise. Of course the ultimate concepts will not be altogether different unless some goals are assigned weights or importance values of 0 (meaning that they are irrelevant to the final concept, and simply contribute nothing at all).

Creating a Line of Partially Overlapping Concepts. Quite often marketers are interested in a line of products, not just a single product. Or, in many instances, marketers, product developers and advertising agencies are interested in a line of concepts with the same basic umbrella statements, but with different elements to satisfy various objectives. Can we create a series of concepts with the property that they are maximally acceptable (in terms of fitting the defined objectives or goals), but have a subset of elements in common (umbrella set), combined with elements that vary (for variety)? That is, the first concept will be maximally acceptable and will contain a certain number of elements, and the second and third concepts will be highly acceptable, but will have some different elements. This approach generates a set of concepts that differ from each other, but which possess the same overall "umbrella." Table 22 shows the series of concepts for the line.

An Overview to Basic Concept Optimization

Simple experimental designs can be used profitably to create winning concepts. These concepts can either be statements about product features, statements about positioning, or both. In contrast to conventional concept development, concept optimization identifies those specific elements which motivate purchase or which communicate a desired "tonality."

What is needed for concept optimization is a more relaxed, open atmosphere than has traditionally been the case. Typically, marketers, product developers and advertising agencies have developed concepts while "under the gun" to come

Table 20 Optimal Concepts—Different Communication Objectives

(a) Overall liking, but tuned to a younger image

Attribute	Expected	Goal	Wt
*Total Purchase	65	100	1
*Old vs. Young	67	100	1
Facial vs. Body	45	100	0

R1   Feel invigorated
F10   A fresh, clean scent
I4   With pure organic ingredients
U9   Designed for after sports
B4   Deeply cleans your pores

(b) Overall liking, but tuned to an older image

Attribute	Expected	Goal	Wt
*Total Purchase	50	100	1
*Old vs. Young	10	1	1
Facial vs. Body	23	100	0

R10   Take your skin back to a younger tomorrow
U2   Made especially for skin that is dry to slightly dry
B8   Soothes dry, cracked skin
P5   Never feels greasy
C5   Keeps your skin young-looking

(c) Overall liking, but tuned to body use

Attribute	Expected	Goal	Wt
Total Purchase	62	100	1
Old vs. Young	50	100	0
Facial vs. Body	70	100	1

R1   Feel invigorated
F13   A sweet scent
I3   Totally free of skin irritants
U7   For body care
L6   Available in a solid, hard bar that doesn't melt quickly

(d) Overall liking, but tuned to facial use

Attribute	Expected	Goal	Wt
*Total Purchase	49	100	1
Old vs. Young	26	100	0
*Facial vs. Body	2	1	1

R10   Take your skin back to a younger tomorrow
U6   For complexion care
B4   Deeply cleans your pores
P8   Non-allergenic

Table 20 Continued

C9   No dry, flaky patches after use

(e) Overall liking, but tuned to a younger image, body use

Attribute	Expected	Goal	Wt
*Total Purchase	58	100	1
*Old vs. Young	58	100	1
*Facial vs. Body	70	100	1

R1   Feel invigorated
F13 A sweet scent
I3   Totally free of skin irritants
U9   Designed for after sports
L6   Available in a solid, hard bar that doesn't melt quickly

(f) Overall liking, but tuned to older image, body use

Attribute	Expected	Goal	Wt
*Total Purchase	69	100	1
*Old vs. Young	40	1	0
*Facial vs. Body	33	100	1

R10 Take your skin back to a younger tomorrow
U2   Made especially for skin that is dry to slightly dry
B8   Soothes dry, cracked skin
P5   Never feels greasy
C5   Keeps your skin young-looking

Note: Wt = Weight, a measure of relative importance
0 = Not relevant for determining the concept
* = Denotes relevant for determining the concept

up with a breakthrough. The intense competitive pressures often produce tunnel vision, rather than openness. Concept optimization reduces tunnel vision. Concerned parties forget, at least for the moment, the demand to create winning concepts, and instead generate elements. These snippets of ideas are then combined by experimental design to create small combinations (test concepts). Only after the test concepts are evaluated does the creative group recombine the elements into winning combinations. This time, however, they work with a road map rather than a guess.

3. Segmenting Consumers Based on Responses to Concept Elements

Groups defined by brand usage, gender, etc. represent the typical way by which marketers divide up the total population. Although these groups differ in age, products used, etc., their response patterns are often similar, albeit not identi-

Table 21 Set of Concepts with the Same Communication Profile but with Different Weights (Importance) Attached to the Objectives

Communication objective—Develop a concept which communicates "everyday use," "facial use," "uniqueness," and "elegant"

(a) All objectives have equal weight

Attribute	Expected	Goal	Wt
Total Purchase	49	100	0
Everyday vs. Special	31	1	1
Facial vs. Body	7	1	1
Common vs. Unique	58	100	1
Ordinary vs. Elegant	35	100	1

F8	A fresh scent
U6	For complexion care
B6	Helps control excess oil
P4	Dermatologist-tested
C4	Firms and tightens skin

(b) Everyday use is much more important than the other goals

Attribute	Expected	Goal	Wt
Total Purchase	55	100	0
Everyday vs. Special	26	1	10
Facial vs. Body	31	1	1
Common vs. Unique	44	100	1
Ordinary vs. Elegant	35	100	1

R2	Feel fresh all day
F8	A fresh scent
U6	For complexion care
C4	Firms and tightens skin
L7	Available in pure white

(c) Uniqueness is much more important than the other objectives

Attribute	Expected	Goal	Wt
Total Purchase	48	100	0
Everyday vs. Special	37	1	1
Facial vs. Body	5	1	1
Common vs. Unique	68	100	10
Ordinary vs. Elegant	48	100	1

I4	With pure organic ingredients
U6	For complexion care
B6	Helps control excess oil
P4	Dermatologist-tested
C10	Minimizes dry skin, fine lines, and wrinkles

Concept Development, Testing, Optimization

Table 21 Continued

(d) Elegance is much more important than the other objectives			
Attribute	Expected	Goal	Wt
Total Purchase	44	100	0
Everyday vs. Special	39	1	1
Facial vs. Body	3	1	1
Common vs. Unique	62	100	1
Ordinary vs. Elegant	52	100	10

U6	For complexion care
B6	Helps control excess oil
P8	Non-allergenic
C10	Minimizes dry skin, fine lines, and wrinkles
L5	Available as a soft, creamy soap

(e) Everyday use and elegance are much more important than the other objectives			
Attribute	Expected	Goal	Wt
Total Purchase	55	100	0
Everyday vs. Special	26	1	10
Facial vs. Body	31	1	1
Common vs. Unique	44	100	1
Ordinary vs. Elegant	35	100	10

R2	Feel fresh all day
F8	A fresh scent
U6	For complexion care
C4	Firms and tightens skin
L7	Available in pure white

Note: Wt = Weight, a measure of relative importance
 0 = Not relevant for determining the concept

cal. Yet, any casual observation of consumer behavior quickly reveals that the consumers show reliable differences in terms of what appeals to them. These individual differences, while obvious on an observational basis, do not emerge as clearly when we divide the consumers according to conventional demographic breakouts.

There may be a more productive way to divide the consumer population. This method uses the pattern of consumer reactions to concepts. The original approach is known as sensory segmentation. Briefly, sensory segmentation assumes that in the population there exist basic groups of individuals showing reliable, yet quite different patterns of product acceptability. For instance, one segment of individuals may like flowery fragrances whereas another segment may like more vanilla-like or powdery fragrances (Moskowitz, 1986). These segments do not emerge from conventional ways of dividing the population, yet

Table 22 Set of Concepts That Have a Specific Umbrella But Differ in Other Objectives as Well

Forced In

F	Fragrance Mention – one item from the category
R	Emotion Reward – one item from the category
*P8	Non-allergenic
*C8	For Year-round skin care

(a) Objectives: maximize acceptance, daily usage, facial, ordinary

Attribute	Expected	Goal	Wt
Total Purchase	31	100	1
Old vs. Young	28	100	0
Everyday vs. Special	40	1	1
Facial vs. Body	16	1	1
Common vs. Unique	45	100	0
Ordinary vs. Elegant	36	1	1

R10	Take your skin back to a younger tomorrow
F1	Unscented
U6	For complexion care
*P8	Non-allergenic
*C8	For year-round skin care

(b) Objectives: maximize acceptance, special usage, facial, ordinary

Attribute	Expected	Goal	Wt
Total Purchase	40	100	1
Old vs. Young	27	100	0
Everyday vs. Special	44	100	1
Facial vs. Body	16	1	1
Common vs. Unique	47	100	0
Ordinary vs. Elegant	35	1	1

R10	Take your skin back to a younger tomorrow
F1	Unscented
B4	Deeply cleans your pores
*P8	Non-allergenic
*C8	For year-round skin care

(c) Objectives: maximize acceptance, younger (critical), daily usage, facial, elegant

Attribute	Expected	Goal	Wt
Total Purchase	41	100	1
Old vs. Young	48	100	10
Everyday vs. Special	38	1	1
Facial vs. Body	28	1	1
Common vs. Unique	53	100	0
Ordinary vs. Elegant	33	100	1

Concept Development, Testing, Optimization

Table 22 Continued

R1	Feel invigorated
F1	Unscented
B6	Helps control excess oil
*P8	Non-allergenic
*C8	For year-round skin care

(d) Objectives: maximize acceptance, older (critical), daily usage, facial, elegant

Attribute	Expected	Goal	Wt
Total Purchase	35	100	1
Old vs. Young	19	1	10
Everyday vs. Special	41	1	1
Facial vs. Body	25	1	1
Common vs. Unique	50	100	0
Ordinary vs. Elegant	38	100	1

R10	Take your skin back to a younger tomorrow
F1	Unscented
B8	Soothes dry, cracked skin
*P8	Non-allergenic
*C8	For year-round skin care

Wt = Weight, a measure of relative importance
0 = Not relevant for determining the concept
*Denotes relevant for determining the concept

the segment differences in sensory preferences are reliable and dramatic, even on an individual-by-individual basis.

The specifics for concept response segmentation appear in Table 23A. For concepts there are two major segments of consumers based upon their response patterns. One group (Segment 1) comprises consumers who consider soap as more functional—primarily, for cleaning. These consumers respond more to concepts that deal with the function of the soap as a cleaning agent. The other group (Segment 2) comprises consumers who look at soap as part of skin repair. Although cleaning is important to them, it is the total experience which attracts them, and for which the soap is a vehicle (see Table 16).

Segmenting consumers on the basis of their response patterns provides the marketer with a tool to create better concepts. If in the population there really do exist groups of individuals who respond differently to the same message, then a single concept may not appeal to all individuals. The market may require a line of concepts. These concepts may be slight modifications of each other (changed so that some concepts appeal to one group, and other concepts appeal to the second group). Or, as is often the case, if the segments radically differ

Table 23A Sequence of Steps to Segment Consumers on the Basis of Responses to Concepts

1. Each consumer rates a set of concepts on both purchase intent and communication attributes. The communication attributes are descriptive (e.g., for male vs. female), and not evaluative. That is, the communication attributes are not restatements of liking or purchase intent.
2. On a total basis, compute the average of each concept on every communication attribute. Each concept now has a profile of average ratings, one per communication attribute.
3. On an individual-by-individual basis, plot that person's purchase intent rating for each concept (1–5 scale) on the Y axis of a graph, versus the communication value of that concept on the X axis.
4. Step 3 generates a separate graph for each individual for each communication attribute. The graph is a scattergram of points.
5. Fit a quadratic function to the data, of the form:

Purchase Intent = $k_0 + k_1$(Communication Attribute) + k_2(Communication Attribute)^2

6. Find the communication attribute level at which the person's purchase intent peaks.
7. Steps 3–6 generate a data matrix. The rows correspond to the panelists, the columns to the communication attributes. The numbers in the body of the table correspond to the optimal communication level at which the individual's rating of purchase intent reaches its highest point.
8. Factor analyze the matrix to reduce redundancy, and save the factor scores. Each consumer panelist generates a set of factor scores.
9. Cluster the panelists based upon the factor scores, using "K means clustering" (divide the population into a small set of homogeneous groups).
10. Panelists falling into the same cluster are assumed to have similar values or points of view.
11. The segmentation is independent of the consumer panelist's level of interest in the concept. The segmentation is a function of the optimal communication patterns.

from each other, then the marketer may need two entirely different, non-overlapping concepts in order to maximize the appeal to these groups and to achieve the greatest consumer interest.

That the segments differ from each other can be seen in Table 23B, which compares the two segments on a number of different concept elements. Table 23B shows that on a total panel basis some concept elements evoke moderate to high consumer interest. There are, however, other concept elements which

Concept Development, Testing, Optimization

Table 23B Concept Response Segments: Soap Elements Which Are Highly Acceptable to Each Segment

	Elements appealing to segment 1 (fresh/clean)				
No.	Element	Total	Seg1	Seg2	Diff
27	Does not contain harsh chemicals or heavy perfumes	10	16	4	12
20	A fresh, clean scent	10	15	5	10
46	Deeply cleans your pores	12	14	9	5
18	A fresh scent	9	13	5	8
28	Totally free of skin irritants	10	12	7	5
60	Leaves your skin feeling clean and fresh	2	11	−4	15
29	With pure organic ingredients	8	10	6	4
61	Leaves your skin beautifully clean, soft, and healthy	4	10	0	10
1	Feel invigorated	5	10	1	9
19	A clean scent	7	10	3	7
	Elements appealing to segment 2 (skin repair/skin healthy)				
No.	Element	Total	Seg1	Seg2	Diff
68	Minimizes dry skin, fine lines, and wrinkles	9	2	16	−14
62	Firms and tightens skin	9	2	15	−13
47	Tones and smoothes your skin	12	8	15	−7
65	Doesn't dry like most soaps	9	6	14	−8
49	Moisturizes while it cleans your skin	9	4	14	−10
67	No dry, flaky patches after use	11	8	13	−5
50	Soothes dry, cracked skin	7	2	13	−11
35	Made especially for skin that is dry to slightly dry	9	4	13	−9
63	Keeps your skin young-looking	10	8	12	−4
24	The scent of freshly cut roses	6	1	12	−11
31	With essential moisturizing ingredients	7	2	12	−10
23	A sweet scent	8	6	11	−5
25	A garden-fresh scent	5	1	10	−9
22	A soft romantic scent	7	5	10	−5

Diff = difference between the segments

truly interest one group, but not the other. If we were to only look at the total panel data we might miss some of these winning elements, because for the total panel they do not look particularly impressive. It is only when we divide the consumers into these subgroups (based upon response patterns) that the differences truly emerge.

Strategies to Optimize Concepts in the Face of Segmented Preferences

Based upon the data in Table 23B, we can develop concepts which maximize consumer purchase intent for either one segment, the other segment, or both.

If we maximize overall purchase intent, then we can estimate the likely purchase intent to be assigned by the segments. Table 24A (first concept) shows this result. Although we have maximized total purchase, we have not separately insured high acceptance from both segments, respectively. Rather, in pursuing the total panel we have let the segments "fall as they may." An alternative strategy is to optimize purchase intent for each segment. Table 24A (second and third concepts) presents these results.

Table 24B shows what happens when we design a concept that both persuades the consumer to buy, and communicates "body soap." Again, because the two segments differ we can create two concepts, one for each segment.

4. Advertising Concepts and the Effect of Graphics

The previous concepts for soap dealt with verbal elements, and intermixed both functional and positioning statements. This next case history deals with a concept for a fine fragrance. The Vaughn Marketing Company recently purchased a group of three small fragrance companies. These boutique companies were known in the industry as marketers of fine but inexpensive fragrances, sold in department stores. Under Amanda Vaughn's guidance, marketers in these three fragrance companies planned to change their image slightly to create an upscale fragrance.

The first step developed a strategy for marketing and distribution. The second step provided a brief to the fragrance houses (which supplied the actual perfume). The fragrance brief or description of the product was "general," outlining marketing's goal to penetrate the fine fragrance field. It was not clear what to do. With guidance from the fragrance house the marketers were able to home in on a basic strategy that called for an evening fragrance, romantic yet light. This type of fragrance profile was familiar to the marketers. They felt confident that they could create an upscale version of the fragrance.

A fine fragrance is not a soap. Fragrance advertising concepts are much more ethereal, far less rooted in specific benefits. Furthermore, much of the advertising concept is conveyed in subtle graphics, words and shades of color. Can the process of developing a fragrance advertising concept be systematized by concept optimization research? The research does not replace the creative process, rather, the research identifies the "hot buttons" or the strong executional elements in different test concepts. That is, research shows what works and what doesn't in terms of attracting and intriguing consumers.

Despite the fact that concepts for fine fragrances are less tangible than soap concepts, the researcher can still bring some order to the problem and uncover the elements which work. In fact, the procedure is quite similar to that followed for the soap concepts:

1. Develop elements
2. Identify restrictions

Concept Development, Testing, Optimization

Table 24 Concepts Designed Specifically for Total Panel and the Two Preference Segments:

A. Optimal Concepts for Purchase Intent

(a) Optimal concept for total panel

Attribute	Expected	Goal	Wt
Total Purchase	79	100	1
Segment 1	76	100	0
Segment 2	78	100	0

F10	A fresh, clean scent
I3	Totally free of skin irritants
U2	Made especially for skin that is dry to slightly dry
B5	Tones and smoothes your skin
C9	No dry, flaky patches after use

(b) Optimal concept for segment 1 (fresh/clean)

Attribute	Expected	Goal	Wt
Total Purchase	60	100	0
Segment 1	88	100	1
Segment 2	36	100	0

R1	Feel invigorated
I2	Does not contain harsh chemicals or heavy perfumes
B4	Deeply cleans your pores
P8	Non-allergenic
C2	Leaves your skin feeling clean and fresh

(c) Optimal concept for segment 2 (skin healthy)

Attribute	Expected	Goal	Wt
Total Purchase	70	100	0
Segment 1	46	100	0
Segment 2	93	100	1

F14	The scent of freshly cut roses
I6	With essential moisturizing ingredients
U2	Made especially for skin that is dry to slightly dry
B5	Tones and smoothes your skin
C10	Minimizes dry skin, fine lines, and wrinkles

B. Optimal Concepts for a Body Soap, as Well as for Purchase

(a) A body soap for total panel

Attribute	Expected	Goal	Wt
Total Purchase	66	100	3
Segment 1	77	100	0
Segment 2	54	100	0
Facial vs. Body	61	100	2

(continued)

Table 24 Continued

R1	Feel invigorated
I2	Does not contain harsh chemicals or heavy perfumes
U7	For body care
B5	Tones and smoothes your skin
L6	Available in a solid, hard bar that doesn't melt quickly

(b) A body soap for segment 1 (fresh/clean)

Attribute	Expected	Goal	Wt
Total Purchase	66	100	0
Segment 1	83	100	3
Segment 2	48	100	0
Facial vs. Body	57	100	2

R1	Feel invigorated
I2	Does not contain harsh chemicals or heavy perfumes
U7	For body care
B4	Deeply cleans your pores
L6	Available in a solid, hard bar that doesn't melt quickly

(c) A body soap for segment 2 (skin healthy)

Attribute	Expected	Goal	Wt
Total Purchase	65	100	0
Segment 1	48	100	0
Segment 2	83	100	3
Facial vs. Body	46	100	2

F13	A sweet scent
I6	With essential moisturizing ingredients
U7	For body care
B8	Soothes dry, cracked skin
C10	Minimizes dry skin, fine lines, and wrinkles

3. Create combinations
4. Test with consumers

The key difference here is that most of the elements are pure positioning elements. In addition, a number of the elements are graphics.

Running a concept study for fragrances is similar to running a concept study for soap. The panelists are given a set of concepts, and rate these concepts on a variety of attributes (see Table 25 for questionnaire). Again, the panelists rate overall persuasion (purchase intent) as well as "communication" (bipolar attributes, such as inexpensive vs. expensive). In the fragrance study the concepts comprise both pictures and text. The pictures are taken from "scrap art" (from

Concept Development, Testing, Optimization

Table 25 Questionnaire for Concept Evaluation—Fine Fragrance

Instructions

You will be evaluating a set of concepts today about fine fragrances—the type that you purchase in the better department stores or boutiques.

For each concept, please read the concept and look at the pictures before you assign any rating.

Once you have read the concept, please answer these five questions:

1. How interested are you in buying this fragrance?
   5 = definitely would buy
   4 = probably would buy
   3 = might/might not buy
   2 = probably not buy
   1 = definitely not buy
2. How expensive is this fragrance?
   0 = cheap → 100 = expensive
3. How frequently would you wear this fragrance?
   0 = never → 100 = daily
4. For whom is this fragrance designed?
   0 = for a younger woman → 100 = for an older woman
5. How similar is this fragrance to others on the market?
   0 = very different → 100 = very similar

---

magazines, stock pictures, etc.). Pictures are treated exactly like other elements. They are not selected to "go with" the text, although they are represented among the restrictions so that illogical combinations of text and pictures cannot occur.

The data is treated in exactly the same way as the soap concept data. Table 26 shows the additive model for the fragrance concepts. Of key interest in Table 26 is the performance of the various pictures. Pictures by themselves do not drive purchase intent. It is the *subject* and *execution* of the picture that are important (just as the content and execution of the verbal portion of the soap concept is critical, not just the fact that the concept has words). Thus, pictures by themselves do not guarantee success. In fact, some pictures detract from acceptability.

The key results from this study on concept elements for fine fragrances follow:

1. Names make a difference. In many other product categories names are irrelevant unless they have been previously supported by advertising. In other product categories new names (those developed specifically for the product) do not have much impact on purchase intent. In this regard fine fragrances differ from more functional products.

Table 26 Contribution of Elements to the Fragrance Concept

		Negative coefficient → Positive coefficient →	Purchase	Cheap vs. expensive	Never vs. daily	Younger vs. older	Unique vs. similar
Element	Text						
	Additive Constant		34	51	64	37	61
	Name						
A1	Elixir of Love		4	3	−3	2	2
A2	Charmoir		−6	5	2	−4	3
A3	Softest Night		6	6	3	−2	−2
A4	Dreamy Silk		2	2	−1	−3	3
A5	Sammy		−3	−5	4	−4	3
A6	Apassionata		4	2	−2	3	−6
A7	Sweethearts		−1	−6	3	−7	2
	Heritage						
B1	A glamorous fragrance from France		5	4	1	2	3
B2	From the finest perfumers of the world		−2	2	−4	4	−4
B3	A generation of perfume science		6	−3	0	4	−6
B4	A unique perfume, just for you		2	3	1	−3	4
B5	The purest blend of the finest fragrance oils		−5	0	−2	−1	3

Concept Development, Testing, Optimization

	Description Of Fragrance				
C1	Sweet yet very romantic	2	1	3	-3	3
C2	Soft, scented, spicy	3	-2	3	2	4
C3	A flower garden	5	-3	4	-2	5
C4	The freshest flowers and spices	4	-5	2	-3	4
C5	Sweet, light, heavenly	-2	-2	2	-2	2
C6	A blend of woods and spices and flowers too	-4	-3	-3	-1	0
	Usage				
D1	For your most romantic moments	4	-3	4	-2	4
D2	Designed especially for you	-5	-5	2	-1	5
D3	Wear it when you want to feel special	-2	-2	-2	2	-2
D4	For romantic moments, or just for every day	3	-4	3	-3	3
D5	It becomes you for evening	5	2	2	4	-3
	Picture/Visual				
E1	A young man kissing a young woman	2	-3	2	-2	2
E2	A smiling young woman at a store	5	2	-1	1	3
E3	A romantic evening meal for two	-3	1	1	3	-2
E4	A woman, the pyramids, a camel and moonlight	4	5	0	3	-4
E5	A picture of the Caribbean in the evening	2	3	-2	3	-3
E6	A man and woman in a Rolls Royce convertible	6	9	-3	5	-7

2. Pictures differ in their ability to promote purchase intent. There is no single rule governing which particular pictures win or lose. However, there is a clear range of contributions for pictures.
3. There are differences in communication between the elements, but the differences are not large. Consumers do differentiate the various elements within the framework of the concept. Some elements better fit an expensive fragrance than other elements (e.g., "a glamorous fragrance from France"). Furthermore, some elements are clearly more youth-oriented and others have a more mature orientation.

Finally, the optimization exercise can lead to improved concepts, as shown in Table 27. This exercise generates different concepts based upon the marketing objectives (e.g., maximize overall purchase intent, maximize purchase intent but insure that the concept has an image more appropriate for inexpensive and younger, rather than for older consumers, etc.).

5. Creating Product Concepts for R&D

Concepts can be aimed strictly towards the product developer in order to give the developer a direction to follow. To illustrate a "product concept" let us consider a hand lotion.

The Nesbitt Corporation plans to develop a line of hand lotions. Nesbitt had been in the category several years before, and had marketed a line of successful lotions with the added benefits of specific ingredients (e.g., aloe vera). Over the years, however, Nesbitt's share of the lotion market steadily declined due to a combination of product problems, marketing inefficiencies, poor advertising, and a revolving door of product managers, each of whom changed the overall lotion strategy until the strategy simply fell apart. In 1992, however, top management at Nesbitt decided to review its poor performance in the hand lotion category and improve it. Management recognized that with an improved line of lotions Nesbitt might recapture a respectable share of the market despite the widespread competition that was emerging from multinational corporations.

The immediate problem was to design the line of lotions with the property that the lotions would be attractive to consumers on a conceptual basis. Once the appropriate line of items was defined, the product developers could then develop the optimal products.

Here the focus lies on developing "product concepts" or specifications. The concept is not a technical specification, but rather a consumer statement of what functional properties consumers would like to see in the product (in non-technical consumer terminology). That is, what particular features should the lotion have? Should the lotion be light or heavy, colored with a pearlized appearance

Table 27 Optimal Concepts for Fine Fragrances with Different Marketing Objectives

Maximize total panel

Attribute	Expected	Goal	Wt
Total Purchase	61	100	1
Cheap vs. Expensive	55	100	0
Younger vs. Older	42	100	0

A3 Softest Night
B3 A generation of perfume science
C3 A flower garden
D5 It becomes you for evening
E2 A smiling young woman at a store

Total panel, yet expensive

Attribute	Expected	Goal	Wt
Total Purchase	56	100	1
Cheap vs. Expensive	69	100	1
Younger vs. Older	41	100	0

A3 Softest Night
B1 A glamorous fragrance from France
C1 Sweet yet very romantic
D5 It becomes you for evening
E4 A woman, the pyramids, a camel and moonlight

Total panel—younger, inexpensive image

Attribute	Expected	Goal	Wt
Total Purchase	40	100	1
Cheap vs. Expensive	29	1	1
Younger vs. Older	28	100	1

A7 Sweethearts
B3 A generation of perfume science
C4 The freshest flowers and spices
D2 Designed especially for you
E1 A young man kissing a young woman

---

or simply white? What type of fragrance should the lotion have or should it be unscented?

Typically, product concept studies comprise substantially fewer elements than do advertising concepts. Product concept studies are significantly simpler because a product can have relatively few physical factors, in contrast to the many ways that a product can be "romanced" in advertising. The concept study for lotion involved two aspects of the product: aesthetic or sensory aspects of the product itself (including the package) and the end benefits of the product (what the lotion does for the skin in terms of smoothing, making it softer, etc.). Product concepts designed for development do not "romance" features. The objective is to guide development, not to persuade consumers to purchase the product by force of positioning. This approach develops one concept or set of concepts which describes the product most likely to be purchased by a consumer. The product development group then uses the information to guide the creation of the actual lotion product.

The concept study for product design was executed in the same way as that for fragrance and soap concepts. Since some of the elements involved packaging, the issue arose as to how to present information about the container and the nozzle, respectively. Some individuals at Nesbitt wanted to describe the packaging verbally. Others wanted to present the packaging information graphically, as pictures (actual representations of what the package would look like). This second approach won out.

The generation of elements was straightforward. R&D presented a variety of alternative elements to the marketing and marketing research groups. After a two-hour ideation session, the group emerged with 140 different elements, including pictures, descriptions of product characteristics, ingredients, etc. Many of these elements were unfamiliar to consumers, who would have to be the ultimate judge of the concepts. The group culled the elements down to a manageable list of 34, including pictures of caps and containers, respectively. The group also provided a list of restrictions regarding elements that could not go together in a concept, in this case primarily because the elements represented incompatible ingredients or incompatible end results. Table 28 shows the elements.

The study was conducted among 120 consumers who said that they would be interested in purchasing a hand and body lotion. These were women ages 20–64, representing a wide distribution of potential consumers. Each consumer rated 45 randomly selected concepts from the full set of 68 combinations. Each concept was rated on purchase intent and 3 image attributes; namely, functional use, masculine vs. feminine, and younger vs. older. The study was conducted in three markets across the U.S. (New York; Dallas; Portland, Oregon).

The analysis of the data showed substantial discrimination between the elements in terms of consumer acceptance. Table 28 shows the part-worth contributions of the various elements to purchase intent. Additionally, the consumers fell into two segments. The segmentation suggested two clear groups. One group (Segment 1) was "sensory" oriented. They reacted strongly to color and to the statement about the container and color. They were indifferent to the end benefits. What is interesting about this group is that they appeared to relish the visual

Table 28 Additive Model for an R&D-Oriented Hand Lotion Concept

Code		Total	Sensory seg1	Benefit seg2	Cosmetic(−) functional(+)	Masculine(−) feminine(+)	Younger(−) older(+)
	Additive constant	41	36	45	42	51	43
	Color						
A1	White	2	2	2	3	0	5
A2	Pink	−3	6	−15	−4	4	5
A3	Pink pearlized	−6	7	−15	−7	7	7
A4	Light blue	3	5	2	−2	−3	−3
A5	Light blue pearlized	−2	6	−8	−5	1	−4
A6	Light green	−3	3	−7	2	−4	−3
A7	Light green pearlized	−6	6	−15	−3	0	−6
A8	Ivory	4	2	5	3	2	3
A9	Cream	6	4	7	2	2	4
	Fragrance type						
B1	Sweetish	−2	−1	−3	5	4	6
B2	Flowery	3	4	2	3	6	−3
B3	Powdery	5	2	7	−3	7	0
B4	Spicy/sweet	−6	3	−13	−3	2	6
	Skin feel						
C1	Thin, oily	−2	−1	−3	−2	−3	−3
C2	Thick, oily	−6	−4	−7	4	4	4
C3	Thin, creamy	4	5	3	−4	−1	−2
C4	Thick, creamy	6	3	8	2	6	4

(continued)

Table 28 Additive Model for an R&D-Oriented Hand Lotion Concept

Code		Total	Sensory seg1	Benefit seg2	Cosmetic(−) functional(+)	Masculine(−) feminine(+)	Younger(−) older(+)
	Container (from picture)						
D1	Long cylinder	−2	1	−4	5	−3	1
D2	Short cylinder	−4	2	−8	2	−1	−2
D3	Long oval bottle	3	6	1	−2	2	3
D4	Short oval bottle	3	7	0	−3	3	−1
	Cap (shown in picture)						
E1	Type 1	2	1	3	1	1	1
E2	Type 2	−1	−2	0	0	0	0
E3	Type 3	2	−2	5	1	0	1
	Color of packaging						
F1	White/oval label	4	3	5	2	3	4
F2	White/rectangular label	2	4	1	5	−4	−1
F3	Pink/oval label	−2	1	−4	−2	6	3
F4	Pink/rectangular label	−2	4	−6	1	3	−2
F5	Light blue/oval label	2	2	2	3	−3	4
F6	Light blue/rectangular label	3	3	3	4	−5	−3
	End benefit						
G1	Restores skin moisture	3	1	4	2	6	5
G2	Softens skin	4	1	6	−4	6	4
G3	Use it for daily skin care	2	−1	4	4	−1	2
G4	Use it before and after kitchen work	4	−2	8	−6	5	−1

Concept Development, Testing, Optimization 61

experience of the product. They did not seem to be interested in the end benefits. This segment comprised 42% of the consumers.

The remaining 58% of consumers fell into Segment 2. These consumers also discriminated between the different colors, packages, etc. Consumers in Segment 2 were more interested in the end benefits. They were most "turned on" by the ability of the product to help the skin. This group can be labeled "benefit oriented." According to the demographics, they were older and tended to have dry or problem skin.

Creating Optimal Concepts for the Total Panel and Segments

The element model in Table 28 enables the developer to create concepts for the total panel and the two segments. Table 29 shows these concepts. Keep in mind that the concepts created are product concepts—actual descriptions of the product from the consumer point of view. Table 29 shows seven concepts, as follows:

1. Optimal concepts for total panel (A) and for each of the two segments separately (B, C).
2. Optimal concepts attempting to satisfy the two segments simultaneously with a single product (D). These two segments of consumers want different sensory characteristics in their product. Consequently, what appeals to one segment will not necessarily appeal to the other, as Table 29 (B, C) shows. If we require that both segments be at least minimally satisfied with the single product, then we can create a compromise product. It does not optimize either group, but it creates a product not adverse to either segment. (It is not clear whether this is a good strategy or not. It is, however, a viable strategy if the marketing objective is to go after a single product that appeals to multiple consumer segments).
3. A line of products designed to be perceived as ultimately cosmetic-oriented, with the products differing from each other (E, F, G). These products are designed to appeal to the total panel, with an accent on communicating for "younger" versus "older" images.

6. Creating And Optimizing Concepts In "Real Time"

Up to now we have been dealing with paper and pencil research. Consumers are presented with stimuli on white sheets of paper, or on concept boards, and rate their impressions. The data is analyzed after the field work is finished, and the result is presented some time later.

Is there a way to accelerate the process so that the experimental design, concept evaluation, analysis and report are done more quickly, or even instantaneously? If so, then the marketer may develop and test concepts at a rate

Table 29 Set of Product Features for a Lotion (Product Concept)

	(a) Total panel		
Attribute	Expected	Goal	Wt
Total	67	100	1
Segment 1–Sensory	55	100	0
Segment 2–Benefit	76	100	0
Cosmetic vs. Functional	44	100	0
Younger vs. Older	70	100	0

A9 Cream
B3 Powdery
C4 Thick, Creamy
D3 Long Oval Bottle
E1 Cap Type 1
F1 White/Oval Label

	(b) Segment 1—sensory oriented		
Attribute	Expected	Goal	Wt
Total	49	100	0
Segment 1–Sensory	64	100	1
Segment 2–Benefit	39	100	0
Cosmetic vs. Functional	37	100	0
Younger vs. Older	55	100	0

A3 Pink Pearlized
B2 Flowery
C3 Thin, Creamy
D4 Short Oval Bottle
E1 Cap Type 1
F2 White/Rectangular Label

	(c) Segment 2—benefit oriented		
Attribute	Expected	Goal	Wt
Total	67	100	0
Segment 1–Sensory	52	100	0
Segment 2–Benefit	78	100	1
Cosmetic vs. Functional	44	100	0
Younger vs. Older	70	100	0

A9 Cream
B3 Powdery
C4 Thick, Creamy
D3 Long Oval Bottle
E3 Cap Type 3
F1 White/Oval Label

Concept Development, Testing, Optimization

Table 29 Continued

	(d) Segments 1 and 2 simultaneously		
Attribute	Expected	Goal	Wt
Total	63	100	0
Segment 1–Sensory	60	100	4
Segment 2–Benefit	65	100	1
Cosmetic vs. Functional	43	100	0
Younger vs. Older	57	100	0

A9 Cream
B2 Flowery
C3 Thin, Creamy
D4 Short Oval Bottle
E1 Cap Type 1
F1 White/Oval Label

	(e) Ultimately cosmetic		
Attribute	Expected	Goal	Wt
Total	44	100	0
Segment 1–Sensory	56	100	0
Segment 2–Benefit	36	100	0
Cosmetic vs. Functional	23	1	1
Younger vs. Older	61	100	0

A3 Pink Pearlized
B3 Powdery
C3 Thin, Creamy
D4 Short Oval Bottle
E2 Cap Type 2
F3 Pink/Oval Label

	(f) Ultimately cosmetic, older		
Attribute	Expected	Goal	Wt
Total	33	100	0
Segment 1–Sensory	57	100	0
Segment 2–Benefit	16	100	0
Cosmetic vs. Functional	23	1	1
Younger vs. Older	67	100	1

A3 Pink Pearlized
B4 Spicy/Sweet
C3 Thin, Creamy
D4 Short Oval Bottle
E2 Cap Type 2
F3 Pink/Oval Label

(continued)

Table 29 Continued

Attribute	(g) Ultimately cosmetic, younger		
	Expected	Goal	Wt
Total	48	100	0
Segment 1-Sensory	58	100	0
Segment 2-Benefit	41	100	0
Cosmetic vs. Functional	28	1	1
Younger vs. Older	45	1	1
A5 Light Blue Pearlized			
B3 Powdery			
C3 Thin, Creamy			
D4 Short Oval Bottle			
E2 Cap Type 2			
F4 Pink/Rectangular Label			

hitherto only dreamed of. The speed of data acquisition and analysis also reduces the risk because the marketer and the product developer can test alternative hypotheses in "real time" with consumers on an ongoing basis, until the concept is "just right."

Current "real time" research is done in focus groups. The moderator presents the concept to the consumers who react to one or several test concepts. The consumers in the group discuss their feelings about what they see and read. The marketer and marketing researcher are separated from consumers by an intervening "one way mirror." The corporate representatives can see and hear what the consumers are doing, how consumers react to the concepts (or products), as well as watch facial and body expressions. The company representatives form their opinion of what drives the consumer reactions based upon what they see and hear.

What is missing from this "real time" research is a way to quantify consumer responses (although this can be done in focus groups), followed by an instantaneous data analysis, and the ability to instantaneously reconfigure a new concept based upon the consumer's reactions.

The author has developed a "real time" concept development and optimization procedure known as "IdeaMap"™ (Moskowitz, 1994). IdeaMap enables the researcher to investigate many different concept options, with these options varied by experimental design. Consumers react to these concepts and to "static concepts" comprising a fixed set of elements. (Static concepts may be derived from the advertisements of competitors which measure competitive performance.)

Concept Development, Testing, Optimization

The case history for IdeaMap involves a sanitary napkin. The manufacturer wanted to identify the hot buttons for a sanitary napkin concept. Additionally, the manufacturer wanted to determine how the competitor's positioning elements fared. It was vital to measure the performance of the competitive elements to see whether competitors were using any "magic bullets" that attracted consumers. Finally, in order to gain additional insights, the manufacturer wanted to create concepts comprising winning elements on an individual-by-individual basis at the time of the evaluation.

Setting Up an Interactive Concept Development System

In order to create an interactive concept system we must satisfy the following four prerequisites:

Individualized Design. The stimuli for each panelist must vary along a self-contained experimental design. Since each consumer can only evaluate a limited number of concepts in a test session, the experimental design for each individual must be relatively small. For instance, each panelist may evaluate 20 elements combined in 25 combinations. (Compare this to more conventional designs. Conventional concept optimization uses the larger experimental designs. There may be 300 elements combined into 600 combinations. Each consumer evaluates a randomized subset of concepts, and assesses only a small portion of the set of concepts. Therefore, each consumer never sees a full experimental design, but only a portion of it.)

Interactive Stimulus Presentation and Data Acquisition. The concepts must be presented rapidly, and the data acquired quickly. This speed requirement necessitates a computer interview. With a computer the task of presenting concepts and acquiring data can be accomplished rapidly, without having the panelist go through page after page of concepts, rating each by paper and pencil. Furthermore, the computer can present pictures along with words.

Individualized Data Analysis. The data must be analyzable for a single individual in a relatively short time (e.g., 30 seconds or less) in order to develop and use the model for that individual consumer (while the consumer is still at the test session). The individual regression model is easy to compute in "real time," given current powerful microcomputers, with regression packages and fast "mathematical co-processors."

Real Time Concept Optimization. During the postevaluation "discussion phase" the researcher must be able to redesign concepts "on the fly" using the individual's data (or any data set previously developed). That is, the concept optimizer must be able to create actual new concepts in "real time" (including pictures), present them to the consumers, and solicit their reactions. These are not test concepts *per se* (which have just been presented and analyzed), but rather new concepts (combinations of existing elements) based upon the data just obtained. In this way the researcher can present newly restructured concepts to the consumer and discuss the consumer's reaction to these concepts.

The IdeaMap System™

IdeaMap enables the researcher to accomplish the foregoing objectives. IdeaMap is an outgrowth of the concept optimization technology. Here are its 12 major components:

1. *Experimental Design*: Each consumer is presented with concepts created by a small experimental design (screening design—see Table 30). The screening design in Table 30 is efficient because it combines many elements into a relatively small number of combinations. Since each consumer evaluates 25 concepts (comprising 20 elements arranged in different combinations), a given set of concepts for a consumer can be analyzed and modeled for that consumer alone. We do not need a large number of consumers to obtain data for analysis. One consumer suffices because the design is tailored to that consumer.

2. *Variation of Categories and Elements Across Consumers*: The categories and elements in the design vary from consumer to consumer. That is, if there are ten categories for a product, then each run or iteration of 25 concepts (as shown in Table 30) corresponds to a randomized set of five categories, comprising four elements per category. If there are 20 consumers, each consumer will be presented with a unique set of five categories (from the set of ten), and a unique set of four elements for each category. (There will of course be overlap, since there are only ten categories. However, for each consumer, IdeaMap chooses the five categories anew.) The major constraint in the design is that all elements must eventually appear approximately equally across the full set of panelists.

3. *Limited Rating Scales*: Each consumer only assigns one rating per concept, namely, acceptance (see below). However, it will be important for subsequent analyses to determine the "tonality" of the elements (i.e., their connotation or meaning). In order to do so, prior to the evaluation, a small separate group of matched consumers (n = 5 to 15) rates each element on a variety of bipolar non-evaluative attributes (a semantic profile scale). The scale is –9 to +9 (–9 = all of one characteristic → +9 = all of the other characteristics). For instance, one semantic differential scale is "older → younger." A –9 is "only for older," a +9 is "only for younger." (The separate panel of consumers rates each ele-

Table 30 Experimental Design for IdeaMap (5-Level Plackett Burman Screening Design)

Concept	\multicolumn{5}{c}{Category}				
	A	B	C	D	E
1	4	1	3	1	1
2	0	4	1	3	1
3	3	0	4	1	3
4	3	3	0	4	1
5	2	3	3	0	4
6	3	2	3	3	0
7	4	3	2	3	3
8	1	4	3	2	3
9	2	1	4	3	2
10	2	2	1	4	3
11	0	2	2	1	4
12	2	0	2	2	1
13	4	2	0	2	2
14	3	4	2	0	2
15	0	3	4	2	0
16	0	0	3	4	2
17	1	0	0	3	4
18	0	1	0	0	3
19	4	0	1	0	0
20	2	4	0	1	0
21	1	2	4	0	1
22	1	1	2	4	0
23	3	1	1	2	4
24	1	3	1	1	2
25	4	4	4	4	4

Note: Each number in the body of the table corresponds to an element. The "0" element is the "null element" (i.e., element from that category is not present in that concept).

ment in advance of the study on the semantic differential scale, so each element has a location or value on the "older → younger" dimension.)

For the sanitary napkin, a small matched group of panelists rated each napkin element on the following five bipolar semantic scales:
1. older (−) → younger (+)
2. traditional (−) → contemporary (+)
3. bulky (−) → comfortable (+)
4. risky (−) → safe (+)
5. ordinary (−) → unusual (+)

4. *Multiple Iterations to Create a Larger Design*: Each consumer may go through several "iterations" or "runs" of 25 concepts each. A given individual may go through as many as five runs. Across the five runs the consumers will see each of the categories several times. A single element may appear in several designs for one panelist.
5. *Evaluation of Many Different Test Concepts*: The system is sufficiently randomized so that the odds are that each consumer will see totally different combinations, with very few combinations repeated. The randomization insures that IdeaMap will not be subjected to biases engendered by testing only a limited number of concept combinations. For 100 panelists, each of whom rates 100 combinations (four iterations of 25 concepts each), the total number of concepts is 10,000 (100 × 100). For reasonably large numbers of elements and categories, it is highly unlikely that there will be any repeated concepts, unless there is a "built-in" combination that all panelists evaluate.
6. *Any Type of Scale Accepted*: The consumer can use virtually any type of rating scale. The simplest scales are one of three types: accept (coded in the computer as the value 100), or reject (coded in the computer as the value 0); degree of liking (e.g., on a 9-point scale, from 1 = hate → 9 = love); or purchase intent (e.g., 1 = definitely not purchase → 5 = definitely would purchase).
7. *Iteration*: Each iteration lasts approximately 5 minutes during which time the consumer evaluates 25 combinations. Thus, the iteration is fairly short.
8. *Regression Analysis*: IdeaMap uses "dummy variable" regression to develop the model. Each panelist generates his own unique model. The independent variables correspond to the elements, and either take on the value 0 (if the element is absent) or 1 (if the element is present). The dependent variable is the individual consumer's response to the concept.
9. *Estimating Responses to Untested Elements*: A single consumer cannot evaluate all elements during a 30-minute interview, especially when there are 150 elements or more. Consequently, there must be a way to estimate how the consumer would have responded to elements that were not evaluated. Table 31 presents the "smoothing" algorithm which shows how to estimate an individual consumer's additive value for elements not tested, given elements that were tested. The approach is based upon finite element analysis and computational grids. The greater the number of elements tested, the fewer the number of untested elements whose additive value the algorithm must estimate.
10. *Individual Model*: On an individual-by-individual basis, IdeaMap creates a model showing the part-worth contribution of that element to consumer purchase intent.

Concept Development, Testing, Optimization

Table 31 Algorithm for Estimating Individual "Part-Worth" Contributions for Elements Not Tested by that Consumer

1. Locate all elements in a semantic space.
2. Find the eight "nearest neighbors" of each element.
3. For each element tested, "flag" the element to show that it was tested.
4. Starting with the first element, determine whether the element was tested.
   • If yes, skip to step 6
   • Otherwise go to step 5
5. Compute the average of the nearest neighbors for this element (that was not tested), and replace the value of the element with the average of the eight nearest neighbors. Proceed to step 6.
6. Go to the next element on the list.

Note: Continue to perform steps 4–6 again and again, until there is no change in the "replacement values" of the untested elements.

11. *Aggregate Data*: The data from a group of consumers can then be combined to create a total panel model or subgroup model (by averaging together the part-worths or contributions of the elements across panelists).
12. *Optimize Concepts*: The Concept Optimizer creates new concepts for the individual (or for a group of individuals) based upon that individual's additive model. The Concept Optimizer has available to it both the additive values for purchase intent (from the individual or group of consumers) and communication values (from the estimates created by a smaller group of consumers, and discussed in Step 3 above).

7. Putting IdeaMap into Action with Sanitary Napkins

The Lindberg Company had been looking into the sanitary napkin market for some years as a potential category for growth. Lindberg Inc. is a small midwestern U.S. company with a limited budget for consumer research. Any market introduction by Lindberg must have a high degree of success. Fortunately, management at Lindberg is research-oriented. For the past ten years they have conducted focus groups as a way to get consumer reactions to new ideas, and to determine what consumers want in the way of new products. For the most part management has been happy with the performance of their focus groups. Since the marketing objectives of Lindberg were not particularly aggressive, and their marketing and financial expenditures were low, the focus group approach was entirely appropriate.

Recently, however, with their new focus on sanitary napkins, marketing recognized that they were about to enter a significantly more competitive category than the one in which they had previously competed. Most of their previous products had been niche products, with relatively little competition. The potential for the company in the sanitary napkin category was significantly greater than it was for any of the previous categories. The only problem was finding a positioning, or point of difference, either in product features or in communication of benefits to the consumer.

The director of marketing research suggested using IdeaMap for the following four reasons:

1. The conventional focus group approach could provide elements for the concepts, but could not easily develop concepts.
2. The marketing consultants used by Lindberg were at a loss for new product or positioning ideas in the sanitary napkin category. These consultants came from large-scale companies with sales in the billions of dollars, and were not used to focusing on niche products.
3. An ideation session held with R&D, marketing, marketing research, and a group of consumers suggested a large number of elements to use. It was important to screen these elements in order to discover the true "hot buttons."
4. Lindberg was a small company and could not afford the very large projects that would have been needed to truly cover all fronts in the category. Cost of research was a significant issue. IdeaMap was very cost effective.

These considerations dictated using the IdeaMap approach with 75 consumers. Marketing felt that with 75 consumers they could obtain the necessary information to help them design the product. They recognized that the base size of 75 was not nationally representative, but was still acceptable in terms of stability of data.

Basic Elements

Table 32 shows a partial list of the sanitary napkin elements, and their location on the set of five semantic differential scales (done by a small consumer panel before the actual evaluation). Marketing wanted to satisfy three objectives by choosing the elements:

1. Determine the relative strength of a competitor's advertising positions. Consequently, some of the elements in the set comprised selling messages used by competitors.
2. Discover consumer reaction to various features (presented in a "low tech" fashion).
3. Discover the consumer reaction to various realistic benefits.

Concept Development, Testing, Optimization

Table 32 Some of the Elements for Sanitary Napkins and their Ratings on Semantic Scales

Elements	Dimensions				
	A	B	C	D	E
Feel as feminine and pretty as you are	0	-6	8	0	4
As unmistakably feminine as you are	4	-6	8	0	4
We understand your need to feel secure	0	-4	0	-2	-8
When you want that feeling of confidence	0	0	0	-2	-8
Trust what works for you	0	0	0	8	2
Nature made you a woman . . . we make it easier	-8	-4	8	0	2
We understand a woman's needs	-8	0	8	2	-8
Confidence is priceless	0	0	0	8	8
So dependable, even a teenager will feel more confident!	8	4	4	8	4
Feel cleaner, drier, better	0	0	0	2	-8
Feel confident during your busy day	0	0	0	4	-2
You have plenty to worry about—let us ease things for you	0	0	0	4	2
"Those days" don't have to be days of dread!	4	-4	0	4	8
Absorbs your worries	0	0	0	4	8
With a million things on your mind—this doesn't have to be one of them	-4	0	0	8	8
All you need to feel secure	0	0	0	8	0
Meets your protection needs best	0	0	-4	6	-8
Designed to stop accidents before they start	0	0	0	8	8
Gives added protection	0	4	-4	4	-8

Codes For Dimensions - Sanitary Napkin
A. Older (-) → Younger (+)
B. Traditional (-) → Contemporary (+)
C. Bulky (-) → Comfortable (+)
D. Risky (-) → Safe (+)
E. Ordinary (-) → Unusual (+)

Restriction and Concept Size

IdeaMap works in a slightly different manner than conventional concept optimization. Both use pairwise constraints (create concepts, making sure that elements which can go together, do go together). However, with IdeaMap no concept can have more than five elements, nor fewer than two elements. (See Table 30 for the schematic design of concepts.) For conventional concept optimization, in contrast, the number of elements in a concept is, theoretically, unrestricted although in practice no concepts have more than seven or eight elements.

The Actual Stimulus Set

Each consumer evaluated three sets of 25 concepts each during a 25-minute test session. Each set of 25 concepts followed the same schematic experimental design shown in Table 30. For each of the three sets of 25 concepts (iterations), the computer selected a new set of five categories. The probability of a category appearing in the second iteration was a function of the number of elements in that category (more elements in a category increased the probability of choosing the category), and whether or not the category had just been chosen in the previous iteration (thus decreasing the probability of choosing the category again).

The concepts appeared on the color computer screen. If there was a picture connected with the concept, then the picture appeared at the top of the concept, much as it might appear in a printed advertisement. The consumer read the concept, looked at the picture, and then rated her interest in purchasing the product on a 9-point interest scale (1 = definitely would not purchase → 9 = definitely would purchase). The computer program recorded the panelist's rating, as well as the speed with which the panelist rated the concept. (Speed was defined as tenths of seconds elapsed from the time the concept appeared on the screen to the time the panelist rated the concept.) Table 33 shows the data from three concepts rated by one panelist.

Analysis of Data from Each Individual

The stimuli for each panelist comprise three experimental designs, with the elements statistically independent of each other. The Plackett Burman screening design assures this statistical independence. The data matrix for each panelist comprises 75 rows (one per test concept) and up to 60 columns (one column for each element, assuming no repeated elements). The dependent variable comprises both the rating of acceptance and the speed of reaction. If the panelist rated the concept as 7, 8, or 9, then the dependent variable "interest" is coded as 100, whereas if the panelist rated the concept as 1-6, then the dependent variable is coded as 0. Speed is a continuous variable equal to the number of tenths of seconds elapsing between the appearance of the concept on the computer screen and the panelist's rating.

After the individual model was created for the empirical set of elements tested, the smoothing algorithm estimated the likely ratings to be assigned to elements not tested (see Table 31 for the algorithm).

Each panelist evaluated different subsets of elements. Therefore, from panelist to panelist the smoothing algorithm was used to "fill in" a different set of elements.

Table 33 Concepts for One Panelist Concept, Rating of Interest, "Speed of Response"

Run # 1 Concept # 1

P6 The ultimate in feminine protection
S3 Available in a super-maxi size
C1 A style shaped to fit YOUR shape
F5 Feel fresh in summer's heat
V4 Time exposure (visual)
Rating = 7   Speed = 98 (9.8 seconds)

Run # 1 Concept # 2

P5 Stay cleaner and drier
S4 Available in a super-thin size
C2 Protection that moves with you
F5 Feel fresh in summer's heat
Rating = 2   Speed = 87

Run # 1 Concept # 3

P2 Designed to stop accidents before they start
S2 Available in mini-pads for light days
C4 Wider adhesive strips prevent bunching and twisting
F5 Feel fresh in summer's heat
V5 Mother/daughter (visual)
Rating = 6   Speed = 83

Speed = tenths of seconds

Combining the Data and Developing a Group Model

The individual data showed a variety of different patterns for sanitary napkins. As one might expect in any cosmetic or health and beauty aid category, individuals differ in what they like. It is generally not productive to look at the data from one individual, but rather to average the data from many individuals. Table 34 shows part of the "average" model for the total panel. This average or group model is simply the average of the component models.

8. Concept Development Incorporating Video— Applying IdeaMap to Package Design

The previous sections dealt with the use of pictures and words in concept development. In some cases (especially packaging), person-to-product interaction

Table 34 Part-worth Contributions of Elements for Sanitary Napkins—Results of Full Panel of 75 Consumers on a Subset of Elements (Results Ranked by Interest)

Interest scale	Speed*	Element text
51	56	Additive constant
10	−2	Neutralizes odor-causing bacteria
9	−2	Small enough to carry discreetly in your purse
9	31	When you need extra hours of protection
8	−21	Finally a product that delivers on its promises
8	10	The cleaner, drier way to a fresh day
8	22	Extra long for extra protection
8	5	Yoga position (visual)
8	23	Environmentally friendly materials
0	50	Available in a super-thin size
0	13	Extra absorbency
0	−24	Mother/daughter (visual)
−1	13	Pads (visual)
−1	31	Personalized choice for every woman
−1	22	Lets you feel fresh and safe
−1	−11	We understand a woman's needs
−1	−14	In shapes curved like you are
−1	−25	All you need to feel secure
−1	−20	Feel confident during your busy day
−1	20	With a million things on your mind—this doesn't have to be one of them
−1	13	Blond at mirror (visual)
−1	12	Nature made you a woman...we make it easier
−2	32	Available in ultra-thin size
−2	−22	Feel cleaner, drier, better
−6	13	When you don't want anything to get in your way
−7	−25	Six styles make it a matter of your personal choice
−7	−14	Be as active as you want to be
−8	22	Compact and contoured
−9	14	Prevents odor
−17	23	Protection that moves with you

* Tenths of second (e.g., Total speed = Additive constant + Contribution of each element)

is critical. We can apply the IdeaMap approach, this time however incorporating video graphics instead of static, unmoving pictures. The approach is straightforward, and follows the IdeaMap procedure. The only difference is that the concepts now comprise words and either video clips or static pictures.

Concept Development, Testing, Optimization

Specifics of the Case History—"Baby Wipes"

The case history illustrating the principle is for a "baby wipe" or "pre-moistened towelette." This product is a well-known item both for babies and for general use by adults. In this particular study a panel of 52 consumers each rated different sets of concepts. The concepts comprised different statements about the "baby wipe." Many concepts contained visuals. Some visuals were static pictures, others were video clips demonstrating a specific action involving the user and the product. The objective was to identify the optimal way to communicate different benefits (e.g., easy to use, effective in doing the job), as well as the contribution of different visuals to the communication.

Each panelist evaluated 75 different combinations, rating every combination on four attributes (purchase intent, uniqueness, effective in doing the job, easy to use). These represent four different attributes which cover different aspects of the product and of the use experience.

Panelist Impressions and Resulting Model

Panelists reported that the task was considerably more "fun" to do, and that the pictures actually brought to life the features of the product. More importantly, however, the panelists felt that the video clips were significant improvements over the pictures because with the video clips the panelists could actually perceive the interaction of the product and the person (albeit on a conceptual or descriptive basis). Panelists reported that the video clips added a sense of reality to an otherwise less-than-real description.

Table 35 shows the winning versus losing elements for the seven different categories. The "interest" value is the percentage of consumers who, having seen the element in a concept, rate the concept as 7–9 on a 9-point scale. Thus, for the first category (opening/closing), the element "The snaps are easy to close and open, even when the baby doesn't hold still" has an interest value of 4. This means that when consumers saw that element in a concept, an additional 4% of the consumers said that they would buy the baby wipe product. These utilities are the part-worth contributions to purchase intent that we have seen previously, and are akin to the measure "% top two box," but done on an individual-by-individual basis. The "interest" is the conditional probability of choosing to buy the baby wipe product, given the presence of the element in the concept.

Visuals vs. Video Clips

A key result from this project is the difference in the consumer responses to different visual representations of the same product. Table 36 shows the different scores for interest, uniqueness, effectiveness and ease-of-use. The table shows

Table 35 Winning and Losing Elements—IdeaMap Applied to Packages of "Baby Wipes"

Category	Element	Interest
Opening/Closing	The snaps are easy to close and open, even when the baby doesn't hold still	4
	Opening these wipes is the easy part	−5
Threading	You're in control of how big or small a wipe you get	3
	Now you decide what size wipe you need	−7
Baby's Comfort	Added aloe makes them so soothing	7
	Moisturizes baby's skin	−3
Sensory Attributes	Soft and moist	2
	Smells like a baby *should* smell	−4
Uses	Keep some in the car for the inevitable	4
	One of Mommy's essentials	−4
Disposability	Biodegradable because we care about your child's future too	1
	Gentle for baby *and* Mother Nature	−4
Packaging	Easy to remove with one hand	4
	Better packaging—for better moisture retention and product sterility	−2
Video Clips	Opening yellow box, and showing towelettes	6
	Opening green lid, taking towelette out	−8

Table 36 How Different Visuals Contribute to Consumer Acceptance and to Communication of End Benefits (All Products Show a Yellow Container Containing Baby Wipes)

Visual	Interest	Unique	Easy	Effective
V13 Showing plastic, plain yellow box	4	−3	0	8
V14 Opening yellow box and showing towelettes	6	2	−2	5
V16 Opening yellow box and showing aluminum seal	2	−2	−1	
V1 Opening and removing top of "snap" style, yellow container with aluminum seal	0	−4	−6	3
V10 Removing cardboard wrap and cellophane from yellow plastic box	1	0	−7	6
V11 Removing cellophane and opening yellow plastic box	−1	−2	−1	1
V15 Removing towelette from yellow box and shutting lid	5	4	−3	6
V18 Removing towelette from yellow box, wiping hands and shutting lid	−2	−3	−7	2
V12 Removing towelette from yellow box and wiping hands	−3	6	−3	4
V17 Tearing aluminum seal and showing towelettes (yellow box)	1	2	−3	18
V2 Tearing aluminum seal from yellow "snap" style container	−6	−1	−3	−4

that adding motion to the package design modifies the consumer's impression of the baby wipe product. Furthermore, the added motion does not necessarily increase purchase interest. It may actually decrease purchase interest, meaning that concepts may promise (or infer) more than they can actually deliver, when the concept information is presented as static information. With video representation of the product, the consumer has new information which more realistically represents the product.

An Overview—Video in Concepts

As researchers seek to develop ever better concepts, and as they try to reduce the risks involved in new product development, we may expect to see much greater use of video and multimedia in concept development. The IdeaMap approach described above allows the researcher to work with consumers in "real time." The use of video and other multimedia tools extends IdeaMap, using more powerful modes of stimulus presentation. The result is a better, more reality-grounded base of data from which to make a decision.

Overview to IdeaMap

IdeaMap provides the product developer and marketer with another way to determine the "hot buttons" in a category. Unlike concept optimization, however, IdeaMap only requires the panelist to assign a measure of interest (from hate to love, or from definitely would not buy to definitely would buy). The data for each panelist is "smoothed" to estimate that panelist's response to elements not tested. Afterwards the data can be analyzed like any other data from concept optimization (i.e., identify preference segments, identify responses of different user groups, optimize the concept). We saw this approach of multiple rating attributes for baby wipes.

If the identification of hot buttons alone is of interest, then IdeaMap may be the best solution to the age-old research problem of identifying the key elements that drive communication in a product category. If, however, it is important to obtain both consumer acceptance and consumer perception of the attributes of a communication element, then the concept optimization method is better because the consumer rates both acceptance and communication aspects of the concept. Alternately, the IdeaMap procedure can be expanded so that the respondent rates multiple attributes, not just interest. (Examples might include appropriate for heavy flow, etc.) Then, the IdeaMap approach would yield several individual models, not just interest.

9. General Overview

Concept development and testing lie at the foundation of product development. Without a good concept the product developer does not necessarily know what to create, nor does the consumer know what to expect. The objective of concept development is to identify hot buttons for product development and product positioning. This chapter presents methods which allow the researcher to delve into the consumer's mind, in an unobtrusive way, and identify both the product and positioning variables. This chapter advocates a designed approach to development, rather than a random, possibly haphazard approach. With designed experiments, even in the realm of concepts and communications, it becomes possible for the developer to increase the speed, scope, and precision of the task, and truly create better ideas for products.

References

Cattin, P. and Wittink, D.R. 1982. Commercial use of conjoint analysis: a survey. Journal of Marketing, *46*, 44-53.

Draper, N.R. and Smith, H. 1981. Applied Regression Analysis. John Wiley & Sons, New York.

Fern, E.F. 1982. The use of focus groups for idea generation: the effects of groups, size, acquaintanceships, and moderator on response quantity and quality. Journal of Marketing Research, *19*, 1-13.

Green, P.E. 1984. Hybrid models for conjoint analysis: an expository review. Journal of Marketing Research, *21*, 155-169.

Green, P.E. and Helsen, K. 1989. Cross validation assessment of alternatives to individual-level conjoint analysis: a case study. Journal of Marketing Research, *26*, 346-350.

Green, P.E. and Krieger, A.M. 1991. Segmenting markets with conjoint analysis. Journal of Marketing, *55*, 20-31.

Green, P.E. and Srinivasan, V. 1978. Conjoint analysis in consumer research: issues and outlook. Journal of Consumer Research, *5*, 103-124.

Green, P.E. and Srinivasan, V. 1978. A general approach to product design optimization via conjoint analysis. Journal of Marketing, *45*, 17-37.

Hagerty, M.R. 1985. Improving the predictive power of conjoint analysis: the use of factor analysis and cluster analysis. Journal of Marketing Research, *22* (May), 169-184.

Hayes, T.J. 1989. The flexible focus group: designing and implementing effective and creative research. *In:* Product Testing with Consumers for Research Guidance, L. Wu and A. Gelinas (eds.), STP 1035, pp. 77-84. American Society for Testing and Materials, Philadelphia.

Johnson, R.M. 1974. Trade-off analysis of consumer values. Journal of Marketing Research, *11*, 121-127.

Moskowitz, H.R. 1986. Sensory segmentation of fragrance preferences. Journal of the Society of Cosmetic Chemistry, *37*, 233-247.

Moskowitz, H.R. 1989. Sensory segmentation and the simultaneous optimization products and concepts for development and marketing of new foods. *In:* Food Acceptability, D.M.H. Thomson (ed.), pp. 311-326. Elsevier Applied Science, Barking, Essex, U.K.

Moskowitz, H.R. 1992. Sensory segmentation: comparison of attitudinal and behavioral methods. *In:* Product Testing with Consumers for Research Guidance, L. Wu and A. Gelinas (eds.), STP 1055, pp. 7–21. American Society for Testing and Materials, Philadelphia.

Moskowitz, H.R. 1994. Food Concepts and Products: Just-In-Time Development. Food and Nutrition Press, Trumbull, CT.

Moskowitz, H.R. and Rabino, S. 1983. The trading of purchase interest for concept believability. International Journal of Advertising, 2, 265–274.

Szabo, B. and Babuska, I. 1991. Finite Element Analysis. John Wiley & Sons, New York.

Wittink, D.R. and Cattin, P. 1989. Commercial use of conjoint analysis: an update. Journal of Marketing, 53, 91–96.

2
Benchmarking a Product Category

1. Introduction

Years ago most researchers in the packaged goods businesses were content to evaluate their products versus the major market brands, and in many instances against the market leader alone. The prevailing thinking at that time was that the market leader was doing the "correct thing," and that to a great extent the success of the leader was due to basic product characteristics. Only in the rarest of instances was a manufacturer inclined to evaluate more than just a few products in the category. This myopic approach spawned research efforts that often simply identified the physical properties of the market leader in order to copy them.

With the development of research techniques to evaluate multiple products, and with the growing importance of "benchmarking" as a valid way to establish one's own competitive strengths, researchers began to test many products in a category rather than concentrating solely on the market leader. Benchmarking produces the following five types of information:

1. Performance of the major and minor products in the competitive frame on acceptance, sensory/aesthetic characteristics, performance, and "image," respectively
2. Effect of branding a product on perceptions of the product
3. Identification of those specific product characteristics driving acceptance versus rejection, respectively

4. Extraction of the quantitative relation between the sensory characteristics of a product and overall acceptance
5. Synthesis of sensory profiles corresponding to development targets.

2. A Case History—Anti-Perspirant/Deodorant

In 1982 Marilyn Henninger became president of Richards Company Inc., a small company in northern Vermont. The Richards Company had been founded some 30 years before by an entrepreneur, Richard Strong, based on patented technology for anti-perspirants. Over the years the Richards Company successfully competed for a niche position in the category, positioning its products as "all natural" and "unisex."

Richards made significant profits in the 1960's and 1970's, although it never developed into a major player in the category. Over those years Richards gained and retained consumer loyalty by providing ecologically responsible products that were effective, reasonably priced, and well-packaged (but not overly so).

The 1980's saw the first erosion of Richards' business as competition began to heat up. What had been a modestly competitive category turned excessively competitive and sometimes cutthroat. With all of the new introductions of products, line extensions, etc., the trade began to charge "slotting allowances" in order to benefit from this intense competition. In order to place its products on the shelf the manufacturer had to pay an up-front fee.

As part of a revised marketing thrust, Henninger commissioned a marketing research study to assess consumer response to the full range of solid anti-perspirants and deodorants. A cursory inspection of the shelves in different stores uncovered dozens of different competitors. These competitors varied in shape, color, and fragrance, even within the same form. In order to make the project feasible and to avoid any undue cost, Henninger recommended evaluating a limited number of products across manufacturers instead of evaluating all of the products currently available. Various manufacturers offered the same products in a variety of scents (including fragrance-free).

In general, when evaluating products in a competitive frame the researcher or marketer must decide exactly what defines the frame. There are no fixed rules. The competitive frame may comprise a general group—e.g., any product used to reduce underarm wetness and ensuing problems. Or, the competitive frame may comprise a very finely tuned group—e.g., all unscented, roll-on products that are classified as being anti-perspirants.

In this study, the marketing group decided to consider a wide set of stick forms, and not define the competitive frame to be any tighter. The marketing group reached its decision because of three key factors:

Learning. Marketing wanted to learn a lot about the category, and was worried that by narrowing the specifications (i.e., to scented products only, or to feminine-oriented, rather than unisex or masculine-oriented), they would overlook potentially valuable information and insights that could help them.

Multiple End Uses of the Data. Marketing needed to test a reasonably wide range of products with different sensory characteristics because they planned to use the information both for positioning guidance (viz., to discover the strengths and weaknesses of the competitors) and for development guidance. For example, if the company wanted to develop a new stick product, then what particular characteristics should the product have, and would those characteristics be identical to any products currently on the market?

Funds Were Limited. Marketing had to confine themselves to one product form, rather than assessing the full array of available product forms. During initial discussions about the project, the opinion was voiced by quite a few participants that Richards might consider launching a roll-on product. Some individuals, feeling that a roll-on product was a definite possibility over the next two-year period, suggested that the study encompass roll-on products as well, so that the research could truly pay out in the "intermediate term." However, it soon became apparent to all concerned that the potential cost of the project would be 45% higher if the study were to assess roll-ons. Consequently, the study was limited to stick products, despite the recognition that a significant amount of valuable information could be learned by testing another product form, and indeed a form that was being considered for launch 2 years down the road.

3. Should the Products Be Tested "Blind" or "Identified"?

Over the past 15 years marketers and researchers alike have come to realize that products comprise more than just the physical stimulus. Products comprise the image brought to the product through the branding (namely, from advertising), as well as the actual physical characteristics of the physical stimulus.

In this study the funds for research were limited. The marketers decided to evaluate the products on a "blind" basis only. It is hard to disguise some of the stick products, especially when some packaging characteristics (e.g., type of "dome") are closely identified with the manufacturer. In order to maximize the effect of masking, R&D contracted with an outside agency to cover the package with a material that could be formed so that all products had similarly (albeit not identically) shaped packages. The final packages were all oval, and just slightly larger than the largest oval-shaped product currently available in the market. "Blinding" or "disguising" the product may or may not be called for in a category appraisal and benchmarking study. Blinding a product can cost a great deal of money, especially when there are many individual containers to disguise. The expense and ultimate biasing effect of branding, if any, should be considered before going further and incurring the expense.

4. Selecting the Set of Products to Evaluate—Content Analysis

When looking at a full range of competition, often the number of potential products to test looms large and the potential expense appears enormous. In categories awash with different sizes, fragrances and other relatively inexpensive (and functionally similar) variations, the researcher must set boundaries on what is to be evaluated. But exactly how does the researcher select which products to test and which ones to eliminate as redundant?

One efficient way to select products is by fiat. The researcher looks at the biggest sellers, selects some or all of these, and then fills in the available test slots with other, less popular brands. The benefit of this fiat-based approach is that it covers the key "players" in a category. However, the drawback is that it may miss some unique aesthetic or physical variations in products which could eventually contribute to a market success. The fact that this feature belongs to an unusual product, not a market leader, may have nothing to do with the feature, but rather may result from poor advertising or distribution.

An alternative approach, ultimately more productive and educational, uses a content analysis of the features. Once the content analysis is done, the researcher can identify a reduced set of products, maximally "dissimilar to each other" in terms of the sensory characteristics.

The content analysis is easy to perform, and can be accomplished in eight simple steps with as few as one or two judges or as many as a full consumer panel. Here are the eight steps:

1. Inspect the full range of products, and record the relevant characteristics
2. Count the number of times each characteristic appears in the set
3. Identify key characteristics which appear only once or twice in the set. These are unique characteristics. Through judgment, keep those unique characteristics for future consideration or discard the characteristics that seem to be irrelevant and just happenstance. (Judgment plays a key role in Step 3 because one does not know ahead of time which unique features may ultimately drive consumer acceptance.)
4. Select product characteristics which appear at least two to three times. For different variations of a common theme (e.g., fragrance type), do not worry about the "fine differences" between two similar fragrances (e.g., two baby powder fragrances that differ qualitatively, albeit slightly). It is unlikely that two products that are colored "green" will have the exact matching shades. At this stage focus on the general feature, not a specific execution.

5. Revisit the products, this time to determine whether each product either possesses or lacks the feature. Under fragrance there will be a variety of general fragrance categories. This will be the same with colors, shapes, etc.
6. Consider all aspects of the product available to the senses. These aspects include appearance (of the container, of the product), fragrance, and feel of the product both through touch and application (feel and dynamics of the product as the product is dragged across the skin).
7. For each product create a profile of 1's and 0's, where the value "1" for a feature denotes that the product possesses that feature, whereas a "0" denotes that the product lacks that feature. If the physical feature is continuous, then divide the continuum into discrete sections. Each section is a discrete feature.
8. Look at the frequency with which each feature appears in the full set of competitive products. By sorting the products according to manufacturer, the researcher may be able to identify a "sensory signature" for that manufacturer if such a signature exists. The "signature" is the commonality of sensory attributes across the various products (e.g., similar appearance of the product, similar type of container, etc.)

Table 1 describes the content analysis and the subsequent data manipulation used by Henninger to reduce the number of products for consideration. Table 2 shows a partial set of five products that the researcher considered and the clusters into which the five in-market products fall. Based upon the clusters identified for the anti-perspirants, the marketing research director suggested that the set of 73 products be reduced to a set of 20 for testing.

The final number of clusters to be used in selecting the products is left to the researcher's judgment. Too few clusters will lump together qualitatively dissimilar products. The clustering exercise will not provide sufficient differentiation among the products. Too many clusters will generate an unduly large number of products to test, and defeat the purpose of reducing the product set.

Selecting the Products for Evaluation

The clustering yielded 20 different products. In addition, the marketing group wanted to evaluate four additional market leaders which were qualitatively similar to some of the products already selected. (It was important for marketing to obtain a "direct read" of these competitors because of their strong position in the marketplace.) The final set of products totaled 24.

Selecting the Relevant Population to Test

Who should participate in the evaluations? There were opinions voiced on this topic:

Table 1 Content Analysis of the Competitive Frame Sequence of Analytical Approaches

Step 1—Purchase key products on the market, and if necessary, create prototypes

Step 2—Identify key categories of sensory attributes (e.g., color, shape, fragrance, texture)

Step 3—Within each sensory category (e.g., color) list the key different subcategories (e.g., pink, white, green)

Step 4—Create a matrix whose rows correspond to products and whose columns correspond to sensory attributes. Each sensory attribute (e.g., pink, white, green) has its own column

Step 5—Populate the matrix (from Step 4) with 0's in every row

Step 6—Go through each of the products. For a specific product, look at each attribute (column). If the product has that specific attribute, then for the specific product/attribute replace the value "0" by the value "1" (to denote that the product possesses that specific feature or sensory attribute)

Step 7—Follow Step 6 until all the products have been categorized

Step 8—Cluster the products in Step 7, using a "K means clustering program." The program will combine products with similar sensory profiles (similar patterns of 1's and 0's). The user can specify the number of clusters to create.

Option: Factor analyze the columns, save factor scores, and then cluster the products on the factor scores, instead of on the features themselves.

General User Cross Section of "Non-Rejecters." Those at the Richards Company voicing this opinion felt that it was vital to measure the responses of consumers who would comprise the current and potential purchasers of the category. The panel would thus comprise both males and females who were current category users, or who expressed an interest in using products in the category but were not currently doing so. The category was defined to be "stick, antiperspirant and deodorant" products. The rationale for this opinion was that the original objective was to design a new product for the relevant consumer population. In this case "relevant" was defined as an individual who would be likely to purchase. As long as the consumer was not averse to purchasing and using a product in the category the consumer satisfied the criteria for inclusion. (Sometimes researchers adopt more stringent criteria for participation. They select only those consumers who state that they would definitely or probably buy and use the product. The researchers eliminate anyone who says maybe/maybe not. Here a non-rejecter was defined as anyone who, on a 5-point purchase intent scale, said that they would definitely, probably, or might/might not buy the product.)

Table 2 Example of Content Analysis of Five Anti-Perspirant Stick Products

Product	201	202	203	204	205	Product	201	202	203	204	205
Cluster	1	3	2	5	19	Cluster	1	3	2	5	19
Colors						**Fragrance**					
White	0	0	0	0	0	**profile**					
Green	0	0	0	1	1	Intensity	7	7	6	6	6
Pink	1	1	1	0	0	Sweet	5	5	3	3	3
						Floral	6	6	3	1	3
Appearance						Rose	4	6	0	0	0
Shiny	1	0	1	1	0	Fruity	0	0	0	0	0
Grainy	1	1	1	1	1	Citrus	0	0	0	0	0
Markings	0	0	1	1	0	Medicinal	0	0	0	1	0
Large Size	0	1	0	0	0	Spicy	0	0	2	3	0
Rectangle	0	0	1	1	1	Woody	0	0	0	0	0
Oval	1	1	0	0	0	Powdery	0	2	3	0	0
Round	0	0	0	0	0						
Top Curves	1	1	1	1	0						
Physical measures						**Texture**					
Length	91	82	89	86	78	Hard	1	1	0	0	1
Width	65	56	57	57	47	Dense	1	0	0	0	1
Depth	26	21	27	25	32	Sticky	1	1	1	2	1

Note: Fragrance notes rated on a 0–9 scale.
Physical measures = Actual measured values by instrument.

Targeted Current User. An equally vociferous opinion was offered by another group at Richards who felt that only current users of the category should be included. These more "purist-minded" individuals felt that the relevant population should comprise only those individuals who actually exhibited the desired purchase behavior. This second group felt that it was a potential waste of research money to assess the reactions of individuals who simply "might buy" the product. It was not clear to this group what could motivate the "aware, non-user" to buy, and it seemed fruitless to spend money on them. There was too little money available and too much to do to fritter away precious dollars on a possible user group.

After much debate, management chose the second (purist) approach to sample the responses of consumers who currently use stick deodorants and anti-perspirants, independent of fragrance type and manufacturer.

Four Key Issues in Running the Product Evaluation

Issues always arise in any large-scale study of this type where there are a variety of products, attributes and underlying questions to answer. Four issues emerged:

How Many Panelists Should Rate Each Product? Clearly the greater the number of panelists the better the data. Increasing the sample size of participating consumers better "fixes" the mean rating of a product and reduces the variability around the estimate. The standard error of the mean, a measure of variability of the mean from study to study, decreases with the square root of the base size. For this study it was vital to test a sufficiently large number of consumers to insure reliable data. It was also necessary to have a high base size per product because the marketing group wanted to analyze the ratings by different user groups (e.g., users of Brand A versus users of Brand B). Given the ultimate cost of doing the project, the marketing and market research groups agreed to secure 50 ratings per product, divided approximately equally across different ages, genders, and brand usage.

How Many Products Should a Panelist Rate, and over What Period of Time? Anti-perspirants differ from simple fragrances and foods which can be evaluated in a single session. (For fragrance evaluation, a panelist can rate 12 or more fragrances over 3 hours without losing sensitivity.) It takes time for anti-perspirants to work. They build up a residual that must be flushed out before consumers can evaluate the next product. Since this test required actual product use, R&D suggested that each panelist test seven products. Furthermore, they suggested that each panelist test one product over 5 days, and then use nothing for days 6 and 7. Afterwards, the panelist would rate the next product. Over the course of the study, therefore, each panelist would participate for 7 weeks (one product for 5 days, and a 2-day washout period each week, respectively).

Based upon the requirement of 24 products, seven products per panelist, and 50 ratings per product, the total base size came to $(50 \times 24)/7 = 1200/7 = 172$. Recognizing that there would be dropouts, the marketing research director suggested recruiting a total of 200 panelists.

How Frequently Should the Panelist Rate the Product, Given That the Panelist Must Live with Each Product for 5 to 7 Days? There are three schools of thought on rating frequency in home use tests.

Last Usage. One school of research believes that the panelist need only rate the product after the last usage period. The rationale is that the objective of the research is to obtain a single, overall perception of the product. This perception is thought to be most easily obtained at the end of the usage period, after the panelist has had a chance to use the product. (When there is a washout period, one should instruct the panelist to rate the product on day 5 rather than on day 7, in order to insure that the panelist is actually rating a product while using it.) The panelist must have the product available, however, in order to rate the attributes most accurately. Henninger's marketing research director opted for this approach because of its simplicity, ease of execution, and data analysis. The panelists did not have to keep daily records of their experience with the product.

During Each Usage. Another faction of researchers believes that the best procedure has the panelist rate the product at each usage, or at least once each day. These researchers believe that there may be changes in a consumer's opinion over time, and that it is vital to capture this change if it exists. The researchers believe that experience with products may affect consumer acceptance and perception, because with repeated product use consumers change their focus and consider new aspects of the product. Daily ratings are probably the most conservative approach, but they generate an extraordinarily large amount of data if the panelists complete the full questionnaire each day. Occasionally these researchers use a scaled-down approach. On the first and the last exposure to the product the panelist rates the product on the full set of attributes, including sensory, liking, performance and image. On intermediate days the panelist rates the product on a more limited set of performance attributes, usually omitting sensory characteristics, which should remain unchanged. Henninger's research director at first had considered this compromise option, but recognized that it would still yield an exceptionally large amount of data to process. The odds were that little additional information would be gained from repeating the questionnaire.

From Memory. The third faction of researchers believes that the panelist should rate the product "from memory" after completing the usage. The rationale was that, in a product evaluation, interest focuses on what people think of the product, given experience with the product, but the evaluation should not be artificial. ("Artificial" to these researchers is a situation that is unnatural.) These researchers felt that the panelist should be familiar with the product, but need not have the product in front of them, as long as their experience with the product had been quite recent. For category appraisal this is probably the least sound method. By relying upon memory, without the product actually being present, the panelist runs the risk of errors in memory and failure to identify key (and unusual) sensory characteristics that are not necessarily salient, but easily recognized when the product is present.

Based upon the cost considerations, the task force at Richards decided on option 1—one rating occasion per product per panelist to be done on day 5 of the usage period.

How Can the Researcher Maintain Consumer Interest? In these types of studies the consumer must participate for an extended period of time. How does the researcher maintain consumer interest? Probably the easiest way is to pay the consumers to participate. If we consider the consumer data to be the foundation of all subsequent decisions, then paying the consumer is akin to insuring the best "raw materials" for the product. Consumers disinterested in the task may not provide reliable and valid data. Paying the consumers a fair but reasonable amount of money (e.g., $120 for 7 weeks of participation) insures that the data will be of high quality. The panelists become and remain "ego-involved" in the task, and do their best to provide accurate data.

Experience with paying panelists to participate has been particularly positive. In long-term product evaluations lasting a month or more, most "dropouts" occur early (e.g., between the orientation session and the return of the first product a week later). Occasionally a panelist will drop out later on in the study, usually due to unforeseen difficulties (e.g., sickness, travel schedules that could not have been known earlier). In large scale studies once the initial dropouts occur (within the first week), there is usually more than a 90% retention rate, especially when the remuneration is reasonable.

Additional motivation can be insured by having an interviewer check the consumer's ratings from time to time. In this study, the consumers took products home on a weekly basis to use, and returned the products on a Monday evening. The return visit provides an opportunity to check the data from the previous week. The interviewer asked the panelist to justify four attribute ratings assigned to the product just rated. This procedure of questioning the panelist after each product insures that the panelist takes the task seriously. The panelist knows that he or she will be questioned, and thus makes every effort to assign accurate ratings.

The Questionnaire

The questionnaire links the consumer with the researcher. In the past, questionnaires were relatively simple because the panelist's task was simple. The panelist simply rated the degree of acceptability of the product. Occasionally, the questionnaire might probe reasons for liking. The panelist could write the answers in phrases or sentences (so-called verbatims or open-ends) or could assign scale values to the product on a variety of diagnostic scales (e.g., "like appearance," "like fragrance," etc.) The diagnostic questions enabled the researcher to ascertain why a specific product was liked and another product was disliked, or at least to guess the reasons, based upon the consumer's own responses. In most cases these diagnostics were helpful, especially when the product had major defects.

As researchers need to address more issues, questionnaires have become increasingly complex. Consumers today are asked many more questions about their responses to the product. Not only are they required to rate liking, but they also must rate many other characteristics of the products. The consumer assigns a profile of attribute ratings to the product in order to show the different perceptual attributes he believes the product to possess.

How Much Information Should the Consumer Provide? A perennial issue that arises in product testing is the amount of information to be obtained from a consumer. There are various opinions. The minimalist end of the continuum comprises practitioners who believe that the consumer cannot validly provide any information other than "like/dislike" and perhaps "image characteristics" (e.g., high quality). (In practical terms, however, the image characteristics are often restatements of "like/dislike.") Practitioners holding this point of view believe that all other data about the product needs to be obtained from an "expert" panel of individuals, trained to provide sensory attribute ratings. In the world of applied product testing the minimalists are in the decidedly shrinking minority.

The maximalist end of the continuum comprises practitioners who believe that the consumer can rate a virtually unlimited number of attributes. Marketing researchers fall into this group. To these practitioners, as long as a consumer can maintain interest, he can rate dozens of attributes. The limiting factor is consumer fatigue and boredom, rather than an innate limitation on his intellectual abilities. This point of view is clearly evident in many product studies conducted by marketing and R&D. A cursory glance at the questionnaires used in marketing research usually reveals several dozen attributes per product. Consumers appear able to complete the questionnaire, although we have no criteria by which to assess the validity of the data obtained with such large questionnaires.

For this particular study the marketing research director recognized that there was a need to obtain as much information as possible about the products. There was also a need to insure that the data was valid. These two objectives generated the following compromise:

a. The size of the questionnaire had to be limited to a reasonable number (here 21 attributes).
b. The questionnaire had to address sensory, liking, image and performance attributes.

Scales. The scales are either unipolar or bipolar. Unipolar characteristics are attributes which extend in one direction (e.g., for fragrance intensity the scale goes from $0 =$ no perceivable fragrance $\rightarrow 100 =$ extremely strong fragrance). Bipolar characteristics are attributes which extend in two "opposite" directions. (The word "opposite" is put in quotes because the directions need not be logical opposites, although they could be construed as opposite. An example of a bipolar attribute is "type of fragrance," $0 =$ cosmetic fragrance $\rightarrow 100 =$ functional fragrance.)

The 0–100 point scale is easy to use. Consumers are familiar with the scale from school, and find the attribute scales easy to understand. Anchoring terms at both ends of the scale also helps the panelist by reducing potential ambiguity. Table 3 shows the questionnaire for the study.

Table 3 Questionnaire for Stick Anti-Perspirant Product

Please look at the product, but do not apply it yet

A. How much do you like the way the product looks in the container?
  (0 = hate → 100 = love)
B. How much do you like the way the product smells in the container?
  (0 = hate → 100 = love)
C. How easy is it to hold the product in your hand?
  (0 = very hard → 100 = very easy)

Now, please apply the product as you normally apply stick products

D. How much do you like the product overall?
  (0 = hate → 100 = love)
E. Which one of these five statements best describes how you would feel about buying the product if it were offered at the price you ordinarily pay?
  01 = definitely not buy
  02 = probably not buy
  03 = might or might not buy
  04 = probably would buy
  05 = definitely would buy
F. How much do you like the way the product goes on your skin?
  (0 = hate → 100 = love)
G. How much do you like the way the fragrance smells on your skin, now that you have applied it?
  (0 = hate → 100 = love)

Now look, smell, and feel the product again

H. How strong is the fragrance of the product in the container?
  (0 = very weak, almost without fragrance → 100 = very strong)
I. How hard does the product feel?
  (0 = very soft → 100 = very hard)
J. How creamy does the product feel?
  (0 = not at all creamy → 100 = very creamy)
K. How smooth does the product feel?
  (0 = very rough → 100 = very smooth)
L. How difficult was it to apply the product?
  (0 = very easy → 100 = very difficult)
M. How much residue did the product leave on your skin?
  (0 = none → 100 = a lot)
N. How much product did you use on each application?
  (0 = none → 100 = a lot)

(continued)

Table 3 Continued

Now, thinking about this product overall

O. How would you describe the product in terms of who it is meant for?
(0 = strictly for men → 100 = strictly for women)
P. How would you describe the age of the person for whom the product has been designed?
(0 = strictly for younger people → 100 = strictly for older people)
Q. How well does this product stop underarm wetness?
(0 = very poorly → 100 = very well)
R. How well does this product stop underarm odor?
(0 = very poorly → 100 = very well)
S. How refreshed do you feel with this product?
(0 = not at all refreshed → 100 = extremely refreshed)
T. How long-lasting is this product in terms of its action?
(0 = very short-acting → 100 = very long lasting)
U. How unique is this product?
(0 = identical to others on the market → 100 = very unique)

5. Analyzing the Data

The 8-Step Analysis

When evaluating multiple products for benchmarking and for category appraisal the researcher follows the eight steps listed below:

1. Create a matrix of product × attribute.
2. Adjoin to this matrix another matrix of product × key subgroups × liking ratings.
3. Map the category by spatial representation to show how products cluster together in terms of sensory similarity.
4. Determine the basic sensory attributes in the category which best differentiate products.
5. Identify what sensory likings "drive" overall liking.
6. Identify how specific sensory characteristics drive overall liking.
7. Develop a sensory model which integrates all of the sensory attributes and overall liking.
8. Use this sensory model to identify how to change products in order to maximize acceptance and/or image characteristics.

Summary Data of Product × Attribute

Of all the analyses that one might perform, this basic summary analysis is probably the most useful. There are many things that the marketer and product developer learn by summarizing the data. Some of the insights and learnings are:

1. How well do the products score? What is the range of overall liking?
2. Along what perceptual attributes do products cluster and along what specific attributes do they differ? This analysis shows the "homogeneity" vs. "heterogeneity" of the category. (For instance, the products might differ in both type and strength of fragrance, but be quite similar in the perceived "drag" they exhibit when the products are dragged across the skin.)

Table 4 shows a partial table of the results. The numbers in Table 4 are mean ratings, computed from the panelists who evaluated the particular products. The only number that is not a mean value is the purchase intent value, which shows the proportion of consumers who said that they would definitely or probably buy the product (ratings 4 or 5 on a 5-point scale).

Although Table 4 shows only a partial list of the products, we begin to see a number of patterns emerging based upon the summary statistics computed across the 24 means:

maximum mean achieved by a product
minimum mean achieved by a product
range (maximum mean – minimum mean)
standard deviation of the 24 means.

1. In terms of liking, the products differ more on fragrance in the container than on virtually any other attribute.
2. The products do not evenly distribute along the 0–100 liking scale. There appear to be 3–4 clusters (see Figure 1). One cluster comprises very well-liked products. These score closer to the 50's and 60's. There is a smaller

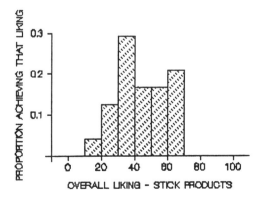

Figure 1 Distribution of the 24 stick anti-perspirant deodorant products on overall liking (0–100) scale.

Table 4 Partial Consumer Data from Anti-Perspirant Study and Summary Statistics for the Means Across 24 Products

	Product				Maximum	Minimum	Range	Standard deviation
	201	202	203	204				
Liking								
Purchase								
(% Top two box)	42	38	30	52	60	19	41	11
Total	38	41	22	59	68	20	48	15
Segment 1	46	47	29	54	78	23	55	16
Segment 2	48	68	83	42	87	14	73	19
Segment 3	28	22	18	71	96	18	78	24
Males	31	48	13	57	76	11	65	19
Females	44	34	30	61	82	18	64	16
Younger	27	48	28	59	74	18	58	18
Older	54	31	12	59	77	12	65	17
Use Brand A	28	48	17	49	78	17	61	17
Use Brand B	45	35	30	51	75	20	55	15
Liking								
Overall	38	41	22	59	68	20	48	15
Appearance/container	49	51	57	61	74	41	33	8
Fragrance/container	73	75	43	80	80	34	46	12
Fragrance/skin	48	53	50	63	74	42	32	8
Feel/application to skin	71	62	68	63	84	42	42	8
Product								
Fragrance strength	29	28	40	52	64	28	36	9
Hardness	72	72	73	57	85	34	51	11
Creamy	51	53	49	45	54	33	21	5
Smooth	50	54	45	48	54	28	26	6
Application								
Easy to apply	65	78	53	76	79	39	40	10
Amount/residue	31	33	35	39	53	28	25	7
Amount used	63	61	68	60	82	41	41	8
Image/performance								
Masculine/feminine	45	48	52	55	68	38	30	7
Younger/older	39	45	29	35	46	16	30	7
Unique	38	36	56	66	66	27	39	10
Stops wetness	59	59	65	60	76	47	29	7
Stops odor	43	43	47	48	52	43	9	2
Refreshes	65	62	61	66	79	55	24	6
Long-lasting	69	68	68	72	81	67	14	5

set of less well-liked products (40's), and quite a number of products performing very poorly (mid-30's and below).
3. The products differ on the sensory dimensions of both fragrance and texture (especially hardness). Consumers perceive differences (especially in terms of the "drag" or effort needed to pull the product across the skin as measured by the attribute "easy to apply").

Information of this type is quite helpful to the marketer who begins to get a sense of the nature of the competitive frame. If all the products score very highly (e.g., 55+ on the overall liking scale), then this may indicate that consumers have a difficult time differentiating the products on liking (even though they have no problem differentiating the products on sensory characteristics). Strategically, the "tightness" of the liking ratings (all products clustering together) means that in order to differentiate one's entry from the pack of competitors it might be necessary to go beyond the product itself and change its positioning and advertising. On the other hand, when the products scatter across the liking scale as they do here, then this is evidence that consumers have innate sensory preferences that translate into acceptance differences. Marketing may be one facet of the strategy, but another facet could be a focused product development effort in order to increase the level of consumer satisfaction.

How Different Subgroups of Consumers Respond to the Products

The average rating glosses over substantial variability of responses among consumers. For instance, a product averaging 50 on the 0–100 point scale may do so because it is truly only a "fair" product (so all the consumers rate the product around 50), or the average may result from combining two groups of consumers, one group who really loves the product and rates it closer to 65, and another group who really dislikes the product and rates it close to 35. We can see these patterns most easily in Figures 2(A–C).

A standard way to determine the homogeneity of the ratings looks at subgroups of consumers. Usually these subgroups are defined in a meaningful way, such as breaks by age, product usage, gender, etc. Researchers and marketers look for differences among subgroups in order to identify opportunities and problems with the product that would not be apparent from the total panel. Table 5 shows the liking ratings assigned by the different subgroups to the products. For the most part the subgroups show similar pattern ratings. Where there are differences, they appear to be random.

Patterns to look for which signal true subgroup differences are:

Scale Usage. One subgroup of consumers consistently overrates or underrates the products. This could be a "response bias" having nothing to do with the

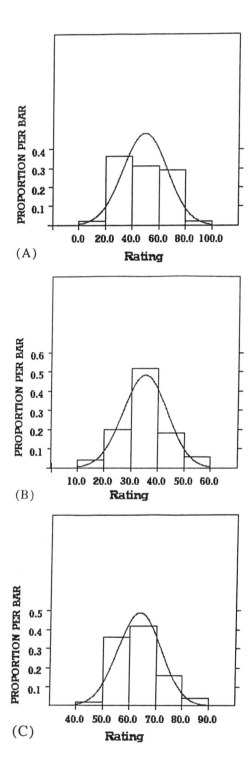

Table 5 Mean of Liking Ratings—Total Panel and Key Subgroups (Ranked by Total Panel)

Product	Total panel	Gender		Age		User	
		Male	Female	Young	Older	Brand A	Brand B
205	68*	53	62*	63*	74*	60*	57
209	66*	76*	56	74*	53	64*	75*
213	65*	59	71*	57	77*	78*	72*
206	65*	72*	58	73*	53	65*	74*
212	60*	75*	46	73*	42	72*	55
211	59	60*	58	59	58	68*	61*
204	59	57	61*	59	59	69*	49
210	56	56	56	66*	41	67*	47
219	56	66*	46	56	54	49	52
208	48	33	62*	44	53	40	41
216	47	58	36	60*	36	36	48
220	44	31	58	59	22	55	43
202	41	48	34	48	31	48	35
223	39	42	37	45	31	38	47
221	38	50	26	33	46	28	39
201	38	31	44	27	54	28	45
222	37	33	41	43	27	40	27
224	37	45	29	29	48	25	30
214	33	24	41	19	52	27	32
207	33	29	36	18	55	38	39
215	29	40	18	27	31	22	20
218	22	13	31	21	23	23	25
203	22	13	30	28	12	17	30
217	20	11	28	22	16	27	26
Maximum	68	76	82	74	77	78	75
Minimum	20	11	18	18	12	22	20
Range	48	65	64	57	65	57	55
Standard deviation	15	19	16	18	17	18	15

*Denotes winner in terms of liking ≥ 60.

Figure 2 Comparison of three distributions for a product where mean rating is 50. Figure 2A shows the distribution for the full panel. Figure 2B shows the distribution for one subgroup (low raters). Figure 2C shows the distribution for another subgroup (high raters).

Table 5 Continued

Mean ratings between product	Total panel	Gender		Age		User	
		Male	Female	Young	Older	Brand A	Brand B
0-29	4	5	4	8	5	8	4
30-39	7	4	6	1	4	3	6
40-49	4	4	5	4	4	4	7
50-59	4	6	5	5	9	1	3
60-69	5	2	3	3	0	6	1
70-79	0	3	1	3	2	2	3
80-89	0	0	0	0	0	0	0
90-99	0	0	0	0	0	0	0

products themselves. Or, this subgroup difference in magnitude of rating could reflect the fact that one subgroup of consumers truly likes the products more, and another subgroup truly likes the products less. The best way to establish the difference between these two hypotheses is to identify the attributes on which the two subgroups score the product similarly. If there is such an attribute, then we know that consumers use the 0–100 scale similarly. Consequently the differences in liking ratings must be attributable to differences in level of liking, not to response bias.

Group Differences. If there exist natural clusters of products based upon commonality of physical features (e.g., fragranced vs. unfragranced, pink vs. white), then do the subgroups respond differently to the different clusters? For example, do males respond less positively to pink products than females? Do users of fragranced products respond negatively to unfragranced products, whereas users of unfragranced products respond positively? This subgroup analysis requires an additional criterion by which to divide the products, in order to produce a two-way table (see Table 6).

In the author's experience there is generally a great deal of similarity between subgroups in the pattern of liking ratings that they assign to "blind" products. Unless the subgroup of consumers is based upon specific acceptance/rejection of certain physical characteristics (e.g., users versus nonusers of fragranced products), it is unlikely that subgroup analyses will reveal easily interpretable differences. Occasionally the liking ratings will vary from subgroup to subgroup, but the pattern underlying that variation will not be obvious. It will appear more like "noise" than like "signal."

Table 6 Example of Liking Ratings by Gender for Stick Anti-Perspirant—Sorted by Product Color

	Liking ratings		
Product	Males	Females	Male-female
White			
211	60	58	2
224	45	29	16
202	48	34	14
218	13	31	−18
213	59	71	−12
201	31	44	−13
205	53	82	−29
212	75	46	29
209	76	56	20
214	24	41	−17
207	29	36	−7
Average	47	48	−1
Green			
210	56	56	0
219	66	46	20
220	31	58	−27
217	11	28	−17
216	58	36	22
208	33	62	−29
223	42	37	5
221	50	26	24
215	40	18	22
Average	43	41	2
Pink			
204	57	61	−4
203	13	30	−17
222	33	41	−8
206	72	58	14
Average	44	48	−4

Summary Table		
	Male	Female
White	47	48
Green	43	41
Pink	44	48

Mapping the Category

During the past decades it has become fashionable to "map" product categories. Mapping is a statistical technique which locates stimuli (e.g., products) in a geometrical space of low dimensionality. Products lying close together are qualitatively "similar," whereas products lying far away are qualitatively "dissimilar." The map may be created on the basis of the sensory or image attributes, or created on the basis of a consumer's judgment of overall "similarity" or "difference." In either case the map shows which products cluster together with other products.

From time to time researchers take the maps even one step further. They try to understand the axes of the map, and give meaning to these axes. For instance, if in a product category the map comprises two dimensions, and based upon the layout of the products one dimension is texture and the other is darkness, then the standard interpretation is to assume that these are the two fundamental dimensions of the category. In the main, however, the mapping technique should simply be considered as a way to locate the products in geometrical space. The coordinates of the map are only suggestions about the basic dimensions underlying the category.

Mapping can be done with different sets of attributes. The attributes in this study are the non-evaluative sensory and usage characteristics, such as perceived darkness, drag of the product as it was being applied, etc. The attributes are "non-evaluative" because there is no location on the scale which is "better" than any other. The issue is "How much of an attribute is there?" rather than "How much do I like the product attribute?" [Liking ratings are evaluative, because the panelist rates his own feelings about the product, rather than rating the more "objective" product characteristics.]

For the mapping, Henninger's research director used factor analysis. Factor analysis (in this case principal components) reduces the number of attributes (all appearance, fragrance, texture characteristics) to a more limited number. Factor analysis develops a reduced set of factors (or dimensions, axes). The dimensions are statistically independent of each other, and are substantially fewer than the original set of attributes. The statistical procedure does not name the factors—naming is left up to the researcher. However, the procedure correlates each attribute from the original stick evaluation test with the functions in order to help the researcher label the factors.

Tables 7A and 7B show the results. Factor analysis revealed three major factors operating in the category. These factors are texture, application/smoothness, and fragrance, respectively (based upon the correlation of the factors with the original attributes). Color did not emerge as a factor because it was not asked as a sensory attribute. Factor analysis also "rotates" the axes in such a way as to make each axis simpler to understand. The rotation of the axes does not

Table 7 Factor Scores for Stick Anti-Perspirants and Correlations of the Factors with the Sensory and Usage Attributes

Product	F1 (Texture)	F2 (Application)	F3 (Fragrance)
A. Factor Scores of Products			
201=A	-0.2	-1.1	-1.5
202=B	0.3	-1.2	-2.5
203=C	-0.2	-1.0	-0.5
204=D	0.2	0.2	-0.2
205=E	1.3	-0.4	0.6
206=F	1.4	-0.6	1.1
207=G	-0.6	-0.1	-1.0
208=H	0.2	-0.5	-1.3
209=I	1.5	-0.1	0.2
210=J	0.6	1.2	0.9
211=K	-0.6	0.7	-0.2
212=L	1.6	0.2	0.1
213=M	1.0	-1.2	1.0
214=N	-1.2	0.2	-0.1
215=O	-0.7	-0.9	0.5
216=P	0.1	-0.6	0.2
217=Q	-1.6	-1.8	2.4
218=R	-2.4	-0.2	-0.3
219=S	0.7	0.7	-0.1
220=T	0.1	0.8	-0.1
221=U	-0.3	0.5	1.1
222=V	0.1	1.5	-0.7
223=W	-1.0	1.1	0.7
224=X	-0.5	2.4	0.0
B. Factor Structure of Attributes			
Soft/hard	[0.91]	0.01	0.06
Creamy	[0.82]	-0.53	-0.03
Residue	0.79	0.09	0.31
Amount used	0.76	-0.15	0.49
Smooth	0.26	[-0.90]	-0.12
Fragrance	0.14	0.10	[0.90]
% Of total variance accounted for	48	17	16

change the data nor the relation between the axes. Rotation attempts to maximize the correlation of a limited number of attributes with each axis. If only a few attributes correlate with each axis, then the researcher has an easier time interpreting the meaning of the axis or factor.

Factor analysis is a simplifying technique which reduces the mass of data to more easily understood dimensions. Factor analysis glosses over many of the nuances in the data. It must perforce wash out some of the fine-grained detail. Therefore, factor analysis (and mapping) should be used as a general tool or "heuristic" to get an overview of the data, to identify the basic dimensions of perception, and to locate stimuli in a geometrical space. Factor analysis should not be used to eliminate questions from a questionnaire because the different attributes tap fine nuances that factor analysis overlooks. We need the finer-grained attributes in order to have a deeper sense of the product. Factor analysis is only a summary, similar to a picture of a city taken from a mile above the ground.

Mapping—Locating the Products in the Factor Space

A key benefit of factor analysis is its ability to locate products in the reduced factor space. Each product generates its own set of "factor scores." A factor score is the projection (or location) of the product on a specific dimension. Since the stick deodorant project generated three factors, each product has a corresponding set of three factor scores (one per factor). We can either show these factor scores in table form (see Table 7A) or we can plot the factor scores on a set of two-dimensional tables (e.g., Factor 1 vs. Factor 2, with each point in the table corresponding to a product). Figure 3 shows a plot of Factor 1 × Factor 2.

By themselves the factor scores are simply a boiled down version of the attribute profile. They don't tell much. But once the factor scores are plotted

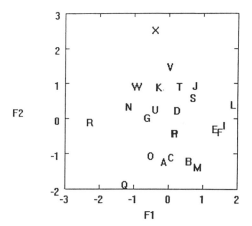

Figure 3 Two-dimensional plot of the 24 stick products, plotted in factor space. The coordinates are the factor scores.

on a graph we can learn a great deal. Specifically, here are the three questions we should ask:

1. Do all of the products cluster together in the map, or do the products spread apart? If most of the products cluster together, but one or two products lie at the extremes, then we conclude that the sensory profiles of the products are quite similar to each other. (There may be one or two quite different products, or "outliers," but most of the products bear remarkable sensory similarity to each other.) Since we can inspect only two dimensions of a plot at any one time, we must create several pairwise plots, which show the various factors plotted against each other (e.g., Factor 1 vs. Factor 2, Factor 1 vs. Factor 3, etc.). Inspecting the graphs can become quite tedious, since there are N(N − 1)/2 graphs plotting pairs of dimensions. [Thus, for 4 factors we would have (4 × 3)/2, or 6 plots. For 5 factors we would have (5 × 4)/2, or 10 plots, etc.]

2. What region of the map looks promising in terms of product acceptance? Using the map we can overlay numbers to show degree of liking or, even better, overlay circles whose area is proportional to the degree of liking. Thus, for 24 products mapped in a two-dimensional plot (e.g., Factor 1 versus Factor 2) we would have 24 points on the map (one point per product). Each point lies in the center of a circle or star. The area of that circle (or star) is proportional to overall liking. We look for promising locations corresponding to regions with large circles. If the large circles scatter across the map we know that on the dimensions being plotted there is no single promising region. If, however, all of the large circles cluster together in one region, then we know that in that region (of products) lie the products that are most highly accepted. Figure 4A shows Factors 1 and 2 plotted against each other (as we saw in Figure 3), but this time each point lies at the center of a star. The area of the star/circle is proportional to overall liking. Figure 4B shows the same type of plot, but this time in three dimensions.

3. Are there holes in the map, and do these holes lie near a region of products that are highly acceptable? For example, we might find that the products distribute to yield three clusters of products. Products in one cluster might be unacceptable, whereas products in the second and third clusters might be highly acceptable. If there is an empty region between cluster 2 and cluster 3 (the two clusters of highly acceptable products), then does this signal an opportunity for a product which has potential for good performance in the marketplace?

Keep in mind that the plots or maps, like the factor analysis, are a heuristic to help the researchers understand the data. We display the data to gain insight. The insight comes from organizing the data in a way that makes it easy to identify clusters of products.

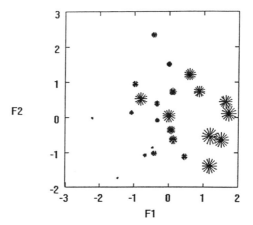

Figure 4A Two-dimensional plot of the 24 stick products, with the area of the star proportional to liking.

Going into Greater Depth—What Sensory Inputs Drive the Category?

Anti-perspirant stick products are complex, exciting perceptions of appearance, fragrance, feel, and end benefits. To which sensory attributes do consumers attend when judging overall liking? Can we say that fragrance drives the product? Appearance? Texture? How can we determine which specific attributes are critical and which are irrelevant?

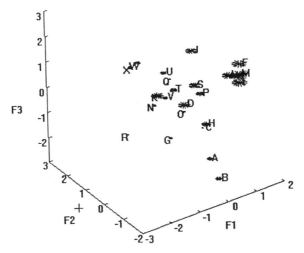

Figure 4B Three-dimensional plot of the 24 stick products, with the area of the star proportional to liking.

Benchmarking a Product Category

One of the easiest ways to find out what drives the category is to look at the liking ratings (see Table 8). From the profile of liking ratings shown in Table 8, we can identify the "strong" and "weak" points of each product. For example, product 211 achieves a rating of 33 on "like appearance." We know that appearance is a weakness of product 211. Conversely, product 205 shows a 72 for "like the way the product applies." We know that feel on skin is a strength of product 205. Thus, at the very simplest level we can find out what "drives" individual products by looking at the liking and disliking scores assigned to the attributes.

We can identify which products are liked or disliked on an absolute basis. We can even develop category "norms" and learn whether a product scores

Table 8 Likes and Dislikes for Each Stick Anti-Perspirant Product (0–100 Liking Scale)

Product	Like overall	Like appearance	Like fragrance	Like fragrance on skin	Like the way it applies
205	68	44	47	42	72
209	66	*39	53	42	65
213	65	54	53	46	70
206	65	*40	44	48	70
212	60	54	56	43	71
211	59	*33	46	*35	62
204	59	48	56	42	59
210	56	47	50	*40	61
219	56	48	49	43	64
208	48	*37	47	*39	62
216	47	*29	51	*40	59
220	44	*34	47	47	57
202	41	*27	51	*35	53
223	39	48	49	*40	57
221	38	*33	47	*35	58
201	38	45	45	*32	64
222	37	*31	50	41	59
224	37	44	54	*37	50
214	33	*32	41	*29	53
207	33	*40	*38	*37	61
215	29	41	*28	*40	60
218	22	*37	*34	*32	*36
203	22	*36	*30	*34	41
217	20	*25	*28	*31	*40

Note: * denotes a product with a low attribute liking (≤40).

higher or lower than the norm. The norm can best be defined as the average of the best selling products. Table 9 shows how we deal with the problem of performance against "category norms" instead of on an absolute basis. We see how each of the 20 products fares against the four best sellers on every attribute. The greater the number of "stars," the more category leaders the product beats. Tables 8 and 9 help the marketer and product developer because they reveal where a specific product is weak and where it is strong, and whether that strength or weakness is a property of the category or just the particular product. Most researchers assume that the high liking attribute drives overall acceptance, and a low liking attribute drives overall rejection.

Leverage Analysis

The foregoing approach may be too simplistic, and may not teach us enough about the category. Although we know the strengths and weaknesses of a product on a product-by-product basis, we really do not know the pattern. That is, if a product scores poorly on acceptance of fragrance, then does it automatically mean that the product will score poorly overall? And if a product scores well on liking of fragrance then does this substantially increase the odds that the product will be highly accepted overall? There must be a better way to uncover the pattern driving overall liking by looking at the pattern of ratings assigned to the full set of products.

Another way to identify the important drivers of liking is based upon "Leverage Analysis." Here are the three assumptions underlying the approach:

1. We know that as attribute liking increases (and holding all other attributes constant), overall liking should increase as well. Overall liking should not decrease if one aspect of the product improves and the rest of the product remains unchanged. Overall liking and attribute liking should move in the same direction.
2. We also know that overall liking often highly and positively correlates with attribute liking.
3. Suppose now that we fit a straight line relation to the data. The independent variable is "attribute liking." The dependent variable is overall liking. We are interested in the slope of the linear relation which shows how overall liking changes with attribute liking. Is the slope steep? If yes, then small changes in attribute liking covary with large changes in overall liking. We conclude that the attribute is important, because a little increase goes a long way to increase overall liking. Conversely, if the slope of the line is flat, then the attribute is unimportant. Small changes in attribute liking correspond to smaller changes in overall liking.

Benchmarking a Product Category

Table 9A Liking Ratings of the Four "Market Leaders"

Product	Like overall	Like appearance	Like fragrance	Like fragrance on skin	Like the way it applies
205	68	44	47	42	64
210	56	47	50	40	61
220	44	34	47	47	57
215	29	41	28	40	60

Table 9B How 20 Anti-Perspirant Stick Products Fare Against the Four "Market Leaders" (Attributes = Overall Liking and Attribute Liking)

Product	Like overall	Like appearance	Like fragrance	Like fragrance on skin	Like the way it applies
201	*	***	*	0	****
202	*	0	****	0	0
203	0	**	*	0	*
204	***	***	****	***	*
206	***	**	*	****	****
207	*	**	*	0	**
208	**	**	****	0	***
209	***	**	****	***	****
211	***	0	*	0	***
212	***	***	****	***	****
213	***	***	****	***	****
214	*	0	*	0	0
216	**	0	****	**	*
217	0	0	*	0	*
218	0	**	*	0	0
219	***	***	****	***	****
221	*	0	****	0	*
222	*	0	****	**	*
223	*	***	****	**	*
224	*	***	****	0	0

Notes: Each star (* → ****) denotes the number of market leaders that the product "beats" in terms of the attribute liking. Thus, product 219 beats three of the four market leaders on overall liking, and beats all four market leaders on liking of fragrance.

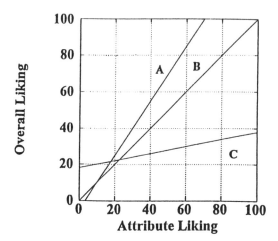

Figure 5 How attribute liking (abscissa) drives overall liking (ordinate). The lines are fitted. The steeper the line the more important the attribute.

Figure 5 shows the schematic case for three slopes. For our 24 products, the key drivers are application, then fragrance on skin, then appearance, and then fragrance (see Table 10A).

We can do the same analysis on a subgroup-by-subgroup basis. We saw above that the subgroups were fairly similar in terms of the ratings. However, if we delve more deeply into the data then do we find that there are any subgroup differences? For example, if we look at males versus females, do we find that they attend to different aspects of the product?

To answer this question we must analyze each subgroup separately. Each subgroup generates a set of ratings for overall liking and attribute liking. By analyzing the subgroups separately, we use each subgroup as its own control. If a subgroup tends to uprate the products on overall liking, then it will tend to uprate the attribute liking. Furthermore, if consumers in the subgroup tend to "stretch" the range of liking ratings from the bottom to the top of the scale, then they should exhibit the same stretching behavior for attribute liking. By using the subgroup as its own control we identify meaningful subgroup differences. Table 10B shows the results, and reveals that there are differences. These are shown by asterisks posted next to the numbers. (The asterisks identify "exceptions" to the general pattern.)

Comparisons to Another Category—Facial and Baby Soap. Do the same rules of attribute importance hold for other categories? We can perform the same analysis for the other categories, looking at the total panel to see whether, across product categories, consumers pay equal attention to the same attribute. Table 11A shows the results for body and facial soap. The results differ from our stick

Table 10A Slope (M) of the Straight Line Relating Overall Liking to Attribute Rating—Total Panel—Equation: Overall Liking = M(Attribute Rating) + B

Attribute	Slope	r^2
Attribute Liking		
Way It Applies	1.32	0.5
Fragrance/Skin	1.31	0.5
Appearance	1.13	0.4
Fragrance	0.83	0.5
Product Image		
Feminine	1.53	0.5
Unique	0.06	0.1
Younger	−0.19	0.2
Product Performance		
Long-Lasting	1.71	0.6
Refreshes	2.27	0.6
Stops Odor	0.63	0.3
Stops Wetness	0.46	0.3

Note: r^2 = Goodness-of-fit (100 × r^2 = percent of variability accounted for by the line).

Table 10B Slopes Relating Attribute Liking or Image to Overall Liking: Key Subgroups

		Key subgroups tested in the study					
		Sex		Age		Use	
Attribute	Total	Male	Female	Young	Older	Brand A	Brand B
Attribute Liking							
1. Way It Applies	1.32	1.34	1.38	* 1.10	* 1.06	1.14	* 1.53
2. Fragrance/Skin	1.31	1.17	1.39	1.26	1.45	1.50	** 0.60
3. Appearance	1.13	1.24	* 0.89	* 0.95	1.25	* 1.37	* 0.90
4. Fragrance	0.83	0.85	* 1.09	0.86	* 1.07	** 1.20	* 0.63
Image							
1. Feminine	1.53	** −0.20	1.50	1.50	* 1.24	* 1.21	* 1.27
2. Unique	0.06	0.05	* 0.34	** −0.21	−0.14	0.15	0.20
3. Younger	−0.19	* −0.44	−0.17	** 0.35	* −0.42	* −0.40	−0.12
Product Performance							
1. Long-Lasting	2.27	2.33	2.51	2.12	2.47	* 2.48	* 2.47
2. Refreshes	1.71	1.96	* 1.46	1.70	* 1.45	1.87	1.83
3. Stops Odor	0.63	* 0.90	0.53	0.73	* 0.83	* 0.89	0.61
4. Stops Wetness	0.46	0.29	0.54	0.30	* 0.65	* 0.68	* 0.65

* Denotes a difference between the subgroup slope and the total panel slope.
** Denotes a major difference between the subgroup slope and the total panel slope (change in sign, very large difference, etc.).

Table 11A How Attribute Liking and Image Ratings Drive Overall Liking (Category = Body and Facial Soaps)

Attribute	Slope	r²
Sensory Liking		
Fragrance/Skin	1.09	0.6
Fragrance/Wet	0.97	0.6
Size	0.82	0.5
Shape	0.77	0.5
Usage Properties		
Ability To Clean	2.06	0.7
Rinses Off Completely	1.78	0.4
Skinfeel While Using	1.61	0.8
Lather Amount	1.23	0.6
Comfortable	0.76	0.3
Image/End Benefits		
Refreshes	1.66	0.7
For Special Uses	0.18	0.5
For Body Care	0.16	0.5
Unique	0.14	0.3
For A Younger Person	−0.49	0.2

anti-perspirant deodorant study. Appearance is least important; fragrance is most important. Table 11B shows the same analysis for sanitary napkins.

How Sensory Attributes Drive Overall Liking

We know from scientific literature that as a sensory characteristic increases, liking changes. Although the prototypical relation may follow an inverted U shaped curve (see Figure 6A), in actuality the relation between sensory attribute level and liking can follow any of a number of different curves, such as those

Table 11B How Attribute Liking and Image Ratings Drive Overall Liking (Category = Sanitary Napkins)

Sensory liking	Slope	r²
Size	0.71	0.6
Inside Feel	0.47	0.3
Inside Contents	0.43	0.4
Color Shade	0.38	0.4
Outside Feel	0.11	0.4

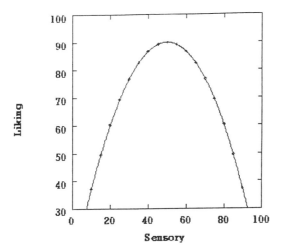

Figure 6A Inverted curve relating non-evaluative sensory attribute level (x axis) to liking (y axis).

shown in Figure 6B. The relation may be steep or flat, an inverted U shaped curve, a straight line up, a straight line down, or even a U shape. The relation between liking and a sensory attribute is developed in a straightforward manner, using regression analysis. The independent variable is the sensory attribute (e.g., amount of effort involved). The dependent variable is overall liking.

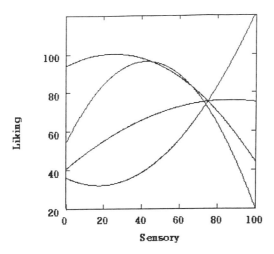

Figure 6B Individual curves relating overall liking to sensory attribute level.

In basic scientific research the investigator can systematically vary the physical characteristics of the product to generate nicely spaced gradations of a physical attribute. Here, the analysis is more difficult because we deal with a complex set of products, using what is currently in the marketplace. In-market products vary on a variety of features. Furthermore, there are often interactions among features. However, in order to be able to discern patterns in the data and understand how sensory attributes drive liking, it is vital to simplify the analysis and assume that only one single sensory attribute drives liking, even though we know that such is not the case.

The sensory attributes for analysis are those texture and appearance attributes which vary in a reasonably continuous fashion (e.g., hardness). Fragrance intensity is a relevant sensory attribute, although fragrance quality encompasses so many variations that we really must break out the products by fragrance types and do the analysis by type.

To develop the relation between liking and sensory attribute we use the following quadratic equation:

Liking = A + B(Sensory Attribute) + C(Sensory Attribute)^2

The quadratic equation above can be estimated with statistical confidence as long as there are at least six products in the database.

To get a sense of the relations we look at plots such as those shown in Figures 7A and 7B. From these two figures we can learn the following:

1. In a category, how does a sensory attribute drive liking? The data shows clearly that a straight line is not appropriate. The relation is a curve.

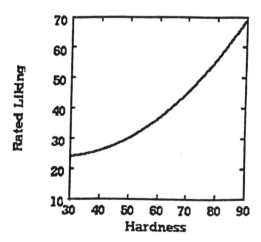

Figure 7A How perceived hardness drives overall liking of stick products.

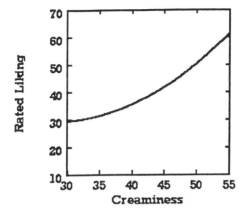

Figure 7B How perceived creaminess drives overall liking of stick products.

Often, researchers use simple linear relations between a sensory attribute and overall liking to find out which are the important sensory attributes. If the data follows a curve, then this straight line analysis is invalid, and will lead to the wrong conclusions.

2. On a sensory basis, where do our products lie versus the optimal sensory level at which liking peaks? Are we close to the optimal level? Are we far from the optimal level? What products lie at the optimal level?
3. If we were to change the product, is it critical to move towards the optimum? For instance, if the relation between sensory level and liking is fairly flat, then it really makes little or no difference where the product lies on the sensory curve. All sensory levels appear to drive liking equally. Changing the sensory level would not necessarily dramatically affect liking, although it might provide a sensory profile that marketers can use to differentiate the product from competitors, or to reinforce specific benefits.
4. If we want to move toward the optimal sensory level, then what products should we use as reference guides? The product developer does not necessarily know the physical formulation of the competition, but at least now has a landmark for development (see Table 12).

6. Creating a Category Model—Using Discrete Features

In this section we create a model relating the presence/absence of physical features to attribute ratings. The objective of the model is to quantify the relation so that a product developer or marketer learns how specific physical features

Table 12 Representative Products Which Have "Optimal Sensory Levels" (Based upon the Relation Between Sensory Intensity and Overall Liking)

Fragrance		Hardness		Creaminess of feel	
Product	Sensory fragrance	Product	Sensory hardness	Product	Sensory creaminess
213	57	205	85	205	54
206	54	206	84	202	53
205	53	209	83	206	52
		212	82	209	52
		219	78	212	52
		222	76	213	51
				201	51

"drive" ratings. The model is "discrete" because it deals with features as either "present" in the product or "absent" from the product. For continuous variables such as hardness, we can divide the continuum into different sections, so that a product has either a "low," "medium" or "high" degree of hardness (the product has only one of the three levels). The model shows the effect of changing the product from the base condition to a new condition by changing features (e.g., what effect does switching the color from white to pink have on perceived "ability to stop odor"?). The seven steps to create the category model follow:

1. Identify the features of the product by a content analysis
2. Assess the degree to which the features are statistically independent of each other
3. Select a limited, yet meaningful number of features to act as predictors in the model
4. Eliminate redundant products if necessary
5. Create the model by regression analysis
6. Use the model to identify the key discrete features which drive the consumer ratings
7. Where relevant, use the Concept Optimizer (see Chapter 1) to identify a set of compatible physical variables which, in concert, produce a defined response profile.

Content Analysis

A content analysis is simply the list of features possessed by a product. Earlier in the chapter we saw an example of content analysis (Table 2). Anyone can do the content analysis. However, the richness and utility of the analysis de-

Benchmarking a Product Category

pends on the descriptive abilities of the person who does it. A typical consumer will describe a product in one way. A trained observer will describe the same product in another way. Each person uses his own attributes and language.

A simple list of features possessed by each product is instructive but not particularly useful. It shows the developer the range of characteristics possessed by products in the category. However the list is not disciplined, nor is it in a tractable form. We can make the content analysis more "user friendly" by creating a matrix comprising feature × product. The columns of the matrix correspond to the products. The rows of the matrix correspond to the features. (The format can be transposed as well, with features as columns.) If a product possesses the feature then we code that fact as "1," whereas if a product lacks the feature then we code that fact as "0." Most of the data in the matrix will be 0's because any single product will possess very few features. However, across the rows (corresponding to the product) there will be a scattering of 1's. If the feature naturally comprises a continuum, then we divide the continuum into exhaustive, non-overlapping regions coded as 0 or 1.

Regularizing the Content Analysis Matrix

If we inspect a full content analysis matrix, then we will see that the features are not present an equal number of times. Some features are present more often, whereas other features are present in only a few products or even in only one product. From a purely statistical point of view, this starting content matrix is unacceptable because the elements do not appear as equally often, and they are more than likely to correlate with each other. Intercorrelation among the features means that it will be quite difficult, if not impossible, to tease out the effect of each of the features in any clear fashion. For instance, if two elements appear together consistently, then it is impossible to determine the effects of these two product features.

Since the benchmarking study is predicated on using in-market products, it is difficult to create an experimental design *a priori*, where all of the elements are statistically independent of each other (the condition known as "orthogonality"). However, the researcher can identify features which are not correlated with each other (or at most minimally correlated), and use these as predictors. Some of the "fine nuances" of the data will be lost, but that is the inevitable price to pay when using existing products. Furthermore, it is not necessary to create a single model using all of the features of predictors. We will develop "micro-models" using only some of the features.

In order to identify a more limited set of features (reduce the features from the larger set identified in the content analysis), we factor analyze the product features. We reduce the number of features to a more limited and usable set by the following three steps:

1. Identify the factors underlying the content analysis (see Table 13A for the results of factor analysis).
2. Identify the single feature most highly correlated with each factor identified in Table 13A (see Table 13B).
3. Use that single feature as a surrogate for all other features correlated with that factor. Sometimes it will be necessary to use several features "loading on" a factor, especially when they are qualitatively independent.

We require the factor analysis to pull out no more than 8 to 12 different features. We purposely want to pull out as many features as we can, up to 12. We want 12 features or fewer, but not more, because we tested 24 products. Ideally we would like to have at least two times and preferably three times as many products (observations in statistical terms) as we have features (independent variables in statistical terms). We need more observations than products because we will use regression analysis to create the model.

Building the Model

There are 18 different features embodied in our eight factors. Our model can only incorporate 12 of them. We select the 12 features (from the 18), and then run the regression analysis in order to develop the category model. The regression analysis estimates coefficients of the following equation:

$$\text{Attribute rating} = k_0 + k_1 + k_2 + k_3 \ldots k_{12}$$

The additive constant k_0 shows the expected rating if none of the specific features are present. It can be thought of as the baseline. In turn, the coefficients ($k_1, k_2, k_3 \ldots k_{12}$) correspond to the 12 features. If the coefficient is positive, then when the feature switches from state "0" (i.e., not present in the stick product) to state "1" (viz., present in the stick product), the consumer's rating for the attribute increases. The higher the coefficient the more impact the feature has as a "driver" of the consumer rating. For instance, among males, switching from a smooth to a "grainy" appearance increases liking by 5 points. If the coefficient is negative, then this means that when the feature switches from state "0" to state "1" the rating drops.

The coefficients for the "macro-model" appear in Table 14. By inspecting the table the marketer and researcher learn what feature specifically drives consumer responses. For instance, what feature drives the perception of "stops odor?" From a relatively complex array of products it has become possible to identify some key features as drivers. Two key drivers are "green color" and "medicinal fragrance."

Getting More "Richness" Out of the Data by Micro-Modeling

The previous analysis showed us how to understand the driving features in a category. In reality, we lose a lot of information by using the limited set of 12

Table 13A Factor Structure for Content Analysis of Stick Anti-Perspirant Products (Number = Correlation of Each Product Feature with the Factor at the Top)

	Factor							
	1	2	3	4	5	6	7	8
Color								
White	0.2	0.0	−0.2	0.0	0.2	−0.1	0.1	−0.9
Green	0.4	−0.1	0.4	0.1	−0.2	−0.1	0.2	0.7
Pink	0.8	0.1	−0.1	−0.1	−0.1	0.2	−0.4	0.3
Appearance								
Shiny	0.7	0.2	0.0	0.1	0.2	−0.2	0.2	−0.2
Grainy	−0.1	0.5	0.1	0.2	0.0	0.3	0.5	−0.5
Marking	0.1	0.3	0.6	−0.4	−0.2	0.2	0.1	0.1
Large Size	0.2	0.1	0.1	0.2	0.8	0.0	0.1	−0.3
Rectangle	0.2	0.0	0.1	−0.8	0.1	0.0	−0.2	−0.2
Oval	0.1	0.2	0.0	0.9	0.0	0.1	0.1	0.0
Round	−0.7	−0.2	−0.1	0.0	−0.4	0.0	0.2	0.0
Top Curve	0.0	0.9	−0.1	0.0	0.2	−0.1	−0.2	−0.2
Feel								
Hard	0.2	−0.3	0.0	0.7	0.1	0.3	0.1	−0.2
Dense	0.2	0.0	0.3	0.4	0.6	−0.2	0.0	0.4
Sticky	−0.1	0.0	0.0	0.0	0.9	0.0	0.2	0.0
Physical								
Length	0.3	0.0	0.0	0.0	0.9	−0.1	−0.1	−0.1
Width	−0.1	0.2	0.2	0.6	0.4	0.2	0.0	−0.4
Depth	0.3	0.0	−0.2	−0.3	0.6	−0.5	−0.1	0.3
Fragrance								
Intensity	0.1	−0.5	0.6	−0.1	0.3	0.3	−0.1	0.1
Sweet	−0.1	−0.1	−0.1	0.0	0.0	0.9	0.0	−0.2
Floral	−0.3	0.0	0.1	0.1	−0.1	0.9	0.2	0.1
Rose	0.7	0.0	−0.1	0.1	−0.2	0.3	0.1	0.2
Fruity	0.0	0.1	0.0	−0.1	−0.1	−0.2	0.9	−0.1
Citrus	0.0	0.0	−0.1	−0.1	0.0	−0.1	0.9	0.1
Medicinal	0.1	−0.3	0.8	0.1	0.1	−0.2	0.1	0.0
Spicy	0.0	0.1	0.9	0.0	0.0	0.0	0.1	0.1
Woody	0.0	−0.1	0.6	0.5	0.0	−0.2	0.0	0.1
Powdery	0.1	0.0	−0.5	0.0	0.0	0.6	0.2	0.0

features as predictors of all attributes. Because we have so few products and so many predictors, we must jettison many predictors for the sake of statistical validity. We may have thrown out too much. Is there a way to look at the effect of many more features on consumer responses?

Table 13B Representative Product Features (from the Content Analysis) Which Best Represent Each of the Eight Factors

Factor 1—Pink, shiny, rose
Factor 2—Top curve
Factor 3—Marking, medicinal, spicy
Factor 4—Oval, hard
Factor 5—Long, sticky
Factor 6—Sweet, powdery, floral
Factor 7—Grainy, fruity, citrus
Factor 8—Green

Note: Several factors exhibit 2–3 seemingly unrelated features which correlate highly with them. In those cases all the relevant features are considered.

One valid but essentially non-productive way uses individual data rather than means. If we obtain 50 ratings per product and if we test 24 products, then we have a large matrix of $50 \times 24 = 1200$ rows (one row per product). This is certainly more than 24, and allows us to estimate the effects of many more features. However, in actuality this is a poor procedure despite its statistical validity. The real information is contained in the product variation, not in the consumer variation. Individuals differ in what they like and what they dislike. But right now this variation between consumers is not of interest. We only used the 50 consumer ratings per product to get a better "fix" or "estimate" of the average for that product. Indeed, were we to have a perfectly reliable, 100% representative sample obtained from just one person, then we would need only that person's data for a valid study. The model that we create considers product differences, not people differences.

A better way to glean more information is to recognize that we can decompose the model into a set of "submodels" or "micro-models." We can break apart the set of physical features into appearance features, fragrance features, and tactile features, etc. Each separate set of features can then be used as a set of predictors, but only for the relevant consumer attributes linked to these features.

Every consumer-rated attribute has several possible models: an appearance feature model, a tactile feature model, etc. These are "micro-models," not "macro-models," because each model deals with a limited set of features. The micro-models would not necessarily fit as well as a general model because they are limited to a few predictors, all in the same general category. However, the micro-models would reveal the finer-grained details of the product and how different physical features in the same general set are able to drive those relations.

Table 14 Additive Model Showing How the Presence of 12 Sensory Features in the Stick Anti-Perspirant Product Drives Consumer Ratings

	Like overall	Purchase	Like seg 1	Like seg 2	Like seg 3	Like male	Like female
Constant	24	29	43	61	32	22	27
Pink	6	6	5	-7	11	5	4
Green	2	6	7	-3	-8	5	-2
Shiny	-4	-2	-4	2	-2	-4	-3
Grainy	4	4	4	-7	4	5	5
Marking	3	5	6	-6	7	3	7
Large size	-1	-5	4	1	4	-4	-1
Sticky	6	4	4	-6	5	5	4
Sweet	0	-3	-3	3	-1	4	-4
Rose	-6	-6	-7	6	-8	-9	-7
Fruity	5	3	3	-5	-2	0	4
Medicinal	-8	-7	-6	4	-5	-4	-4
Powdery	-4	1	-1	2	-4	-5	3
Multiple r²	0.56	0.45	0.47	0.43	0.47	0.55	0.54

(continued)

Table 14 Continued

	Like overall	Like young	Like old	Like/use Brand A	Like/use Brand B
Constant	24	19	32	21	19
Pink	6	6	6	5	6
Green	2	4	-2	7	-3
Shiny	-4	-3	-2	-1	-4
Grainy	4	5	4	4	6
Marking	3	6	0	7	4
Large size	-1	-4	2	-3	-4
Sticky	6	5	2	6	5
Sweet	0	3	-2	-3	0
Rose	-6	-5	-7	-5	-6
Fruity	5	3	1	5	-1
Medicinal	-8	-3	-1	-2	-5
Powdery	-4	-1	-3	2	-2
Multiple r^2	0.56	0.59	0.67	0.45	0.56

	Like overall	Like appearance	Like fragrance	Like fragrance on skin	Like as it applies
Constant	24	51	37	46	59
Pink	6	8	4	6	3
Green	2	5	5	1	-1
Shiny	-4	-5	1	-4	3
Grainy	4	6	3	6	12
Marking	3	9	3	2	4
Large size	-1	-7	1	-2	-2
Sticky	6	3	-2	3	12
Sweet	0	7	2	3	3
Rose	-6	-4	-12	-8	-5
Fruity	5	-7	14	-6	-3
Medicinal	-8	-9	-2	-5	2
Powdery	-4	-2	-3	-4	5
Multiple r^2	0.56	0.55	0.68	0.47	0.52

	Sensory fragrance	Sensory hardness	Sensory creaminess	Sensory smoothness
Constant	38	53	42	46
Pink	3	7	1	-2
Green	4	2	-4	3
Shiny	-2	7	4	2
Grainy	-5	9	5	-2
Marking	6	4	0	-2
Large size	-11	-6	-6	-1
Sticky	13	5	9	3
Sweet	10	3	2	1
Rose	-10	2	-3	-2
Fruity	-11	-1	0	6
Medicinal	11	2	-4	-3
Powdery	1	4	-4	-3
Multiple r^2	0.45	0.57	0.63	0.57

Table 14 Continued

	Easy to apply	Amount residue	Amount used	Feminine	Young	Unique
Constant	39	31	57	46	42	48
Pink	4	7	3	11	-8	15
Green	2	-1	2	-6	7	
Shiny	0	-2	4	-3	6	-5
Grainy	23	4	7	-5	-2	-2
Marking	12	3	9	2	4	-16
Large size	-11	-1	-2	-1	-2	-13
Sticky	5	7	7	4	2	9
Sweet	6	7	2	7	-2	9
Rose	-3	-8	-3	-6	0	0
Fruity	2	-8	2	-5	5	-10
Medicinal	-2	-6	-1	-10	3	5
Powdery	-1	-5	-2	11	7	-10
Multiple r^2	0.45	0.56	0.26	0.57	0.66	0.46

Table 14 Continued

	Stops wetness	Stops odor	Refresh	Long last
Constant	57	45	57	69
Pink	−1	2	3	3
Green	2	4	8	3
Shiny	3	0	−1	−3
Grainy	−4	−2	2	4
Marking	5	2	8	3
Large size	−4	−3	−8	−5
Sticky	7	3	7	5
Sweet	1	2	5	4
Rose	4	−1	−7	−2
Fruity	2	−2	−11	−7
Medicinal	12	8	−4	6
Powdery	3	1	−7	−1
Multiple r^2	0.35	0.39	0.56	0.24

Some of these partial models appear in Table 15 (A–C). From these models we can quickly discern the key drivers in a category, looking at appearance, fragrance, and texture (tactile). Table 15 also shows the additive constant. The additive constant is the expected attribute rating assuming that the feature has the zero condition (defined at the bottom of the table) present. The other features have positive or negative coefficients. These coefficients correspond to the expected change in the attribute rating when we switch from the zero condition to the specific feature. To estimate the attribute rating we simply add together the additive constant and the additive value (coefficient) of each feature. The sum is the best estimate we have for the expected value of the attribute, considering only this limited set of features. The "goodness-of-fit" statistic lies around 0.3–0.5. The r^2 value corresponds to the proportion of the variation in the attribute that can be explained by using the features to predict the attribute. The r^2 values are relatively low for the simple reason that the consumer attribute is often driven by features not considered by the model.

Another way to create a partial model for each attribute is to use the method of stepwise regression. The stepwise regression allows any feature to "enter the equation," as long as it accounts for a significant proportion of the variability. The stepwise regression method is often used to identify the key predictors when there are a large number of possible independent variables from which to choose. Although the stepwise regression is statistically sound and widely used, it often comes up with the same set of variables, again and again. We are not interested in the limited set of variables, but rather in the nuances that would be overlooked by simply choosing the most powerful set of predictors.

Table 15 Additive "Micro-Models" Relating Specific Categories of Features (Appearance, Tactile, Fragrance) to Attribute Ratings (Each Category of Features Yields One Set of Models)

	A. Appearance Features as Predictors										
	Constant	White	Green	Pink	Shiny	Grainy	Marking	Large size	Oval	Rect-angle	Top curve
LIKING											
Purchase	28	-2	-4	4	-1	-4	5	4	8	2	13
OVERALL											
Segment 1	22	-6	-3	1	-4	1	4	8	17	10	15
Segment 2	27	6	5	8	-4	-7	0	17	5	-4	19
Segment 3	59	-3	0	3	6	-2	-4	-14	17	27	-19
Males	22	4	2	13	-9	-15	13	24	11	3	26
Females	14	-5	0	2	-4	-1	4	9	22	19	17
Younger	25	-2	-2	5	-2	-3	3	9	18	8	12
Older	25	-14	-4	0	-1	-1	9	8	15	11	21
Use/Brand A	13	11	1	7	-6	-1	-4	9	26	15	6
Use/Brand B	36	-8	-12	2	-6	-6	20	12	15	8	14
	21	-1	2	9	-8	2	-4	16	12	5	13
OVERALL											
Appearance	22	-6	-3	1	-4	1	4	8	17	10	15
Fragrance	50	-7	-1	5	-4	1	0	5	9	8	4
Fragrance/skin	NA	NA	NA	NA	NA	NA	NA	NA	NA	NA	NA
Applies	NA	NA	NA	NA	NA	NA	NA	NA	NA	NA	NA
	83	-23	-17	-18	-1	20	5	2	-15	-21	4
SENSORY											
Fragrance	NA	NA	NA	NA	NA	NA	NA	NA	NA	NA	NA
Hard	NA	NA	NA	NA	NA	NA	NA	NA	NA	NA	NA
Creamy	NA	NA	NA	NA	NA	NA	NA	NA	NA	NA	NA
Smooth	NA	NA	NA	NA	NA	NA	NA	NA	NA	NA	NA

Table 15 Continued

A. Appearance Features as Predictors

	Constant	White	Green	Pink	Shiny	Grainy	Marking	Large size	Oval	Rect-angle	Top curve
APPLICATION											
Easy to apply	70	-30	-29	-26	0	24	13	-8	-2	-3	7
Residue	44	-13	-11	-8	-4	12	3	5	-8	-9	3
Effort	51	-6	-4	-14	0	8	2	-2	-11	-11	-1
Amount used	85	-21	-17	-20	0	17	7	3	-19	-26	5
IMAGE											
Feminine	55	-11	-10	-6	-5	12	1	3	-6	-6	4
Young	38	3	0	-6	6	-7	6	-5	-3	-9	3
Unique	60	-8	6	8	-4	-1	-3	1	-17	-11	3
Stops wetness	85	-17	-11	-17	2	6	12	-1	-17	-19	0
Stops odor	59	-10	-7	-8	-1	5	5	1	-10	-10	1
Refreshes	78	-25	-18	-19	-4	18	10	0	-10	-11	5
Long-lasting	82	-16	-13	-15	-6	18	1	-1	-10	-13	4

NA - Attribute not appropriate for predictors.
Zero conditions (additive constant):
   Color - other than white, green, pink   Marking - none
   Size - small   Shape - round
   Top curve - none (flat)

B. Fragrance Features as Predictors

	Constant	Fragrance intensity	Sweet	Floral	Rose	Fruity	Citrus	Medicinal	Spicy	Woody	Powder
LIKING											
Purchase	39	1	-4	3	-5	-3	6	-1	4	-3	3
Total	39	3	5	-10	-11	-8	16	-2	13	-4	6
Segment 1	46	-5	5	-9	-13	-9	21	-1	11	-7	7
Segment 2	49	-1	10	-2	15	7	-24	4	-10	30	7
Segment 3	35	6	8	-27	-4	-31	47	0	27	-16	11
Males	39	0	7	-11	-14	-5	14	-7	22	3	3
Females	36	5	3	-9	-6	-13	20	5	5	-5	12
Younger	40	0	9	-10	-7	-8	13	-5	18	-2	5
Older	35	7	-1	-10	-14	-9	21	5	6	-2	10
Use/brand A	35	5	6	-10	-8	-3	8	1	19	-12	8
Use/brand B	34	3	14	-15	-10	-17	24	1	12	3	7
OVERALL											
Appearance	39	3	5	-10	-11	-8	16	-2	13	-4	6
Fragrance	NA	NA	NA	NA	NA	NA	NA	NA	NA	NA	NA
Fragrance/skin	67	-5	-1	6	-15	5	0	-1	3	8	2
Applies	47	-7	8	-3	4	-23	23	4	13	39	15
	NA	NA	NA	NA	NA	NA	NA	NA	NA	NA	NA
SENSORY											
Fragrance	44	8	2	1	-4	-12	11	2	4	-17	-9
Hard	NA	NA	NA	NA	NA	NA	NA	NA	NA	NA	NA
Creamy	NA	NA	NA	NA	NA	NA	NA	NA	NA	NA	NA
Smooth	NA	NA	NA	NA	NA	NA	NA	NA	NA	NA	NA

Table 15 Continued

B. Fragrance Features as Predictors

	Constant	Fragrance intensity	Sweet	Floral	Rose	Fruity	Citrus	Medicinal	Spicy	Woody	Powder
APPLICATION											
Easy to apply	71	0	-3	9	-13	15	-16	-8	0	-13	-10
Residue	40	2	1	-1	-4	-5	6	-7	8	-17	-5
Effort	40	5	-3	1	-6	4	-3	1	-6	-4	-2
Amount used	67	6	-3	1	0	5	-9	-3	9	-33	-8
IMAGE											
Feminine	56	2	0	-1	-2	-4	4	-11	6	-11	-6
Young	35	1	-5	2	0	14	-16	-6	5	-10	-7
Unique	61	-1	-6	7	6	-17	15	-8	-2	3	-19
Stops wetness	63	2	-4	6	1	5	-10	6	-1	-17	-9
Stops odor	48	2	-1	2	-2	-1	-1	0	0	-10	-5
Refreshes	72	1	-1	5	-8	-3	3	-7	6	-18	-10
Long-lasting	77	3	0	0	-7	-1	2	-6	4	-20	-5

NA = Attribute not appropriate for predictors.
Zero conditions (Additive Constant):
  No fragrance
  No definable fragrance character

Benchmarking a Product Category

C. Tactile Features as Predictors

	Constant	Hard	Dense	Sticky	Long	Wide	Deep
LIKING							
Purchase	32	−1	0	25	−20	5	6
Overall	34	−1	0	29	−22	6	9
Segment 1	34	−3	3	22	−5	4	6
Segment 2	64	7	−10	−28	13	2	1
Segment 3	24	−7	7	35	−16	8	15
Males	35	2	−3	41	−32	6	9
Females	32	−3	3	19	−13	8	11
Younger	37	−1	0	38	−31	5	10
Older	29	−1	0	18	−9	10	9
Use Brand A	35	−2	2	33	−22	6	7
Use Brand B	30	1	−2	33	−19	8	10
OVERALL	34	−1	0	29	−22	6	9
Appearance	60	1	−2	22	−16	−1	2
Fragrance	NA	NA	NA	NA	NA	NA	NA
Fragrance/skin	NA	NA	NA	NA	NA	NA	NA
Applies	69	−1	0	13	−14	−1	−1
SENSORY							
Fragrance	NA	NA	NA	NA	NA	NA	NA
Hard	66	−3	2	20	−22	6	−2
Creamy	46	−4	4	3	−6	1	−2
Smooth	45	−7	7	−8	5	−2	−2
APPLICATION							
Easy to apply	67	−6	5	2	−12	6	−7
Effort	43	2	−4	−9	7	−4	0
Amount used	67	−1	−1	14	−11	−2	0
IMAGE							
Residue	38	−1	1	8	−4	0	−1
Feminine	51	−2	1	8	−8	2	−1
Young	34	−2	2	3	−6	−4	0
Unique	55	−6	6	8	−10	−6	−6
Stops wetness	70	−2	0	−1	2	−8	−7
Stops odor	51	−2	0	−2	2	−4	−3
Refreshes	70	−2	0	7	−9	−1	−3
Long-lasting	78	−1	−1	1	−5	−3	−2

NA = Attribute not appropriate for predictors.
Zero conditions (Additive Constant):
    Soft
    Low density
    Not sticky
    Short
    Narrow
    Not deep

Looking at the results from Table 15 we can immediately begin to see how the features of the product drive attribute ratings. The models are certainly far from perfect and, like the general model (with only 12 terms, chosen from the many features), our estimates of the attribute levels will be "off," sometimes by a little and sometimes by a lot. But the good news is that one is able to identify the key features quantitatively, to provide insight, even though that insight may only be qualitative. We have used modeling to achieve additional insight into the category.

Using the Macro-Models to Design New Products

The models using the feature are similar to concept models which show how the presence or absence of an element communicates a specific benefit. We can take that analogy one study further by treating the feature model as if it were a concept model. We specify a profile of consumer attributes for the product, and then search for that combination of features which must be present in the product in order to come as close as possible to that desired profile. The consumer-perceived profiles can be changed to accord with different marketing strategies. The physical features will change as well.

We begin this design project with the general model comprising the 12 independent features that we developed and considered in Table 14. (The equations we use appear in Table 14.) Let us now define a variety of strategies. One strategy will create a product that maximally appeals to all consumers (maximizing overall liking) or maximally appeals to males or females, respectively. A second strategy will create a product that communicates a unisex message. A third strategy will have the product communicate a strong efficacy and be highly acceptable to consumers. Table 16 shows the set of features (from the set of 12 identified in Table 14) which in concert best satisfy the specific goals.

Using the Micro-Models to Design New Products

The macro-model above is easy to use. However, it lacks the "fine-tuning," simply because the full gamut of sensory and image attributes cannot be accounted for by 12 "on/off" switches or features. We need to use the set of micro-models that we created above (see Table 15), with each micro-model showing how an attribute can be best predicted from the features.

The approach using the micro-models is described below. We use the micro-models developed in Table 15. For example, the appearance attributes are predicted only by appearance features. The tactile and fragrance features all are assigned values of "NA" in the micro-model because they do not enter the equation. For the image and liking attributes, however, there are three models

Table 16 Recommended Stick Features—Results from Macro-Models

Goal 1:	Total	Males	Females	Unisex tonality	Stops wetness	Stops wetness
Goal 2:	None	None	None	High liking total	Stops odor	Stops odor
Goal 3:	None	None	None	None	None	High liking
Product features						
Color	Pink	Green	Pink	Green	Green	Pink
Shine	No	No	No	No	No	No
Grain	Yes	Yes	Yes	Yes	No	Yes
Marking	Yes	Yes	Yes	Yes	Yes	Yes
Large size	Small	Small	Small	Small	Small	Small
Sticky	Yes	Yes	Yes	Yes	Yes	Yes
Fragrance	Fruity	Fruity	Sweet	Fruity	Sweet	Fruity
Expected ratings						
Liking						
Total	46	42	40	42	32	46
Males	47	47	39	47	34	47
Females	50	44	51	44	39	50
Image						
Male vs. female	66	49	59	49	44	66
Function						
Stops odor	57	59	61	59	68	57
Stops wetness	53	65	69	65	73	53

for each attribute: an appearance feature model, a fragrance feature model, and a tactile feature model, respectively.

Once we have developed the three micro-models (appearance, fragrance, tactile), we proceed as in Table 16. Let us take the same goals as set forth in Table 16. Whereas previously we were limited to 12 features of the model (and only a limited subset of those features for any particular "solution" to the problem), now we have many more features. Since we have many more features incorporated into the model we can create more specific solutions corresponding to the features. Table 17 shows the same six goals, but this time with the more complete set of features, based upon the partial models (shown in Tables 15A-15C, respectively). The expected attribute ratings are derived from averaging the results from the three micro-models.

Table 17 Optimal Combinations of Features for Stick Anti-Perspirant Based upon "Micro-Models": Results for Different Marketing Objectives

	A. Optimizing for Total Panel Acceptance	
Product feature	Expected attribute rating	
Pink	Liking overall	66
No shine	Liking male	71
Grainy	Liking female	68
Marking	Masculine vs. feminine	62
Large size	Stops wetness	59
Rectangle	Stops odor	52
Top curve		
Citrus		
Strong fragrance		
Soft		
Dense		
Sticky		
Short		
Wide		
Deep		

	B. Optimizing for Male Acceptance	
Product feature	Expected attribute rating	
Green	Liking overall	66
No shine	Liking male	74
No grain	Liking female	61
Marking	Masculine vs. feminine	58
Large size	Stops wetness	59
Rectangle	Stops odor	46
Top curve		
Strong fragrance		
Spicy		
Hard		
Not dense		
Short		
Wide		
Deep		

	C. Optimizing for Female Acceptance	
Product feature	Expected attribute rating	
Pink	Liking overall	68
No shine	Liking male	70
No grain	Liking female	68
Marking	Masculine vs. feminine	55
Large size	Stops wetness	57
Rectangle	Stops odor	47
Top curve		
Strong fragrance		
Citrus		
Soft		
Dense		
Sticky		
Short		
Wide		
Deep		

Table 17 Continued

Product feature	D. Optimizing for a Unisex Communication	
	Expected attribute rating	
Pink	Liking overall	37
No shine	Liking male	34
No grain	Liking female	40
No marking	Masculine vs. feminine	50
Small size	Stops wetness	62
Round	Stops odor	47
No curve		
Weak fragrance		
Powder		
Soft		
Not dense		
Not sticky		
Short		
Narrow		
Deep		

Product feature	E. Optimizing for an Efficacy Communication (Stops Wetness, Stops Odor)	
	Expected attribute rating	
Off white	Liking overall	22
Shiny	Liking male	14
No grain	Liking female	27
Marking	Feminine	51
Small size	Stops wetness	80
Round	Stops odor	56
No curve		
Strong fragrance		
Floral		
Soft		
Dense		
Not sticky		
Long		
Narrow		
Shallow		

Product feature	F. Optimizing for Efficacy (Stops Wetness) and Total Panel Liking	
	Expected attribute rating	
Off white	Liking overall	65
No shine	Liking male	70
Grainy	Liking female	58
Marking	Feminine	64
Large size	Stops wetness	70
Rectangle	Stops odor	53
Top curve		
Strong fragrance		
Spicy		
Soft		
Dense		
Sticky		
Short		
Narrow		
Deep		

An Overview to Modeling Products by Discrete Features

Rather than considering the product category to comprise, in the words of William James, a "blooming, buzzing confusion," we can wrest structure from the variability in the marketplace. The sheer range of competitive products available from different manufacturers guarantees that these products will differ from each other. We can use the naturally occurring variation to our advantage. As long as consumers consistently differentiate among the products in terms of liking, or in terms of communicating desired benefits, then we know that the consumer response must be traceable to the physical features, and to nothing else. Feature modeling as we have done here formalizes the process of discovering which physical features drive consumer responses. The analytic process enables the researcher to identify quite rapidly the salient features of the product for next stage development. Of course the researcher must realize that the information should be used for general guidance rather than for precise formulation direction. The researcher did not set up the study in accordance with a statistical experimental design, but rather used the products that were currently available as a means to identify key features in the category.

7. Finding and Filling Holes in an Existing Category

Background

In product development and marketing, manufacturers continually look for market opportunities. Some opportunities may require product improvement. Most opportunities involve finding new products. New products must have some defined advantage, whether they be better, or possess features or benefits not possessed by current products. This never-ending search for new and fulfillable opportunities challenges R&D and marketing alike.

We can consider the benchmarking category appraisal as a way to identify where the current products "lie" in either the world of product features or the world of perceived consumer benefits. The maps that we developed above in Figures 3 and 4 show us how closely products lie to each other. A list of features, on a product-by-product basis, shows the composition of products. What remains to do is to turn the problem around. Rather than mapping the category to locate products, map the category to find holes. Holes suggest opportunities. Going one step further, can we identify opportunities (or holes in the map) corresponding to consumer-perceived benefits? If so, then we can go one step further and identify the features which correspond to that product, and estimate potential acceptance.

How do we find holes in a category corresponding to features? The basic data comprises a matrix of product (row) by feature (column). The numbers in the matrix are either 0 (feature is absent) or 1 (feature is present). Let us assume that there is no other information available.

With data of this type, we ordinarily cluster the rows together. A cluster comprises products (rows) that are similar to each other, based upon the profile of features. Similarity is measured by some measure of association between pairs of rows. Rows that look similar (have a similar profile of features that they comprise) fall into the same cluster. Rows that look different (have a dissimilar profile of features) fall into different clusters. Researchers often do this clustering by inspection. The subjective judgment can be replaced by more objective and standard clustering procedures. Standard clustering defines the characteristics of each of the clusters. The center of the cluster has an expected profile. The profile corresponds to the features defining the cluster.

We now turn the problem around 180°. Clustering places products together based upon the similarity of their features. We are, however, interested in the *empty space*—the regions where there are no products. Clustering "empty space" (rather than clustering features, or locations of "filled space") finds the center of the region where there are *no products*. Figure 8 shows the two plots. The squares (in black, against a white background) correspond to products (Figure 8A). We are not interested in the products. Instead, we are interested in where the products are missing (squares in Figure 8B—the reversed matrix).

The approach is straightforward. We begin with the matrix of products (rows) by features (columns). The beginning matrix has "1" corresponding to the presence of a feature, and "0" corresponding to the absence of a feature. Let us reverse the matrix. For each number in the matrix, compute its inverse. We subtract each number from a constant. Let us call the constant "1." Thus

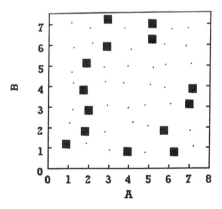

Figure 8A Location of products (squares) in a two-dimensional space. Products are dark, holes or empty spaces are light.

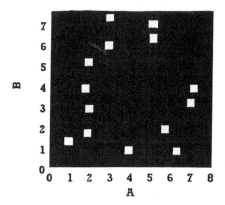

Figure 8B Reversed matrix: location of holes in a two-dimensional space. Holes are dark, products are light squares.

a 0 corresponds to a 1 (1 − 0 = 1), and a "1" corresponds to a 0 (1 − 1 = 0). This is the basic "reversed matrix," and appears in part in Table 18. Table 18A shows the original matrix, Table 18B shows the reversed matrix.

The reverse matrix is now ready for clustering. When we cluster the reverse matrix we cluster "empty space." The centroids of the clusters are the centroids of the empty space. Each cluster of empty space represents a potential product. The clustering program provides an estimate of what features are present in each cluster.

To illustrate this approach, let us consider the case of five clusters in the empty space. The features present in each cluster appear in Table 19A–B. Since the clusters comprise a variety of different cases in the space, the features that appear in each cluster do not average to 1.0. (Some of the cases have the feature, other cases do not have the feature.) We define each cluster in terms of the strongest feature present in that cluster. No cluster can have more than one feature from any one category (e.g., a given cluster can have only one fragrance, one color, etc.)

Once we have the features in each cluster we can go back to the model and estimate the likely profile of consumer reactions corresponding to those features. We can do that using the equations shown in Tables 15A–15C. We can also estimate the likely acceptance rating corresponding to that combination of features. These estimates appear in Table 20.

Identifying Opportunities in Terms of Perceived Consumer Sensory Characteristics and Benefits

The previous section dealt with identifying combinations of features not present. Can we accomplish the same objective in terms of consumer-perceived sensory

Table 18 Example of Basic Feature Matrix (Top) and "Reversed Feature Matrix" (Bottom) for Five Products

	A. Original Matrix (1 = Present, 0 = Absent)				
			Product		
Feature	A	B	C	D	E
Shiny	1	1	1	0	1
Grainy	1	1	1	0	1
Marking	0	0	0	0	0
Large size	0	0	0	0	1
Top curve	1	1	0	1	1
Hard	0	0	0	0	0
Dense	0	0	0	0	0
Sticky	0	0	0	0	1
	B. "Reversed Matrix" (1 = Absent, 0 = Present)				
			Product		
Feature	A'	B'	C'	D'	E'
Shiny	0	0	0	1	0
Grainy	0	0	0	1	0
Marking	1	1	1	1	1
Large size	1	1	1	1	0
Top curve	0	0	1	0	0
Hard	1	1	1	1	1
Dense	1	1	1	1	1
Sticky	1	1	1	1	0

characteristics and benefits? We clustered "holes" in a feature space to yield combinations of features not currently provided. Can we do the same with "holes" in a benefit space? That is, can we identify combinations of consumer-perceived benefits that are not currently being satisfied? And, once given those benefits, then can we use the Concept Optimizer procedure to identify a limited set of features that in concert delivers those consumer-perceived benefits?

Approach

The approach is similar to that for the features. In the work on features, each feature was either present or absent. That allowed us to find the inverse of the feature. When it comes to consumer benefits, however, the panelists rated each product on a 0–100 scale. Consequently each product possesses a set of benefits to different degrees. No product typically achieves a value of 100 or 0 on a specific benefit, so we are unable to apply the strict inverse operation (where

Table 19 Features of Five "Clusters"—Each Cluster Corresponds to a "Hole" in the Feature Space of Stick Products

	A. Averages in Each of the Five Clusters				
Feature	Cluster 1	Cluster 2	Cluster 3	Cluster 4	Cluster 5
White	0.5	1.0	1.0	1.0	0.3
Green	0.5	1.0	1.0	1.0	0.7
Pink	1.0	0.0	1.0	1.0	1.0
Shine	0.5	0.3	1.0	1.0	0.7
Grain	0.0	0.3	1.0	1.0	0.0
Marking	1.0	1.0	1.0	1.0	1.0
Large size	0.0	0.8	1.0	1.0	0.6
Rectangle	1.0	0.5	1.0	1.0	0.4
Oval	0.0	0.3	1.0	1.0	0.2
Curve	0.0	0.7	0.3	0.2	0.7
Top curve	0.2	0.3	1.0	1.0	0.3
Hard	0.0	0.8	1.0	1.0	0.5
Dense	0.0	1.0	1.0	1.0	0.9
Sticky	0.0	0.8	1.0	1.0	0.8
Long	0.6	0.4	1.0	0.4	1.0
Wide	0.2	0.7	0.0	0.8	0.0
Deep	0.8	1.0	0.8	1.0	0.2

	B. Features that Must Be/Can Be Present in the New Products				
	Cluster = product opportunity				
Feature	Cluster 1	Cluster 2	Cluster 3	Cluster 4	Cluster 5
White	No	Yes	Yes	Yes	No
Green	No	Yes	Yes	Yes	Yes
Pink	Yes	No	Yes	Yes	Yes
Shine	No	No	Yes	Yes	No
Grain	No	No	Yes	Yes	No
Marking	Yes	Yes	Yes	Yes	Yes
Large size	Small	Large	Large	Large	Large
Rectangle	No	No	Yes	Yes	No
Oval	Yes	No	Yes	Yes	No
Circle	No	Yes	No	No	Yes
Top curve	No	No	Yes	Yes	No
Hard	No	No	Yes	Yes	No
Dense	No	Yes	Yes	Yes	Yes
Sticky	No	Yes	Yes	Yes	Yes
Long	Yes	No	Yes	No	Yes
Wide	No	Yes	No	Yes	No
Deep	Yes	Yes	Yes	Yes	No

Note: A "1" means that the "hole" should have the feature.
A "0" means that the "hole" should not have the feature.

Table 20 Expected Profile of Key Attribute Ratings For Five Different Proposed Products Corresponding to "Holes" in the Category

Category	Cluster (hole)	Liking total	Liking males	Liking females	Image female	Stops wet	Stops odor
Appearance	1	68	67	69	62	68	53
Tactile		22	12	30	42	64	50
Average		45	39	50	52	66	52
Appearance	2	67	68	72	51	61	48
Tactile		49	48	50	52	53	43
Average		58	58	61	52	57	46
Appearance	3	67	68	72	51	61	48
Tactile		50	52	49	49	61	46
Average		59	60	61	50	61	47
Appearance	4	23	16	30	50	68	51
Tactile		56	58	56	51	53	43
Average		39	37	43	51	61	47
Appearance	5	22	13	26	52	99	63
Tactile		13	0	23	44	72	53
Average		18	7	25	48	86	58

Profiles obtained by:
(a) Using the appearance micro-model (Table 15A) and tactile micro-model (Table 15C). Each model above yields an independent estimate of the attribute rating model.
(b) Averaging the two estimated profiles.

a 1 becomes a 0 and a 0 becomes a 1). We can, however, specify a value at which the feature is considered to be present. This value is arbitrary. For our stick anti-perspirant deodorant case history, let us consider the value of 60 or more (on the average) to represent that a product possesses the benefit, and a value of less than 60 to reflect that a product lacks the benefit. We also reverse certain attributes if desired.

Table 21 shows a part of the consumer benefits matrix, first in terms of the original numbers (on the 0–100 scale), and then in terms of the categorization of the scale values with a mean of 60+ replaced by the value 1 and a mean of 60 or below replaced by the value 0. Finally, Table 21 shows the matrix "reversed" as we did for the feature. Benefits that are absent from the original product are denoted by the value "1," whereas benefits that are present in the original product are denoted by the value "0."

The benefit matrix can be clustered in the same way that we clustered the feature matrix. This time the clusters correspond to the centroid or center of gravity of "empty space" (where there is a specific absence of benefits). The clusters show combinations of current consumer benefits that are not provided by the products currently in the market. The clusters do not, however, provide an indication of benefits that were missing from the original set of products. That is, we cannot discover benefits which did not previously exist. To understand these missing benefits we need to work on concept development, rather than category analysis. Table 22 shows the clusters of current benefits missing in the category.

The final analysis creates combinations of physical features corresponding to these clusters. We can use the Concept Optimizer to "reverse engineer" and thus discover the combination of physical features that corresponds to the desired combination of consumer benefits. The goal to be satisfied is the combination of benefits that are missing. The independent variables are the physical features. Concept optimization determines the relevant combination of physical features which in concert generate the desired combination of benefits (see Table 23).

An Overview

This section presented an approach to creating new products by identifying empty holes in a benefit space. The logic is similar to that followed with identifying combinations of physical features absent from the competitive frame. The approach provides the marketer with an opportunity to create new products with defined benefits, and to rapidly identify the features corresponding to those benefits.

Table 21 Part of the Matrix of Products × Benefits, Presented in Three Ways

Product	a) Original product × attribute matrix			
	201	202	203	204
Easy To Use	65	78	53	76
No Residual*	70	68	65	61
Use A Little*	37	39	32	40
Feminine	45	48	52	55
Young	39	45	29	35
Unique	38	36	56	66
Refreshing	65	62	61	66
Long-Lasting	69	68	68	72
Product	b) Attributes converted to 1 if attribute mean > 60			
	201	202	203	204
Easy To Use	1	1	0	1
No Residual*	1	1	1	1
Use A Little*	0	0	0	0
Feminine	0	0	0	0
Young	0	0	0	0
Unique	0	0	0	1
Refreshing	1	1	1	1
Long-Lasting	1	1	1	1
Product	c) Attribute reversed (1 becomes 0, 0 becomes 1)			
	201	202	203	204
Easy To Use*	0	0	1	0
No Residual*	0	0	0	0
Use A Little*	1	1	1	1
Feminine	1	1	1	1
Young	1	1	1	1
Unique	1	1	1	0
Refreshing	0	0	0	0
Long-Lasting	0	0	0	0

* = Original scale revised to make analysis easier

Table 22 Four Clusters of Benefits not Currently Present in the Category Benefits (where appropriate the scales are reversed)

	Means			Description		
	Easy to use	No residual*	Feminine image	Easy to use	No residual*	Feminine image
Cluster 1	0.27	0.00	1.00	No	No	Yes
Cluster 2	0.00	1.00	0.00	No	Yes	No
Cluster 3	1.00	0.00	1.00	Yes	No	Yes
Cluster 4	1.00	1.00	1.00	Yes	Yes	Yes

* Attribute is reversed

Table 23 Final Profile of Physical Features (Appearance, Touch) Corresponding to the Four Different Clusters of Benefits Missing in the Category

Feature	Cluster 1	Cluster 2	Cluster 3	Cluster 4
Color	White	White	White	Pink
Shiny	No	Yes	No	Shiny
Grainy	No	No	No	No
Marking	No	No	No	No
Large size	Small	Large	Small	Small
Shape	Oval	Oval	Oval	Oval
Top curve	Yes	No	No	No
Hard	Hard	Hard	Soft	Hard
Dense	No	No	Yes	Yes
Sticky	Yes	No	Yes	No
Length	Long	Long	Long	Long
Width	Narrow	Narrow	Wide	Wide
Depth	Deep	Deep	Deep	Deep

8. Creating a Category Model Using Continuous Variables: An Example with Bath Gels

Background

Up to now we have dealt with a product whose features are "discrete" (not continuous), so that the features are either present or absent. (The only exception is fragrance intensity for specific fragrance notes.) Let us extend the notion of a category model to the continuous case. The category is bath gel. The specific problem is to benchmark existing bath gels and a variety of alternative prototypes on overall liking, perceived thickness, perceived amount of lather, etc. The objective of the research is to create a model that allows the product developer to identify optimum sensory levels of the different continuous variables.

Kathleen MacDonnell runs the Hadden Soap Company. Two years before, Hadden Soap had entered the liquid soap business. The unique selling proposition was that their liquid soap entries would be made from the finest ingredients, just as their bar soaps were. After the initial successes, MacDonnell was encouraged by the line extension strategy, and investigated the feasibility of entering the bath gel category. Bath gels were a growing category, and a research study suggested that the Hadden Soap Company might well profit from entering this category. Furthermore, in the minds of consumers, Hadden Soap Company could very well manufacture a bath gel and charge a premium price.

Benchmarking a Product Category

Marketing suggested an immediate push to create and market a line of gels. However, MacDonnell suggested that a more prudent first step would be to identify the "acceptance" drivers in the category, and then fine tune the product to appeal to consumers by "getting the product right." In this case the issue was to identify the correct fragrance type, color (specifically shade because green was selected to be the primary color), viscosity, and amount of lather. Prior to any experimentation, MacDonnell wanted to benchmark the existing products in order to identify the proper sensory levels of products. (This same problem could have been solved by systematic experimental design. For illustrative purposes, however, we are going to consider how the problem can be addressed by benchmarking, and by creating a model based upon the relation between sensory level and liking.)

Research Approach

Researching bath gels is fairly straightforward. Consumers evaluate different products monadically, at home, rating each product on a variety of attributes. Here we concentrate on the attributes of overall liking, depth of color (shades of green), perceived viscosity, perceived oiliness of the gel as it was added, and amount of lathering. The actual questionnaire dealt with many other attributes, including fragrance, feel, and end performance.

A panel of 60 consumers participated. All panelists were category users of bath gel products, having purchased and used a product within the past 2 weeks. Furthermore, panelists were positively disposed to a concept talking about a new line of "high quality, herbal scents, with a note of forest pine and eucalyptus." The criterion for "positive to the concept" was that the concept was to be rated as "definitely would buy" or "probably would buy" when read by an interviewer during a telephone interview.

The product set comprised 18 bath gels, varying in color, fragrance, viscosity and lather. Ten of the 18 bath gels comprised the key competitors in the category. The remaining eight products comprised a variety of selected pine-type fragrances, of green color, created by R&D to represent different types of products that could be made, but were not currently available.

Each respondent used every one of the 18 gels, over a 2-month period, bathing with one gel at least every 3 days, and rating the gel on attributes after the first bathing experience. The questionnaire probed liking, sensory, image, and feeling characteristics (see Table 24). Many of the attributes for image and feeling were "bipolar," with either end of the scale corresponding to an acceptable product. For instance, the attribute of "age" is defined by two end points: "appropriate for younger people" vs. "appropriate for older people." Both sides of the scale are correct—it is only a matter of where the panelist places the rating. Often when the researcher uses a unipolar scale such as "appropriate for an older

Table 24 Questionnaire for Bath Gel

We would like your opinion on each bath gel just before and after using it. Please take a moment to rate each bath gel just before and just after using it.

    Before You Bathe—Please Look at the Gel in the Transparent Container and Smell It:
- A. How much do you like the way it looks? (0 = hate → 100 = love)
- B. How dark is the gel? (0 = very light → 100 = very dark)
- C. How much do you like the fragrance? (0 = hate → 100 = love)
- D. How strong is the fragrance? (0 = cannot smell it → 100 = extremely strong)

    Now, use the Gel in Your Bath the Way You Usually Use a Bath Gel.
    When You are Finished Drying Yourself, Please Take a Moment to Rate the Product:
- E. How thick is the gel as you pour it? (0 = thin → 100 = thick)
- F. How oily is the gel as you pour it? (0 = not oily → 100 = very oily)
- G. How much do you like the feel of the gel as you pour it? (0 = hate → 100 = love)
- H. How much do you like it overall? (0 = hate → 100 = love)
- I. How much do you like the fragrance? (0 = hate → 100 = love)
- J. How strong is the fragrance? (0 = cannot smell it → 100 = extremely strong)
- K. How much lather does the gel yield? (0 = none → 100 = an extreme amount)
- L. How quickly does the gel lather? (0 = very slowly → 100 = very quickly)
- M. How much do you like the lather? (0 = hate → 100 = love)
- N. How oily does the bath feel? (0 = not oily at all → 100 = extremely oily)
- O. How sensuous does the bath feel? (0 = not sensuous at all → 100 = extremely sensuous)
- P. How refreshed do you feel? (0 = not refreshed → 100 = extremely refreshed)
- Q. Describe the bath gel (0 = masculine → 100 = feminine)
- R. Describe the bath gel (0 = for younger people → 100 = for older people)
- S. Describe the bath gel (0 = functional → 100 = cosmetic)*
- T. Describe the bath gel (0 = conventional → 100 = unique)
- U. Describe the bath gel (0 = serious → 100 = fun)

* Defined at the start of the study by examples

person," the scale is set up in such a way that one end of the scale covertly corresponds to a positive attribute and the other end of the scale corresponds to an implicit negative attribute. By anchoring the scale at both ends by equally positive attributes the researcher insures that the bipolar image or performance scale locates the product in a semantic space, rather than restating the attribute of "overall liking."

Results

The basic analysis comprises a product × attribute matrix, similar to the matrices we have seen before for multiple product evaluations. The three higher level analyses are:

Benchmarking a Product Category

1. An analysis of how sensory attributes drive liking and image, from a sensory basis, among the 18 pine-type/green products. By means of a quadratic function, this analysis separately relates each liking and image attribute to every sensory attribute.
2. An integrated model combining all sensory attributes in order to estimate overall liking.
3. An integrated model which discovers the sensory profile of a new product that has specific image characteristics, and use of the product × attribute data to identify reference products or targets for subsequent development.

What Drives Overall Liking?

As a sensory attribute increases, liking first increases, peaks and then drops. The key is to identify the curve relating liking to sensory attributes. The researcher must recognize that many sensory attributes jointly contribute to overall liking, so that a single curve shows just part of the picture. Often purists argue that such unidimensional analysis misleads, given the complexities of what drives product acceptance. However, in the absence of systematic product variation and with relatively few products to study, unidimensional models are both statistically valid (they fit the data, they are parsimonious) and instructive (they show the nature of the relation and provide guidance).

Figure 9A shows the relation between perceived darkness of green of the bath gels and liking of appearance. Figure 9B shows the relation between amount of lather produced by the product in water and liking of lather. Finally, Figure 9C shows the relation for perceived fragrance intensity vs. liking of fragrance. Fragrance was evaluated in the tub, after the product was completely dissolved in the hot water.

From Figures 9A–9C we begin to understand the dynamics of the bath gel. Even though we do not know what ingredients each gel comprises, the products are sufficiently similar to each other to yield a product model.

The attribute × liking curves also show attribute importance. The area under the curve (see Figures 10 A and B) reflects attribute importance. The larger the area subtended by the curve, the more important an attribute becomes to consumers. An attribute can be very important if it spans a moderate sensory range, but within that moderate range it covaries dramatically with overall liking (even if in a curvilinear fashion). For instance, the area under the fragrance intensity curve (vs. overall liking) is quite large. Attribute importance can be measured by the definite integral of the area under the curve.

An "Integrated" Model

One of the biggest shortcomings of category appraisal is that it does not use experimentally designed products, and therefore the analyses often are done on

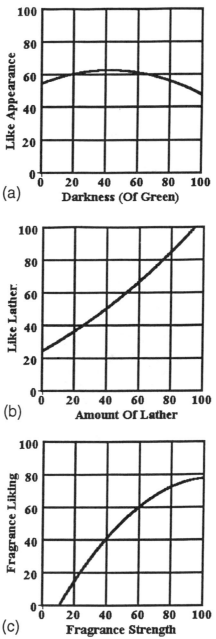

Figure 9 (A) Bath gel: Liking of appearance versus perceived darkness of green. (B) Bath gel: Liking of lather versus overall perceived amount of lather. (C) Bath gel: Liking of fragrance versus fragrance intensity.

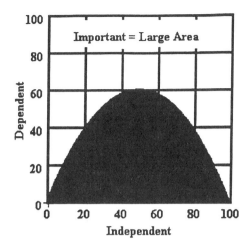

Figure 10A Example of an important attribute. The curve subtends a large area.

an attribute-by-attribute basis (see Figures 9A–9C above for green color, amount of lather and fragrance intensity). Occasionally, researchers develop more complicated models to estimate overall liking using multiple linear regression. We saw an example of multiple linear regression when we dealt with the antiperspirant stick products. For our study here, a multiple linear regression typically generates this type of equation:

Overall liking = $k_0 + k_1$(Green) + k_2(Fragrance) + k_3(Lather)

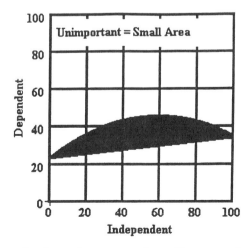

Figure 10B Example of an unimportant attribute. The curve subtends a small area.

The foregoing linear equation requires three predictors. The astute researcher could argue, however, that a simple linear equation doesn't tell the whole story for two reasons:
1. Curvature exists: As a sensory attribute increases, liking first increases, peaks, and then drops. Beyond the optimum there is no further increase in liking with an increasing sensory attribute. As a consequence, the simple linear equation misleads. One might argue for fewer terms, incorporating only those linear terms that add significantly to the goodness-of-fit of the equation. In reality, however, using only statistically significant terms begs the question. The truer relation often is not linear, but quadratic. No number of significant linear terms can compensate for the quadratic nature of the relation, even though the linear equation with many predictor terms appears to "fit the data." Indeed, if the researcher has a large number of linear terms to use, then uncorrelated linear terms do a better job of predicting overall liking than using both linear and quadratic terms for the same independent variable. However, the reality of the situation is that the linear terms will mislead because they miss the true relation, whereas the quadratic equation will not miss the true relation.
2. There are pairwise interactions between variables, whether or not these can be modeled: Appearance and texture/skinfeel may jointly interact in a synergistic or suppressive fashion to determine overall liking. Multiple linear equations as shown above do not take this interaction into account. Furthermore, with only a few products with which to work, and with three independent variables (color, fragrance and texture), there are very few pairwise interactions that can even be considered. Yet with three variables, there are three additional pairwise interactions. (With four independent variables there would be six interactions.) Should any be included? There are too few degrees of freedom left with which to estimate error. The model can quickly become "saturated," perhaps unnecessarily so.

Need for a Reduced Set of Predictors

In order to create a product model of the bath gels we need to reconsider the problem. There are simply too many predictors and too few "degrees of freedom" (defined by the number of cases or products that we test). In the practical world of product testing budgets are limited, so that the statistical power of the study must be compromised by testing a more limited number of samples. Although it is ideal to test many dozens of products to insure that the resulting product model is stable, the practical constraints on procuring or developing these products and testing them with an adequate number of consumers is so great that the more powerful test is simply not feasible.

An alternative method is to work with a reduced set of predictors. Some practitioners like to do a factor analysis on the sensory attributes, identify key factors or dimensions, and then choose one sensory attribute that loads most highly on each of the factors. Table 25 shows the simple correlations between three factors underlying the sensory attributes and the seven sensory attributes. Following this rationale, the researcher would then predict all liking attributes by a combination of three sensory attributes: thick, fragrance, dark.

The approach discussed above and shown in Table 25 is one solution, but not necessarily the best solution. The approach jettisons most of the sensory attributes in favor of a limited number.

There is an alternative method presented here that is far more powerful. The alternative method uses factor scores, following the five steps below:

1. *Use sensory attributes as the base data.* Identify the sensory attributes for the product category.
2. *Perform a principal components analysis.* Perform a principal components factor analysis on the sensory attributes, and rotate the factors (e.g., by quartimax rotation) to achieve a simple structure. Quartimax rotation provides a simple structure of factors. Create the database by appending the factor scores to the product attribute profile (see Table 26).
3. *Use factor scores as the independent variables.* Compute the factor scores, or the location of each of the products on the rotated factors. These factors are statistically independent of each other, and are parsimonious. Whereas there could be as many as 10, 20, 30 or even more sensory attributes considered in the original questionnaire and factor analysis, there are usually no more than 2 to 5 basic factors.
4. *Create models which use factor scores.* Create separate models relating the factors, their squares and their cross terms to every attribute in the study, whether this attribute be liking, sensory, image or performance. Keep in mind that although the factor scores were derived from the sensory attributes, they are now used to predict both the sensory attributes from which they were derived, as well as other attributes.

Table 25 Correlations of Sensory Attributes with Factors

	F1	F2	F3
Thick	[0.93]	0.00	0.21
Lather	0.78	0.25	−0.21
Fragrance/tub	0.06	[0.93]	−0.18
Fragrance/bottle	−0.12	[0.90]	0.29
Dark/bottle	0.00	0.33	[0.86]
Quick to lather	0.39	−0.12	[0.87]
Oiliness	0.69	−0.13	0.64

Table 26 Partial Data Base—Bath Gels

Factor scores	Product		
	101	102	103
F1	−1.00	0.11	−0.03
F2	1.05	0.06	−0.46
F3	0.50	1.23	−1.28
Image Attributes			
Functional-Cosmetic	71	64	70
Masculine-Feminine	62	66	65
Serious-Fun	69	57	64
Younger-Older	66	71	68
Refreshing	74	58	59
Conventional-Unique	74	61	67
Liking			
Appearance	65	59	61
Fragrance-Bottle	63	61	60
Fragrance-Bath	66	64	62
Lather	59	57	62
Sensory Segment 1	66	63	66
Sensory Segment 2	71	64	66
Overall	68	63	66
Sensory			
Darkness-Bottle	63	58	30
Fragrance-Bottle	60	56	52
Fragrance-Bath	57	54	53
Amount Of Lather	62	61	65
Oily Feel	56	64	54
Speed Of Lathering	49	62	43
Thickness Of Gel	55	69	61

5. *Use the product model to optimize.* The equations in Step 4 allow the researcher to identify values of the factor scores that correspond to the most-liked product. Furthermore, once these factor scores are identified, use the product model to estimate the full profile of attributes corresponding to these factor scores. Insure that the factor scores lie within the range tested, and that all of the sensory attributes of the optimum product also lie within the range tested in the study. This is constrained optimization, and is necessary for those category appraisal studies which do not use systematic experimental design of independent variables.

Models for the Bath Gel Study

An example of an equation using factors appears in Table 27A. The equation does not fit the data perfectly. Since there are only three factors, there are at most nine terms and an additive constant. (The more terms we have in the equation, the better will the equation fit the data.) Despite the fact that the equation does not fit the data as well as we would like, the equation is quite useful. When used for optimization the equation can reveal optimum levels of the factors in the middle range, and therefore target product levels in the middle range. This possible outcome is a welcome improvement over the more conventional linear equation which by its very nature forces the optimum to lie at the upper or lower extreme. Table 27B also shows the goodness-of-fit of each attribute equation to the data. The multiple r^2, multiplied by 100, shows the percent variability in the data accounted for by the equation.

Optimizing the Bath Gel Product by the Liking Equation

Our first analysis concerns the optimum product. At what value of the three factor scores does liking reach its peak? Table 28 shows this optimum (#1). Furthermore, the bottom of Table 28 shows the expected ratings for the product on all of the remaining attributes. The equation allows the researcher to estimate the likely rating to be assigned to the product once the levels of the factor scores are established. These levels are established by the optimization procedure, which searches for the specific combination of factor scores maximizing overall liking.

One disturbing finding emerges from this first optimization (#1). Although the factor scores themselves lie within the range that we tested, the sensory and the image attributes occasionally achieve values that can best be described as meaningless. Negative expected values or expected values substantially above 100 are not meaningful when the original scale was bounded by a low of 0 and a high of 100. Furthermore, in some cases there are expected levels of attributes that lie within the 0-100 range, but whose values lie outside the range achieved by the actual test products. There are no reference products for those outlier levels.

We must revise the optimization somewhat to take into account the range of values actually achieved in the empirical part of the study. That is, we can continue to optimize, but we must insure that attributes never achieve levels beyond the highest and lowest levels achieved in the study. This is an example of constrained optimization, where the constraints are provided by the levels in the actual study. Table 28 (#2) also shows the revised levels of the factor scores which satisfy two criteria:

Table 27 Example of Factor Models

A. Factor Model for Liking

Dependent variable—Overall liking
Multiple r^2 = 0.75
Standard error of estimate = 3.77

Regression equation:

Constant	56.70
F1	0.75
F2	4.63
F3	−0.97
F1*F1	0.47
F2*F2	1.39
F3*F3	2.26
F1*F2	−3.61
F1*F3	0.92
F2*F3	2.95

B. Goodness-of-Fit of Factor Models

Attribute	Multiple r^2
Functional-Cosmetic	0.72
Masculine-Feminine	0.44
Serious-Fun	0.79
Younger-Older	0.58
Refreshing	0.75
Conventional-Unique	0.64
Like Appearance	0.58
Like-Fragrance-Bottle	0.82
Like-Fragrance-Bath	0.67
Like Lather	0.86
Like Segment 1	0.82
Like Segment 2	0.69
Like Overall	0.75
Darkness-Bottle	0.93
Fragrance-Bottle	0.92
Thickness	0.97
Fragrance-Bath	0.93
Amount of Lather	0.88
Oily Feel	0.96
Speed of Lathering	0.97
Thickness of Gel	0.94

Table 28 Optimum Factor Scores and Sensory Profiles (Maximize Liking)

	No sensory constraints #1	Sensory constraints #2*
Factors		
F1	−2.3	−1.3
F2	1.9	1.7
F3	1.3	−0.2
Estimated Image		
Functional-Cosmetic	108	80
Masculine-Feminine	88	74
Serious-Fun	102	74
Younger-Older	88	81
Refreshing	108	76
Common-Unique	99	79
Estimated Liking		
Like/Appearance/Bottle	55	61
Like/Fragrance/Bottle	94	71
Like/Fragrance/Tub	88	74
Like/Lather	73	64
Like/Segment 1	85	72
Like/Segment 2	106	81
Like/Overall	94	76
Estimated Sensory		
Sensory/Dark/Bottle	102	70
Sensory/Fragrance/Bottle	63	58
Sensory/Feel	30	51
Sensory/Fragrance	61	60
Sensory/Lather	75	63
Sensory/Oily	64	49
Sensory/Quick Lather	44	39
Sensory/Thick	34	52

* All sensory levels lie within range tested.
Sensory constraints—All sensory attributes constrained to lie within levels achieved by products in the test.

1. Maximize overall liking or another criterion (e.g., feminine image).
2. Insure that no sensory attribute ever achieves a level beyond that achieved by products in the study.

Once the optimization is completed, the researcher now knows the profile of the optimum product. This profile is "synthesized," with the property that all sensory attribute levels by themselves are both achievable (they lie within the range tested) and referenced (some product in the set has a level near the optimum to serve as an exemplar or target for development). It bears repeating at this time that since we are evaluating competitive products that are not linked together by an experimental design, we do not know the actual formulation levels corresponding to the optimum. We know only the products that have the relevant sensory levels.

Optimizing Subject to Constraints

Another way to use the category model consists of optimizing an attribute other than overall liking, either without considering overall liking at all, or by using the overall liking attribute as a constraint. In the product ratings the panelists rated both sensory and image/performance characteristics (emotional feelings as well as image of the product). Table 29 shows the optimum for overall liking (#1), subject to constraints first on the attribute of "feminine" (#2), then on the attribute of "perceived appropriate for cosmetic use" (#3-5). All of the sensory attributes lie within the range tested. Table 29 shows three product profiles satisfying the different levels that the attribute "cosmetic" must achieve.

The researcher can place various constraints on the optimum in order to generate a product with a specific level of acceptance, which is realizable physically, and which generates desired levels of image perceptions. Furthermore, at the end of every optimization exercise the researcher can identify the relevant sensory characteristics corresponding to the product, as well as identify those products that do serve as development targets on an attribute-by-attribute basis.

Turning the Problem Around—Reverse Engineering

Up to now we have looked at the bath gel problem in one direction—going from sensory characteristics to liking via factor scores. Let's turn the problem around 180°, just as we did for concepts. Let us specify the image profile that we would like the product to possess, discover the factor scores, and then determine the sensory characteristics (and thus products) corresponding to the desired profile.

The method used is reverse engineering. The logic is quite simple—adjust the factor scores until a combination of factor scores is reached which yields an expected attribute profile as close as possible to the profile specified by the researcher. At that point we say that a match or fit has been accomplished. The goal or target profile to be matched lies as close as possible to the expected profile that would be obtained for these factor scores. Once the factor scores which generate the match are identified, the researcher can estimate the likely profile of all ratings.

Benchmarking a Product Category

Table 29 Optimum Scores and Sensory Profiles (Sensory Attributes Constrained to Lie Within the Range Tested)

Maximize	#1 Total	#2 Total	#3 Total	#4 Total	#5 Total
Constraint	None	Feminine	Cosmetic	Cosmetic	Cosmetic
Lower		48	50	60	69
Upper		52	55	63	71
Factor					
F1	−1.3	−1.0	−1.3	−1.8	−1.3
F2	1.7	−0.7	−0.6	−0.8	1.2
F3	−0.2	2.3	−0.8	−0.4	−0.4
Estimated Image					
Functional-Cosmetic	80	60	55	63	71
Masculine-Feminine	74	52	59	64	68
Serious-Fun	74	52	58	57	68
Younger-Older	81	56	53	69	74
Refreshing	76	52	51	52	67
Common-Unique	79	56	53	61	72
Estimated Liking					
Like/Appearance/Bottle	61	64	63	48	62
Like/Fragrance/Bottle	71	47	51	61	64
Like/Fragrance/Tub	74	54	54	59	68
Like/Lather	64	56	55	61	62
Like/Segment 1	72	49	55	67	68
Like/Segment 2	81	60	54	57	71
Like/Overall	76	54	56	63	69
Estimated Sensory					
Sensory/Dark/Bottle	70	71	34	41	58
Sensory/Fragrance/Bottle	58	56	52	48	57
Sensory/Feel	51	59	55	79	54
Sensory/Fragrance	60	50	54	54	59
Sensory/Lather	63	55	57	67	59
Sensory/Oily	49	68	48	68	48
Sensory/Quick Lather	39	60	45	54	40
Sensory/Thick	52	63	55	79	54

Table 30 shows four worked examples of this analysis. The goals come from desired image/usage profiles. The results are shown in terms of factor scores and expected sensory profiles. All sensory attributes lie within the range tested (they are constraints in the optimization).

Table 30 Fitting a Pre-Designated Goal Profile

Factor	Product #1		Product #2		Product #3		Product #4	
F1	−2.5		0.0		−0.9		1.9	
F2	−0.2		0.8		−0.8		2.6	
F3	−0.2		0.2		0.4		0.6	
Estimated Image	Est	Goal	Est	Goal	Est	Goal	Est	Goal
Functional-Cosmetic	51	50	64	65	49	50	70	70
Masculine-Feminine	61	60	60	60	50	50	70	70
Serious-Fun	60	65	61	60	50	50	69	70
Younger-Older	49	50	65	65	50	50	69	70
Refreshing	50		62		45		72	
Conventional-Unique	52		65		50		67	
Estimated Liking								
Like/Appearance/Bottle	53		66		65		65	
Like/Fragrance/Bottle	50		58		44		65	
Like/Fragrance/Tub	54		61		48		66	
Like/Lather	55		58		50		64	
Like/Segment 1	53		62		52		34	
Like/Segment 2	55		63		48		94	
Like/Overall	56		62		50		69	
Estimated Sensory								
Sensory/Dark/Bottle	43		58		47		82	
Sensory/Fragrance/Bottle	53		57		53		63	
Sensory/Feel	49		66		63		82	
Sensory/Fragrance	57		58		52		69	
Sensory/Lather	48		62		54		73	
Sensory/Oily	44		57		55		64	
Sensory/Quick Lather	44		50		52		56	
Sensory/Thick	50		67		63		85	

Est = Estimated.
Goal = Goal set by researcher for the product.

Fitting Holes in the Category

One of the benefits of mapping a category is being able to find holes in the map. When the variables are continuous this makes finding the holes very easy. The research locates the factor scores of the products as points on a map. Figure 11 shows the map for the 18 bath gels. Points correspond to products, and conversely empty spaces correspond to opportunities (or "holes" in the category).

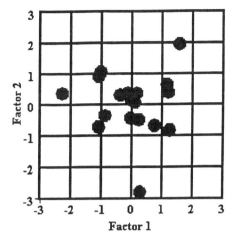

Figure 11 Plot of the 18 gels on the first two factors.

The reverse engineering procedure enables the researcher to locate the hole (e.g., by visual inspection), and then identify the specific factor scores corresponding to these holes. Once these factor scores are identified, the goal fitting algorithm then identifies the sensory profile and the expected image profile of this product. Figure 11 shows many different holes. Table 31 identifies different product profiles, corresponding to three different holes or empty regions in the product map.

An Overview to Modeling and Mapping Continuous Variables in Category Appraisals

Category appraisal with continuous variables as response attributes enables the researcher to identify what drives the product category in terms of sensory attributes, as well as to create an integrated product model. That model, using regression analysis, accounts for interactions among variables, and non-linearities in the response function. The regression model for continuous variables is stronger than regression models created for discrete variables where the attribute can only take on one of a limited number of levels. Furthermore, optimization and reverse engineering procedures add power by identifying factor scores corresponding to optimal products (with or without constraints), as well as identifying factor scores corresponding to a desired image profile. Once the factor scores are discovered, the optimization algorithm estimates the likely profile of all sensory attributes, thus providing the product developer with specific targets or exemplars for development.

Table 31 Factor Scores and Profiles of Three Products Corresponding to "Holes" or "Gaps" in the Category

Factor	Hole #1	Hole #2	Hole #3
F1	0.5	0.5	0.5
F2	2.0	2.0	2.0
F3	−1.0	0.0	1.0
Estimated Image			
Functional-Cosmetic	68	72	84
Masculine-Feminine	63	66	73
Serious-Fun	56	65	79
Younger-Older	80	75	76
Refreshing	58	69	87
Conventional-Unique	70	73	79
Estimated Liking			
Like/Apperarance	70	69	68
Like/Fragrance/Bottle	60	64	73
Like/Fragrance/Tub	65	66	74
Like/Lather	61	61	66
Like/Segment 1	57	56	56
Like/Segment 2	73	80	93
Like/Overall	65	68	76
Estimated Sensory			
Sensory/Dark/Bottle	60	72	85
Sensory/Fragrance/Bottle	58	60	63
Sensory/Feel	74	70	63
Sensory/Fragrance	63	63	62
Sensory/Lather	62	67	76
Sensory/Oily	48	56	66
Sensory/Quick Lather	36	46	55
Sensory/Thick	74	71	65

9. An Overview

In a highly competitive environment it is critical to seize the competitive advantage in the most rapid and cost efficient manner. Benchmarking enables the manufacturer to identify those key features in a category which drive specific perceptions. By evaluating a large number of competitive products, categorizing them in terms of the features that they possess and then identifying combinations of features that are missing in the marketplace, the manufacturer can rapidly identify new combinations of existing features. The foregoing procedure is ideal when it is critical to recombine existing features into new products. The marketing researcher can provide specifications for product development, at least in terms of recommending promising new configurations of product features.

References

Arabie, P. and Hubert, L. Cluster analysis in marketing research. *In:* Advanced Methods in Marketing Research, R. Bagozzi (ed.) Blackwell, Oxford, In Press.

Best, D. 1991. Designing new products from a market perspective. *In:* Food Product Development, E. Graf & I.S. Saguy (eds.), pp. 1–27. Van Nostrand Reinhold/AVI, New York.

Draper, N.R. and Smith, H. 1981. Applied Regression Analysis. John Wiley & Sons, New York.

Johnson, J.L., Dzendolet, E. and Clydesdale, F. 1983. Psychophysical relationship between sweet and redness in strawberry flavored drinks. Journal of Food Protection, 46, 21–25.

Moskowitz, H.R. 1981A. Sensory intensity versus hedonic functions: classical psychophysical approaches. Journal of Food Quality, 5, 109–138.

Moskowitz, H.R. 1981B. Relative importance of perceptual factors to consumer acceptance: linear versus quadratic analysis. Journal of Food Science, 46, 244–248.

Moskowitz, H.R. 1984. Cosmetic Product Testing: A Modern Psychophysical Approach. Marcel Dekker, New York.

Moskowitz, H.R. 1994. Food Concepts And Products: Just-In-Time Development. Food & Nutrition Press Inc., Trumbull, CT.

Moskowitz, H.R. and Jacobs, B.E. 1989. Using in-market products to generate target sensory profiles in early stage development. *In:* Product Testing with Consumers for Research Guidance, L. Wu (ed.), STP 1035, pp. 64–74. American Society for Testing and Materials, Philadelphia.

Piggott, J.R. and Sharman, K. 1986. Methods to aid in the interpretation of multidimensional data. *In:* Statistical Procedures in Food Research, R. Piggott (ed.), pp. 181–232. Elsevier Applied Science, Barking, Essex, U.K.

Schutz, H.G. 1983. Multiple regression approach to optimization. Food Technology, 37, 47–62.

Weisberg, S. 1985. Applied Linear Regression. John Wiley & Sons, New York.

Wilkinson, L. 1992. SYSTAT: The System for Statistics. Systat Inc., Evanston, IL.

3
Practical Considerations for Product Testing

1. General Overview

The discipline of product testing is relatively young. Sixty years ago there was no such discipline as product testing in the cosmetic and food industries. Rather than being a disciplined approach, at that time testing consisted of giving the product to one's friends or relatives and soliciting their opinions about the product (e.g., what was wrong with the product; what could be improved; what were the strengths; what were the weaknesses, etc.) One need only look around in the industry today to see how much things have changed. Testing products for safety, for consumer acceptance, and for claims support has become an established, growing industry, replete with procedures, norms, databases, and the inevitable clash of viewpoints among the experts.

Product testing in personal products came a little later than it did to food science. Consequently there is not much of a solid history in test methods, although over the past decade researchers have made great strides in developing procedures to test personal products. This chapter presents some of the considerations which go into designing a study for a personal product.

Defining the Problem

To an experienced hand in product testing this first step seems to be a truism. So many problems seem self-evident that there doesn't seem to be a need to define the problem. For instance, the marketing manager might simply want to

Practical Considerations

know how well the product performs. The Research and Development (R&D) product developer might wish to learn how well a product delivers against the expectations set up by the concept. An R&D director, in turn, with a more global viewpoint might want to discover specific sensory factors in a product category which make the category "tick" (namely, what is important in a category in terms of the consumer's perceptions).

Defining the problem is harder than just asking questions. The question is often asked in the language of a consumer or marketer. How well a product performs is a colloquially expressed question. The problem may be:

1. How well does the product perform vis-à-vis the current competitors (e.g., is the product as acceptable as the current competition, better, or worse)?
2. How well does the product perform on an absolute basis against relevant norms for the category? (This question can be asked independently of the performance of competitors, and simply concerns the score of the product vis-à-vis the hurdles that must be overcome.)

It is impossible to list every question and provide an approach to answer each. However, it is productive to list a sequence of four steps that one can follow in order to help define the problem:

1. Does the problem/issue involve one product or many products?
2. What will be done with the data (report card or search for patterns)?
3. Can the analysis be done with the data using only one attribute (e.g., liking), or does the analysis require several attributes (e.g., liking, sensory characteristics)?
4. Will the data be used to discover patterns in the data so that the products are simply points used to develop the patterns? Or are the products themselves necessary? That is, in the development of the "pattern," do the many points simply define the curve or pattern? Or is it absolutely vital that we look at a table with the products identified? For instance, in one case we may be interested in the scores of all products made by a single manufacturer, or of a single type, such as a fragrance type. We do a point-by-point or product-by-product analysis, looking at the scores of the individual products tested. In another case we may be interested primarily in the relation between the sensory attribute level of the product and attribute liking (e.g., darkness of color vs. liking of appearance). In order to discover the relation or underlying pattern we require a variety of products exhibiting different attribute and liking levels. We need not know the identity of each point, however. We merely need to know that each product has been rated on the two attributes. We fit a summary curve through the data in order to uncover the pattern.

Selecting the Products to Test

At first glance the selection of products to test seems to be straightforward, but upon deeper reflection it is not as simple as it seems. Should the researcher evaluate his own product alone? Should one include the competitive frame as "benchmarks" against which to test? What if the competitive frame is not clearly defined? How should one select the competitive frame within the constraints of limited amounts of money and time? What about other products besides one's most promising prototype? Can more be learned by including several prototypes, not just that single prototype which looks like it will be a winner?

A decade ago there was an inordinate amount of complacency among researchers. Most researchers used the product test to confirm their wisdom, rather than as an opportunity to learn about the product in terms of consumer-relevant strengths. The historical-minded researcher who delves into the corporate archives will uncover study after study which evaluated only one or two products (typically one's own prototype against the market leader), simply to find a "winner."

As the competitive situation has become increasingly tough for marketers, there has been a slowly dawning recognition that perhaps testing more products rather than fewer is the better use of the research dollar. Although one would never purchase a house without inspecting a variety of houses to get an idea of "what works and what does not," this global view has only slowly penetrated the cosmetics and health and beauty aids fields in particular, and the packaged goods field in general.

There are no fixed answers about what products to test. The researcher can best be guided by asking the following three questions:

1. What is the purpose of the test? Is the test designed to determine whether a product "passes or fails?" If so, then the researcher need not establish a pattern in the data. The research provides a report card, primarily for one stimulus. In this situation the best approach tests several prototypes in order to increase the likelihood that one (or more) of the prototypes will pass the "screen" and achieve the necessary level of acceptability. If there are well-established norms (e.g., overall liking must exceed a certain level, such as 65 on a 0–100 point scale, where 100 = like extremely and 0 = hate), then one might need to test only one product. It is usually better, however, to test several products, simply because the odds of discovering a "winning product" increase with the increasing numbers of prototypes tested.
2. How will the data be used? If the data will be analyzed subsequently to identify directions for product improvement, then it is best to test several products, not just one. In this way the researcher can discover the rela-

Practical Considerations

tion between the sensory characteristics of a product and liking. (That relation cannot emerge when one tests only one product.)

3. Will the data be used to change the product in a gross fashion or a fine fashion? If the researcher is concerned with "gross patterns" in the data, then it makes sense to test both prototypes and competitive products in the same study. (Gross patterns are qualitative in nature—e.g., if the color gets darker, then the consumers like the product more. If the texture gets greasier, then consumers like the product less, etc.) If the researcher is concerned with "fine patterns" dealing with small differences or changes in ingredients, then the researcher should test multiple, and probably systematically varied, prototypes. It is impossible to obtain solid quantitative direction from consumers when the researcher tests only one product or a set of unrelated competitive products (including one's own prototype). The products must be varied on a physical continuum by the developer back in the laboratory, and then tested among consumers.

How Many Questions and What Questions to Ask?

Consumers can respond to many different questions about a product. These include its sensory attributes, patterns of usage, and end benefits. Furthermore, consumers form an impression of the product in their mind that can be best called "image" (e.g., appropriate for males versus appropriate for females).

Given the consumer's desire to please the interviewer and cooperate in a study, it is no wonder that research programs differ so widely in the number of questions asked about a product. Some of the more easygoing (or perhaps less inquiring) researchers are satisfied with a simple series of questions pertaining to product acceptance. These researchers ask questions about overall liking, liking of appearance, of package, of fragrance, etc. There is another school of thought, however, which believes in probing the consumer to obtain as much information as possible. This second school believes that the consumer can validly answer several dozen questions about each product, thus providing a fuller profile of subjective impressions.

There is no right number of questions to ask a consumer. The conservative researcher always runs the risk of asking too few questions. Occasionally the conservative researcher may miss some key factors in a consumer's perception of the product. It is better to be less conservative, therefore, in order not to overlook key characteristics which may be present and which may drive product acceptance.

The more adventurous researcher with an optimistic spirit may also fail because the consumer is not an infinitely patient machine who will simply carry out the orders presented to him. First, the interview may bore the consumer. Even though an interview may be "self-paced" (the consumer fills out a questionnaire rather than responding face-to-face or over the phone to the interviewer), the consumer may simply "burn out." There is a point beyond which the answers are given without any thought, simply to fulfill the task. Second, if the questions are repetitive, then the repetition reduces the panelist's interest. For example, if the panelist has only six questions to answer, and if these questions correlate highly with each other, then the panelist soon loses interest in the task. Researchers are prone to use "laundry lists" of correlated or redundant questions. As an example, questions a–c below are really the same:

a. How much do you like this product?
b. How high quality is this product?
c. How would your friends like this product?

All three questions correlate with each other. Knowing the answer to one question pretty well means knowing the answer to all three questions. There are other examples as well, not so blatant as these, but which inevitably lead to correlated, redundant answers, and thus bore the consumer to tears, or worse, to recalcitrant non-cooperation.

Number of Attributes

Although some researchers believe that consumers have a limited ability to rate products on characteristics, and thus opt to ask only overall liking, in actuality consumers are quite good at assessing the different attributes of perception. If a consumer pays attention to the task, then he can answer well up to 40+ questions. As noted above, this does not necessarily mean that the researcher is investigating 40 different aspects of perception. Rather, it means that the alert and interested consumer can assign 40 (or more) ratings to the same sample, paying attention to each.

Order of Attributes

There is substantial controversy about the order in which one should ask the questions. One school of thought believes that the consumer should rate overall acceptance first, and then follow with "diagnostics." These diagnostics may be attribute liking, such as liking of fragrance, or sensory attributes, such as the perceived intensity of fragrance. Practitioners of this school of thought believe that overall acceptance (or some cognate attribute such as purchase intent) is the key attribute on which to focus first. Only afterward is it appropriate to distract the consumers by asking other questions pertaining to the reason why the product is liked or disliked, or what the salient sensory characteristics are about the product.

Another group of equally vehement and vocal researchers believes that the overall question should appear last after all the other questions have been an-

Practical Considerations 163

swered. They support their belief by stating that when a panelist assigns an overall liking rating first, he will then attempt to justify that initial liking rating as he answers each of the remaining questions. That is, if the panelist dislikes the product intensely, then he will hardly feel confident rating the liking of the appearance high, or the liking of the fragrance high after downrating the product. The panelist will search for reasons to downrate the product, in order to justify the initially low overall rating.

There is no definitive literature to justify either position. Both positions have merit. In this author's experience it is best to have the consumer rate overall liking after experiencing the product, but not necessarily as the last attribute rated in the questionnaire.

1. When a consumer evaluates a product, the attributes should follow the order of appearance. That is, it is best to rate package attributes before appearance attributes, appearance before fragrance, fragrance before feel/texture, and feel/texture before efficacy/performance. There is a natural temporal sequence of attributes which emerges when using a product. The questionnaire should follow this sequence of order of appearance.
2. There is some degree of preference justification. Many personal products have complex characteristics that are important for R&D. It is vital to obtain an accurate measure of these characteristics. A product may be quite acceptable but have some negative characteristics. If preference justification comes into play, even modestly, then it may prevent the consumer from providing a fair assessment of the different characteristics. However, by allowing the consumer to rate the components of the product first, and then having the consumer rate the product overall, the researcher can insure a fairer estimate of the attributes.

Scales for Evaluation

Scales for product testing present a perennial bone of contention among researchers. Virtually nothing sets off controversy among researchers in the way that scales do. For the most part the arguments are empty. Most scales work equally well in differentiating one product from another, as long as the scale is sufficiently wide (e.g., enough categories) to permit a single consumer to differentiate between a group of products.

We can begin with the worst scales and move on to the better, and finally to the best scales.

The absolutely worst scales are those that are meant to differentiate between a large number of products but permit the panelist only a limited number of response categories. For instance, the scale may simply ask the panelist to state whether he would buy the product or not buy the product. The panelist has only two choices. The scale is adequate when used simply to measure the consumer's all-or-none response to the product. However, suppose the panelist must discriminate between six easily discriminable products, but can only use a 2-point scale (e.g., like and dislike) or a 3-point scale (like, neutral, dislike). One single panelist cannot demonstrate discrimination between the six products, because two or more products will be assigned the same scale values. However, products will differ from each other if rated by 100 panelists, not because the panelists on an individual basis can scale the differences, but because panelists do not agree with each other.

The problem of a limited number of allowable scale points can be counteracted by allowing the panelist to use a scale having more points. Two or three scale points are insufficient. But how many points suffice? Should the scale comprise 5 points? Nine points? An odd number of points or an even number of points? An unlimited range? Should the scale be a line scale with no numbers?

Scales, Scale Points, Category Labels. Over the decades researchers have proposed a variety of different scales. Many researchers recognize that the scale must allow the panelist to discriminate between products. There are those who feel that panelists are best off when provided with a limited number of points on the scale. These researchers feel that it is necessary for the panelist to feel in control of the scale. In order for the panelist to fully understand the scale, these researchers feel that there should be no more than nine to eleven categories on the scale. These researchers also feel that the categories should be labeled so that each category or scale point has a fixed meaning. Table 1 lists representative scales suggested by a variety of researchers in different fields. The scales differ both in the number of allowable scale points and the verbal anchors assigned to each point. (During the course of the history of research, the assignment of labels to categories on a scale was a favorite research topic. Researchers weren't comfortable unless they felt absolutely certain that the descriptors or labels reflected the true differences between the category values.)

Line (Continuous) Scale. Some researchers feel that panelists should not be forced to constrain their ratings to lie within a specific number of categories. These researchers feel that the panelists should use a continuous scale. The scale need not be unbounded at the ends. A good example is the line scale. The line scale comprises a line of defined length (e.g., 6 inches). The panelist can place a rating anywhere on the line to define the strength of feeling. The line is anchored at both ends in order to reduce any potential ambiguity. The line can be anchored at the very ends (e.g., for fragrance intensity the anchors can be "absolutely no fragrance detected" to "extremely strong fragrance"). Other researchers feel that the panelist should be given the opportunity to scale perceived magnitudes (or degree of liking) beyond the anchors, because there may be occasions when the stimulus is perceived to be even stronger than the level denoted by the top of the scale. These more conservative researchers locate the

Table 1 A Potpourri of Scales and Adjectives for Liking and Sensory Intensity

Liking scales

2-point scale
　Like
　Dislike

3-point scale
　Like
　Neither like nor dislike
　Dislike

5-point scale
　Like extremely
　Like
　Neither like nor dislike
　Dislike
　Dislike extremely

9-point scale
　Like extremely
　Like very much
　Like moderately
　Like slightly
　Neither like nor dislike
　Dislike slightly
　Dislike moderately
　Dislike very much
　Dislike extremely

Purchase intent
　Definitely would buy
　Probably would buy
　Might/Might not buy
　Probably not buy
　Definitely not buy

Non-Evaluative Scales for Sensory Intensity and Other "Amount" Related Attributes

3 points
　Cannot detect
　Weak
　Strong

5 points
　Cannot detect
　Weak
　Moderate
　Strong
　Very strong

anchors "inside" the line. The outside portion of the scale is available when the stimulus intensity exceeds that denoted by the anchors.

Magnitude Estimation. The scales above (including the line scale) constrain the panelist to assign a rating within the range allowed by the scale. However, nature does not work like that. According to S. S. Stevens of Harvard University, it is important to allow the panelist to assign numbers in the way that is scientifically most appropriate. Stevens believed that the category scales distort the consumer's measurement. A more appropriate method allows the consumers to assign numbers so that the sizes of the numbers match the intensity of the stimulus. This approach, called magnitude estimation (Stevens, 1953), generates ratio scales. The ratios of the ratings match the ratios of perceived intensity. Magnitude estimation requires that the panelist not constrain his or her ratings, but rather use an open, unlimited numerical scale. Magnitude estimation has enjoyed a 40-year reign as one of experimental psychology's key methods for sensory measurement. In applied research, however, magnitude estimation has not been quite as popular because it is difficult to implement in the field (Moskowitz, 1983).

Orientation—Explaining Procedures, Scales and Attributes to Consumers

Many consumers participate in a product evaluation test without knowing what is expected of them. What is the procedure for applying and evaluating the product? The knowledge may vary from knowing exactly what to do with a product to having no idea how to apply it. Most people know what to do with a soap. However, when it comes to a new, complex powder system that must be applied in two or three steps, most consumers will be at a loss the first time they are confronted with the product.

Scales may be difficult as well. To researchers who deal with scales on a daily basis, nothing can be as easy as a scale. But often researchers neglect the consumer. Some scales are extremely easy to understand and to use (e.g., do you like the product, or don't you like the product). Even those easy scales have problems. When consumers want to be "precise" in their ratings they will ask about the case when they feel indifferent to the product. Is indifference to be rated as "like?" As "dislike?" The consumers may wish to utilize an intermediate category (e.g., neutral), and will be confused when that intermediate category is absent. Most consumers don't ask for an explanation, but some do. Even those consumers who don't ask for an explanation may have trouble, however. These consumers may be simply too shy to ask questions.

The situation becomes more complicated when it comes to attributes. Over the past 30 years researchers have become "attribute happy." From a modest beginning with simplistic questionnaires, product evaluation has matured into

Practical Considerations 167

a field where the questionnaire comprises dozens of questions. Not all of these questions are equally comprehensible. Consumers have no problem with simple sensory attributes dealing with visual appearance. The same consumers may have trouble with attributes that deal with fragrance and texture, especially when the attributes are more esoteric. For example, strength of fragrance and even strength of flowery fragrance are understandable to most consumers. But attributes such as "woody" are not equally understood by all consumers. For texture, perception matters are even more complex. Most consumers understand the feeling of "hard" vs. "soft" or "goes on dry." Many consumers do not understand concepts of "rub-in." When they rub a lotion into their skin and are instructed to rate a variety of attributes (e.g., drag, ease of rub-in), consumers often have trouble understanding the exact meaning of some attributes unless the attributes are explained in detail.

Matters become much more complex for "image" attributes. These are the more complex characteristics such as "mature," "feminine," "sophisticated," and the like. Image characteristics require the panelist to evaluate a more complex perception of the stimulus. Image characteristics are at once sensory-based and interpretation-based. There are sensory aspects of the attribute "mature," but there is also an interpretation of the sensory impression which differs from panelist to panelist.

Screening Panelists for Participation

The study begins with the first contact made with the panelist. This first contact takes place when the panelist is qualified and invited to participate. Researchers have devised a variety of ways by which to secure consumer participation, especially in tests requiring extended participation.

Mall Recruitment. The interviewer works at a shopping mall. (These are usually "high traffic" malls, located in an area that possesses the proper demographics.) The interviewer asks passing customers whether they would like to participate in a product evaluation test. If an individual responds "yes," then the interviewer administers a screening questionnaire. Table 2 shows an example of this screening questionnaire. The screener may be short, long, simple or involved, depending upon the qualifications needed. (Keep in mind that the greater the number of qualifications required for participation the greater will be the number of rejected panelists.

Over the past several years marketing researchers have reported that mall recruiting has become increasingly difficult. There are more refusals to participate than ever before (individuals approached who are potential panelists, but who refuse to participate upon request).

In a mall the easiest types of studies are the short ones, such as fragrance evaluation, where the interviewer applies a new fragrance to the consumer's skin

Table 2 Example of a Screener for Participation in a Product or Concept Test: Product = Foot Soak

Hello, my name is _____ from DesignLab. We are conducting a (product/advertising) evaluation test, and would like you to participate. Are you interested?

If panelist says no, terminate and tally.

If panelist says yes, continue.

In order for you to participate, I need to know whether you qualify. Which of the following five categories have you used in the past 3 months?

Nail polish remover (last month)	Y/N
Shampoo	Y/N
Lipstick	Y/N
Foot soak	Y/N
Antacid	Y/N

If the consumer answers yes to foot soak, then continue.
Otherwise, terminate and tally.

You answered yes to "foot soak." Which of these four brands have you used most often:

Mrs. Olney's soak	(1)
Sea Scent soak	(2)
Johnson's foot soak	(3)
Nu Feet	(4)

If Sea Scent soak or Johnson's foot soak not chosen, then terminate and tally. Otherwise, continue:

Now read the concept . . .

Product test

We are holding a product evaluation session on _____, at _____. At this time we will be teaching our consumers how to evaluate six different foot soak products. They will evaluate the foot soaks at home, over a 2-week period, and return to our office with their ratings. The payment for participating is $50.00. Are you interested in participating?

If yes, give the respondent directions to the test site.

Concept test

We are holding a test session on advertising evaluation, on _____. Participants will evaluate the concepts during the course of a 2-hour period, from _____ PM to _____ PM. The session will be held at our office at _____. The payment for participating for the 2-hour session is $30.00. Are you interested in participating?

If yes, give the respondent directions to the test site.

Practical Considerations 169

or to a blotter and instructs the panelist to evaluate the product right then and there. This type of immediate, short-term evaluation is easiest for "recruiting" panelists because it requires a short time commitment of less than 15 minutes. A slight variation of this procedure, with the panelist waiting a total of 20 minutes in order to evaluate both top note and dry down impressions (or at least middle notes) will generate more refusals because of the additional time required. The same test conducted at home will generate an even greater number of refusals because panelists who would ordinarily be amenable to participating in a central location evaluation balk when they feel that their connection with the research will extend beyond the shopping mall. The researcher may have to offer an incentive to the panelist to participate, either in the form of a product or in the form of a cash payment.

Lists of Qualified and Intended Participants. During the past 30 years many consumer research "field services" operating in local areas have built up banks of names of consumers agreeable to participating in a marketing research test. For the most part these consumers are called on a semiregular basis to participate in focus group discussions about products. The field service has the panelists fill out a questionnaire detailing the categories of products that they regularly use, and in some cases the brand used. The questionnaire also contains information about the panelist's age, sex, income (approximate), presence and number of children in the house, etc. Field services using these lists regularly update them to secure new panelists, so that a typical field service might have a list of several thousand potential consumer "volunteers." These individuals are not "professional panelists" (a term that frightens researchers because it means that the individuals are not true consumers). Rather these are simply interested consumers who can be called from time to time to participate in the study.

For extended product tests, consumers who have volunteered their services are the best potential participants for a variety of reasons. First, these consumers have already expressed interest in participating. To some extent the hardest work has been done already. The consumer's resistance to participating has disappeared, although if the study demands are too great, then the consumer may refuse to participate in that particular study. Second, the consumers have been qualified ahead of time. The product usage questionnaire that they fill out at the start of their association with the marketing research firm indicates whether or not they are typical users of the product category. The product usage questionnaire can be updated by a quick phone call to determine whether or not the panelist is a current user. The panelists are quickly reachable to insure that they meet additional qualifications. And, since these individuals are more or less "regulars" as research participants in many studies, they are accustomed to being phoned and asked to participate. (This invitation could be offered once every 2 or 3 months.) Third, these panelists have already participated in one or more focus groups, and perhaps even in one or more long-term product studies. They know the procedure, and are accustomed to participating.

Working with Panelists at the Start of the Test

Although many individuals have participated in product evaluations at some point in their lives, the researcher should always approach the project as if the panelists know absolutely nothing about the product, the category, the procedure, etc. Even though the panelists may be considered "frequent users" of the product, it is best to standardize the interaction between the panelist and the product.

The foregoing viewpoint is not without its detractors. There are many practitioners in the field of product testing who aver (sometimes quite strongly) that the panelist is best "left alone" to interact with the product in the typical fashion. These practitioners believe in maintaining the "ecological validity" of the study—that is, the panelist should be allowed to use the product in the same way that he or she uses the product at home. Only then will the data be valid. Any type of instruction about using the product will be unnatural. However, according to these researchers it is perfectly acceptable to explain to the panelist what the attributes mean. Many people are not accustomed to answering questions about products during the course of usage. These practitioners feel that there will be no bias introduced by a short introduction to both the project and the questionnaire, respectively.

Orientation to product usage can take many forms. At the very simplest stage one merely explains to the panelists what to do during the course of the evaluation. The panelist is told what to do and what to evaluate (or else provided with a sheet of paper listing the various steps in the evaluation). For many product tests this procedure suffices, especially when the product is easy to use (e.g., soap), and the researcher only wants to secure general reactions to the product.

Unfamiliar or Unusual Intellectual Task. Occasionally, however, the researcher wants more than the panelist's general reaction. For instance, with a fragrance the researcher may want the panelist to rate specific attributes (e.g., fit of a fragrance to a concept). The panelist may have to read a concept and rate the degree to which that concept (an advertisement) actually fits the fragrance that he just smelled. To most consumer researchers this task seems fairly simple. There doesn't seem to be much ambiguity about the panelist's task (to rate the degree of "fit" between two stimuli). Yet to a panelist, this task may confuse. When the panelist reads the concept it is tempting to rate the concept or rate the fragrance, but not consider the "match" between the two, even though the panelist is instructed to do so. Thus, the task may be ambiguous, difficult, or simply out of the ordinary. If the researcher suspects that there may be difficulties or misunderstandings encountered when the panelist assigns the ratings, then it is best that the moderator at the orientation session anticipate and solve that potential problem.

Explanation of Unusual (e.g., Technical) Attributes. A detailed orientation should be done when the researcher believes that the attributes are more complex than is usually the case. For a lotion the researcher may want to explore a variety of unusual attributes. These may be drag, rub-in, and other aspects (or simply language) with which the consumer is not familiar. The consumer knows what to do but may not understand precisely what the attribute means. Sometimes practitioners claim that in these situations the best thing to do is work with experts. An alternative is to invest a little time prior to the actual evaluation explaining to the panelist the meaning of the attribute. The explanation may even entail a short orientation exercise giving the panelist a firsthand chance to experience the meaning of the attribute and receive feedback from the interviewer. The benefit of orienting consumers in these attributes versus using experts is that the data now comes from a slightly more "educated" consumer rather than from an expert whose ratings may no longer represent those of the typical consumer.

Insure Understanding of the Order of Evaluation. The third reason for orientation is to demonstrate to the panelist exactly what to do when using the product. For many health and beauty aids products, "what to do" is not an issue. The panelists are familiar with the product and know precisely what they should do. However, in many situations the questionnaire calls for a detailed analysis of consumer responses to product usage in a disciplined fashion (e.g., beginning with appearance, progressing to fragrance, to feel, and then to benefits). Whenever the panelist has to embark on a disciplined series of attribute ratings, it is important to instruct the panelist about what to do, the definition of the attributes, how to apply the product, and how to evaluate it. A short orientation prior to the evaluation insures higher quality data because the panelists now know what to look for, what the attributes mean, and what is expected. Contrary to some points of view, the orientation period and the disciplined evaluation of products do not create an "expert panelist." They simply reduce the ambiguity, and improve the panelist's performance. There is no ongoing feedback from the interviewer to modify the consumer's attitudes or behavior (other than clarifying what to do).

Technical Standardization of Product Usage. Finally, in some situations it is important that the panelists apply the product properly. Although in the "real world" consumers are not instructed what to do (except perhaps by directions on the package), the research may involve specific issues of application and product usage. For instance, in the application of an anti-perspirant pump the researcher may wish to standardize the spray pattern as much as possible because the interest focuses on the degree of "dryness" of the spray. In this case the interviewer can demonstrate to the panelist precisely how to spray the product in order to insure that all the panelists do the same thing.

Maintaining Panelist Cooperation During the Course of the Study

How does the researcher guarantee panelist interest and motivation during the course of the study? In a short, 20-minute mall intercept test (e.g., evaluation of one or two fragrances for liking) motivation can be maintained by having an interviewer interact with the consumer. But what about the extended 6 to 8 week product usage test during which the panelist must evaluate a variety of different products? How does the researcher maintain panelist motivation during this period and prevent the panelist from simply assigning random numbers to the products?

In the majority of studies panelists prefer to cooperate rather than adopt an adversarial position. That is, for the most part panelists prefer to participate in a wholehearted manner, and to assign honest ratings rather than to fill out questionnaires in a desultory, disinterested fashion. Panelists are uncooperative primarily in the following three situations:

1. *Poor Instruction.* Panelists are not properly instructed, and therefore do not know what to do. Improper instruction may result from poor explanation of what to do, or failure to explain the meaning of the attributes. This can be easily remedied by proper orientation at the start of the study.
2. *Poor Monitoring.* Panelists are not properly monitored over the course of the project. Over a 3-week evaluation panelists need to be reminded what to do, and to understand that their performance is being monitored. (Monitoring should not provide feedback about the correctness or incorrectness of ratings, unless that feedback is absolutely necessary.) One way to monitor is to have the panelists take home the product(s) for week 1, and return at the end of the week to "check in," and to receive the second set of products for week 2, and so forth. Even though the same product is to be evaluated over the extended multiweek evaluation, the weekly visits to the research facility maintain panelist involvement and assure motivation. The panelist should return with a completed questionnaire. Furthermore, at each return visit an interviewer should ask the panelist questions about the ratings. The questioning is easy to do—one need only look at the panelist's rating sheet and ask the panelist to substantiate one or two ratings. Neither the panelist nor the interviewer knows the "correct" answer. However, the panelist has to substantiate his or her rating, which insures that the panelist will pay attention and be honest when assigning ratings to the product.

3. *Poor Compensation.* Panelists are poorly paid to participate. Panelists do not need to be richly recompensed for participation. In current dollars, a 3 to 4 hour, supervised central location test (e.g., for fragrance evaluation) should involve approximately $40.00 for consumer participation. An extended 6-week test, with weekly returns, should require approximately $100.00-$140.00 payment. Payments significantly below these levels (and adjusted for market and income level of the participants) will reduce panelist cooperation. If the panelist feels that the money is just too small, then the odds are that the panelist will drop out of the test and fail to return. An even more damaging result will occur if the panelist does return, but "fakes" the data on the day of return, simply in order to have complied with the test requirements. (The panelist may be desperate for the money, but so angry at the low payment as to simply write anything in the questionnaire, just to get it over with.)

If the panelist has been properly instructed, the right amount of money agreed upon for participation, and a weekly return scheduled, then for the most part the researcher has taken the necessary steps to insure panelist interest. During an extended test (e.g., 3+ weeks) it is inevitable that panelist interest will wane, and that some panelists will drop out of the study altogether. Waning interest cannot be further controlled. Dropouts from the study usually occur between the orientation (when the panelist receives the product to take home for the first week) and the first return (when the panelist returns with a completed questionnaire and receives the next product). Depending upon the money paid for participation the drop rate can be very high (not enough money is being paid for participation) or very low (enough money is being paid). Usually, later drops from the study occur for reasons which have nothing to do with the panelist's intrinsic interest in participation. Illness, death in the family or unexpected business crises arise. It is inevitable that with a large number of panelists, something will happen to one or another panelist to prevent completion. In this author's experience, that secondary "dropout rate" should not exceed 5-7% of the panel.

In a long-term study there are other issues to consider. Here are two typical problems:

1. *Partial Data* - During the course of the study a panelist only evaluated half of the products, and dropped out afterwards. Should the ratings be used? This is a frequent occurrence. Unless the study absolutely requires that each panelist evaluate every product assigned to him, there is no problem including the data in the general sample.
2. *Partial Payment* - A panelist has quit after 4 of the 6 weeks. Should the panelist be paid for the ratings that the panelist assigned, and if so how

much? This is a hard question to answer. On the one hand the researcher would like to reward the panelist for participating, especially if the reason for dropping out was beyond the panelist's control. On the other hand the conditions for participation specified that the panelist would be paid only upon completion of the study. The simplest answer to this problem is to pay the panelist if the panelist has completed a significant portion of the evaluations (e.g., more than 2/3) and if the panelist has been forced to drop out by circumstances outside of his control.

Testing Products in Multiple Locations

Most product tests are conducted in a number of geographically dispersed locations. Researchers and marketers want to insure that the responses to the products are not influenced by a provincial or non-representative sample of consumers. In order to insure representativeness, the researcher selects consumers in several markets. The underlying assumption is that by testing in multiple locations the researcher increases the representativeness of the samples.

This is a perfectly correct and appropriate strategy. The issue here is how to insure the same test conditions from market to market. Most researchers who have worked in different markets, even within the same country or state, know that there are differences in the way that local "field services" recruit, orient, and treat panelists during the course of the study.

There are three things to keep in mind, and remedies to employ:

1. There will be significant market-to-market variation in the capabilities of the field services that execute the actual interviews.
2. When the panelists are qualified to participate and then invited to participate in an extended test, the panelist should be "rescreened" at the site, preferably by someone other than the initial recruiter. The screening questions should be the same. Rescreening insures that the panelist who shows up to participate is actually the same individual who was recruited. By repeating the questions, the interviewer can check the answers originally given during recruiting against the answers given at rescreening. If the answers match, then the interviewer can feel assured that the panelist is truly who he says he is.
3. The actual orientation in the product usage can be done two ways. If the orientation is really quite simple then a senior person at the local field service can do it. In order to insure that this person knows what to do the manufacturer should send a videotape of the orientation. The videotape is better than written instructions because it gives the field service employees a sense of how the manufacturer wants the orientation to proceed. If the orientation is more complex and requires demonstration, then the manufacturer might consider sending a representative into the field to

Practical Considerations 175

orient the panelist. The field can be set up in such a way that the different markets are "staggered." That is, the testing at the "lead market" may begin at the start of the week. The employee who does the orientation can then travel to the next market, and conduct the orientation for a new set of panelists the following evening. Although in many ways this seems to be "overkill," it is virtually the only way that the manufacturer can assure that the fieldwork will be done correctly.

The second strategy is better for pre-recruit studies which require technical "know how" (e.g., evaluation of multiple fragrances by a large group of consumers, during a single extended session lasting 4 hours). Most field services do quite well when they have to execute simple studies (e.g., test two fragrances in a 10-minute test, applying one fragrance to the left wrist and the other fragrance to the right wrist). The field services are well-practiced in the procedures demanded by these simple studies. Field services are not as well-versed when it comes to more complex procedures. Even written instructions do not help because the interviewer has no familiarity with these more complex, carefully choreographed procedures. In these cases, having a representative from the manufacturer can assure proper test execution.

Sources of Error in Multiple Product Testing—Field Work

Given the foregoing considerations from orientation to product testing, here are five sources of potential error:

1. Have all of the products been properly labeled?
2. If the products are tested "blind," then are all of the products adequately disguised?
3. Will each panelist follow a "rotation sheet" that insures that every product is tested equally often in every position? There are no standardized randomization patterns for many conditions (e.g., test 7 of 24 products). The researcher must develop a randomization scheme to insure that every product appears equally as often in every test position.
4. Is there an orientation at the start of the test? What is the purpose of the orientation? Will there be an orientation exercise to show the panelist what to do? Who will demonstrate the procedure? Will the procedure also be written down for later reference (at home) by the panelist?
5. If the product evaluation is done at a central location facility (e.g., for fragrance), then has the facility been properly arranged to insure ongoing consumer sensitivity? (For fragrance testing this means installing fans to exhaust the fragrance. For color testing this means knowing ahead of time the nature of the lighting, and insuring that all products are tested under the same light. It is poor research procedure to test the appearance

of a product under incandescent light in one market, and under fluorescent light in another market. The same color will look quite different, and the test will yield discrepant results.)

An Overview

This section has dealt with some of the key operational issues in product evaluations. In the "real world" of applied research rarely is a study perfectly run from beginning to end. Despite the elegance of design up-front and the analysis afterwards, often it is the implementation of the study in the field which distinguishes results that are valid from results that are invalid.

Marketing researchers and sensory analysts alike understand the issues involved in product evaluation. Some of these can be handled by considerations up-front (e.g., relevant attributes, or design of the evaluations) in order to insure that the panelists maintain their motivation and their sensory sensitivity. A host of other unforeseen problems can arise in product testing. Many of these problems have nothing to do with the research, but rather deal with the interface of the consumer and the study. These problems require experience for their solution. This section has outlined some (albeit not all) of the considerations and precautions that the researcher should follow in order to maximize the quality of the data.

2. Developing the Right Language for Questionnaires

Introduction

Most researchers uses attributes in order to measure the private sensory experience of consumers. In many cases attributes are selected simply at random or by guesswork. The researcher may have a "laundry list" of attributes that have been used for years to evaluate products (or concepts) in the category, and so the researcher dutifully includes these attributes in the list to be evaluated. In new categories, the researcher must create a list of attributes on which to measure the product. This creation entails other research steps, such as qualitative research to discuss the nature and meaning of attributes with consumers in an open discussion, and/or a screening of the attribute checklist to determine which attributes are relevant to the consumers.

The Plethora of Attributes

A moderator (discussion leader) in a focus group can elicit many useful attributes from consumers. Most consumers have available to them a large number of descriptive terms for a product. Unfortunately, however, many of these terms are often "buzzwords" and are really nothing more than playbacks from advertising. The researcher need only look through a list of words to discover two categories: simple words that are quite general, and advertising words that are emotional, and not necessarily related to the product characteristics themselves.

Insightful interviewers can elicit attributes that are truly critical, but which do not spontaneously surface in a superficial discussion. Usually these deeper lying attributes are not immediately obvious, but when they emerge during the course of a penetrating discussion with the panelist, both the panelist and the researcher have the "aha" experience of recognizing that they have tapped something very important. For instance, for sanitary napkins, the immediate words might be protection and coverage, but the deeper words or descriptions might be something akin to "a constant reassuring pressure on the skin."

Developing the List of Attributes

To develop a truly good list of attributes requires a series of steps which first create a large list of terms, and then cull that large list to the most relevant ones. The more sources of terms one uses at the start of the process, the better the results will be. The participants may be consumers, in-house product developers, or consumers selected by various psychological tests to be "creative" or "verbal," etc. For the best results the researcher should mix together participants from the foregoing groups into one large group. Up to ten individuals should participate in the discussion. (Too few participants puts an undue burden on each participant to come up with descriptor terms. Too many participants reduces the chance of synergies and interactions between individuals, and stifles the spark which emerges from a group of the right size.)

Here is a straightforward 10-step approach:

1. *Identify Stimuli to Act as "Triggers"*: Within the category identify a range of products which perceptually differ from each other. If the category is a newly developing one, collect products which have relevant features, even if the products are not exactly part of this newly emerging category.
2. *Let Participants Use These "Trigger" Stimuli*: Give the stimuli to consumers, and let the consumers use them (or at least inspect them) at home. The experience with the stimuli will give the panelists a feeling of the main characteristics. Encourage the panelists to use the stimuli, and even take apart the stimuli (if that is possible).
3. *Gather the Panelists for a Discussion Session*: Bring the panelists together for a session that will last between 2 and 4 hours. Tell the panelists ahead of time that during this discussion session they should be prepared to talk about their impressions of the product, its physical characteristics, its benefits, how the product compares to other products that they have used, what the product lacks, etc. Focus the panelist's attention on as many different aspects of the product as possible.

4. *Have Each Panelist (in Turn) Describe One Intrinsic Physical Feature of the Product*: Go around the room, and have each panelist list one characteristic of the product. Do this two times, so that each panelist will have a chance to provide two characteristics. Sometimes panelists will have difficulty because the obvious characteristics will soon be exhausted. However, the group discussion can help the panelist come up with an idea. [The panelist may have to "stretch" a bit, but this stretching is exactly what is needed to reveal other attributes lying below the surface, which are less obvious, but also critical].
5. *Have Each Panelist (in Turn) Describe One Interaction Between the Product and the User (e.g., How the Product Is Used, What Is Actually Being Experienced)*: Follow the same approach as in Step 4 above. Step 5 will elicit attributes which emerge from actual product use. They are not necessarily sensory characteristics of the products per se, but rather perceptions that emerge (e.g., hard to use, goes on roughly, leaves an astringent skin feel, hard to remove, etc.)
6. *Have Panelists (in Turn) Describe One End Benefit or Result That Is to Be Obtained from Using the Product*: The objective here is to uncover consumer-perceived benefits of using the product. Follow the same approach as in Step 4 above.
7. *Open Up the Discussion*: After Steps 4–7 have been followed for at least one round (and in some cases two rounds), open up the discussion to the group. By this time many of the participants will have been sufficiently "energized" and "stretched" so that they will perceive and suggest further sensory, usage, and end benefit attributes.
8. *Polish the Terms*: The exercise will generate a large list of terms. After the exercise has been completed the researcher should collect the terms, "polish them" (if they are phrases, then improve the language to make each phrase easier to comprehend), and then classify the terms into relevant categories.
9. *Cull the List*: A small group of individuals should then go through the list, identifying the redundant elements as well as elements which have little shared meaning. Redundant elements can be stricken from the list. Elements which have little shared meaning may be terms or phrases that are hard to understand, or may have very little generality. These also should be stricken from the list. What remains will be a large list of categorized terms and phrases dealing with the sensory, usage, and end benefits of the product.
10. *Select the Terms for the Questionnaire*: From the final list in Step 9, select the relevant terms to be used for the questionnaire. The terms should be the obvious ones, as well as new terms which will add a new level of depth and insight. (One can never be sure about the utility of

new attributes and terms in the questionnaire until one implements the actual product evaluation with consumers, but at least this step allows the researcher to try new attributes, and may potentially improve the questionnaire.)

An Overview

The typical researcher uses a laundry list of terms in order to assess consumer responses to a product. Over time, many of these laundry lists come to comprise the same types of attributes. (The vast majority of these attributes are really restatements of "liking.") By following the ten steps outlined above the researcher can develop a more comprehensive set of terms, and avoid the ever-present possibility that the terms will simply mirror the buzz words that are broadcast by the media. However, as shown above, the procedure requires an investment of time. There are always terms which lie below the surface, and which need to be "excavated" by an in-depth discussion of the product. It is vital to use the product, to experience its facets and nuances, and to elicit the resultant impressions in a group setting. By doing so the researcher can be assured of having tapped the potentially critical aspects of a product that might otherwise be glossed over in favor of the more common, but less insightful terms which lie at the tip of one's tongue.

3. Going One Step Further in Attributes: Fitting Personalities and Pictures to Products

Introduction

Traditional sensory analysts primarily use sensory, performance and acceptance terms to describe products. Marketing researchers, and especially those with backgrounds from advertising agencies, recognize that there is a substantially greater set of potential descriptions for a product that can be useful for marketing. These descriptions are "personality statements" and "pictures."

When panelists react to a product, they bring many complex perceptions and responses to the study. The conventional attribute scales may be too limiting if the researcher wants to probe more deeply into the panelist's perceptions. It is possible to learn more about the product when the researcher goes outside the conventional scaling procedures, and presents the panelist with complex attributes, defined either by personality statements or by pictures.

Using Personality Descriptors as Attributors

The researcher can instruct the panelist to evaluate a fragrance and rate the degree to which the fragrance embodies different types of personality descrip-

Table 3 Profiles of Four Fragrances, Rated on Liking, Sensory, Image, and "Personality" Attributes*

	Frag A	Frag B	Frag C	Frag D	Range	St Dev
Liking						
Overall	56	53	49	46	10	3.8
Sensory						
Intensity	45	38	57	47	19	6.8
Sweet	56	50	63	49	14	5.6
Fruity	58	52	56	52	6	2.6
Flowery	55	53	60	58	7	2.7
Harsh	23	35	26	25	12	4.6
Spicy	57	52	58	43	15	5.9
Image						
Young-old	57	44	60	42	18	7.9
Male-female	62	60	60	49	13	5.1
Day-evening	58	43	52	58	15	6.1
*Personality**						
Happy-sad	35	38	52	48	17	7.0
Introvert-extrovert	45	65	51	56	20	7.3
Passive-active	47	52	47	53	6	2.8
Immature-mature	62	43	68	51	25	9.7
Shy-inquiring	63	58	52	53	11	4.4
Quiet-loud	57	51	46	50	11	3.9

*Rating of fragrance as if it were a person. First attribute (e.g., introvert) = anchored at 0, rating second attribute (e.g., sad) anchored at 100.
Range = Maximum of four fragrances − minimum rating of four fragrances
St Dev = standard deviation of the four means

tors (or vice versa—how well each personality descriptor fits the fragrance). Table 3 presents the results of one of these evaluations for four fragrances, showing both the sensory/image profiles (from conventional scaling), and the personality profile (if the product were to be a "person," then how would the panelist describe the personality). From the personality profile we can see the large differences between the four fragrances. We might be aware of the differences from the ratings of sensory and image characteristics, but the personality descriptors make the differences more vivid and add a new dimension to the fragrance.

The same type of exercise can be done for physical products and compared to the profile achieved when the panelist knows only the brand name of the product. Do the two agree? Does the physical product convey a personality profile similar to that promised by the brand? Table 4 shows the results of a

Practical Considerations

Table 4 Profiles of Two Eye Shadow Products (A, B), Tested Blind and Branded, and Rated on Various Attributes

	Blind Prod A	Blind Prod B	Branded Prod A	Branded Prod B
Liking				
Overall	50	55	40	60
Sensory/performance				
Dark	35	43	38	45
Shiny on skin	28	31	26	34
Easy to apply	45	58	42	55
Image				
Expensive	26	37	30	37
Unique	35	42	46	51
Personality				
Happy-sad	25	22	18	12
Introvert-extrovert	58	52	63	48
Passive-active	43	45	49	56
Immature-mature	64	68	62	74
Shy-inquiring	45	48	46	55
Quiet-loud	31	28	34	18

study on two eye shadow products, differing both in physical composition and in brand. The differences in the personality associated with the physical product alone are there and obvious. However, the larger, more obvious and more striking differences come when the researcher presents the panelist with the product, coupled with the product brand name and packaging. The differences in product personality emerge even more strongly. There are differences in the sensory attributes and basic image attributes, but they are much smaller than the differences in the personality characteristics ascribed to the products.

Pictures (Images) as Attributes

Pictures, instead of words, comprise an entirely different dimension of an attribute. Psychophysicists know full well that panelists can match the intensity of non-numerical stimuli (e.g., the loudness of tones) to perceived sensory intensity of a criterion stimulus (e.g., the brightness of light, the loudness of sound, or the size of a circle). This is known as "cross modality matching," and forms the foundation of psychophysical measurement (Stevens, 1975).

The applied researcher can go several steps further. The researcher can select a set of pictures varying in content (e.g., scenes vs. people), in activity (active vs. passive activities), and in actual physical construction (e.g., bright colors vs. dark colors, lots of images vs. lots of open space). These are the "attributes." For each stimulus the researcher can show these different pictures to the panelist and instruct the panelist to rate the degree to which the picture matches the stimulus.

To some panelists this task seems absolutely simple. The panelists appear to have little or no problem matching a picture to a product. When asked what rules the panelist follows, these panelists often state that they just "feel" that a picture matches or does not match. They have no idea why the match exists. Other panelists have far more problems with the task. They cannot find a nexus or similarity between a physical product that they try and a set of evocative pictures. It is not clear to these panelists that there should be any connection, except the most obvious ones (e.g., the product has the same color as one or more of the pictures, or the pictures depict the product).

If the researcher does the experiment, selecting pictures of different types and testing these against stimuli, then at least two interesting results occur:

1. Panelists report that certain products go well with "active" or bold colors, whereas other products go well with pastel or "milder" colors.
2. Some products fit with people scenes, other products fit better with landscape scenes.

The results are often idiosyncratic. However, with a reasonable number of panelists initially screened to be able to perform this matching task, the results eventually prove very instructive and often insightful. The panelists can often point out the link between pictures and products.

An Overview

The task of matching personalities and pictures to products transcends the traditional product evaluation mode. Whereas traditional evaluation is intellectual, the personality and picture matching belongs more to the realm of "projective" techniques. The researcher is not interested in the sensory characteristics of the product or brand, but rather the researcher is interested in a deeper, more profound response to the brand. By using personality statements and pictures, the researcher requires that the panelist more clearly express his overall response to the stimulus, using complex stimuli (personality, pictures) as the language of expression.

4. Multinational Product and Concept Evaluation: Some Pragmatic Issues to Resolve

Introduction

In recent years companies have espoused the cause of internationalism and harmonization in product development and testing. Rather than creating products for consumers in one country alone, many companies now emphasize the creation of products which can span multiple countries. Even if one product cannot satisfy consumers in different countries the thinking is that a line of related products may be able to satisfy consumers, and that consumers in each country can be satisfied with a limited number of items from the line.

This section deals with the testing of concepts and products in different countries, with different cultures, different languages, and different ways of using scales. How does a researcher work in different cultures to create a usable, actionable database of responses to products and concepts?

Concept Elements—Problems of Language Across Countries. Both product and advertising concepts are sensitive to nuances. In English, for instance, there are many ways to describe "elegance." These descriptions vary from simple to complex. Most importantly, they tap the essence of the language and the power of that language to describe. However, since languages differ, the same words and phrases may not translate across languages. It is vital that the researcher recognize that statements (and thus basic ideas) which are meaningful and evocative in one language may either work or not work in another language.

Unfortunately there is no way to standardize concept elements across languages. That is, what works in one language may or may not work in another. The researcher must "do the experiment," make the translation (using a native speaker of the language as a translator) and then test the concepts among consumers.

Visuals in Concepts—the Same Across Cultures and Languages? We might believe that visual elements in concepts would remain the same across cultures and languages. This is patently not true. If, in a concept optimization study, panelists from different cultures and countries evaluate the same set of words (in their respective languages) and the exact same set of visual stimuli, and if the panelists rate both acceptance and other impressions (e.g., for whom the product is designed—younger vs. older), then we often find that the same visual performs quite differently in two cultures and languages. Thus, even visuals are not invariant or fixed across cultures.

Differences in Language for Descriptive Attributes and the Need for Local Dialects. Consumers in different countries use words differently. Even a single language, such as Spanish or English, may be used differently in different countries. Consequently, attributes have to be carefully selected and translated by a native in the country where the research will take place. Only then can the researcher be sure to capture the current idiomatic usage of the language. (Ideally it is best to have the translator currently living in the country. That way the translator is *au courant* with colloquial terms.)

It is also advantageous to use a phrase to describe the attribute, rather than a word, and if possible to incorporate one or more concrete products to exemplify the attribute in a concrete fashion. This will anchor consumers from different cultures.

Scale Usage in Different Cultures. In the U.S. we can present consumers with a fixed scale (e.g., 0-100), and assume that consumers will use the scales similarly. As we saw earlier, this may not be the case across countries and cultures. In some countries the consumers feel secure and use the full range of the scale. In other countries, however, there is a covert pressure and a cultural norm not to embarrass the manufacturer by "downrating" the product. The consumers from such a culture thus "uprate" the products, even when they don't like them, in order not to violate unwritten norms of behavior.

The researcher can neither prevent nor overcome the impact of cultural norms associated with scales. However, the researcher can put in a set of reference stimuli (be these standard concepts or products). The test stimuli can then be reported in two ways:

a. Absolute score on the scale
b. Score relative to the reference stimuli.

The researcher should make every effort to incorporate both strong and weak performers in the test to provide wide ranges of stimuli. Furthermore, once the researcher decides to use reference stimuli, these stimuli should be used for a while in order to build up a bank of norms and to increase one's understanding of the scaling behavior in that country.

An Overview

One of the key areas for future development in research methodology will be the creation of common (or at least mutually compatible) questionnaires that can be used across countries, cultures, and languages. The continued internationalization of product marketing will require that researchers accurately measure the performance of products and concepts in markets in which they currently operate, as well as in new markets in which they have little or no experience. The challenge is not to make all countries the same, but rather to create a flexible research program that can be easily and inexpensively implemented in, and calibrated for, various languages and cultures.

5. Paired Comparison vs. Scaling Procedures: Theoretical Differences, Practical Problems

Introduction

Most marketers and marketing researchers are familiar with the conventional paired comparison methods in which the researcher presents the panelist with

Practical Considerations 185

two products and instructs the panelist to select which product of the pair is preferred, has a stronger fragrance, is greasier, etc. The paired comparison methods are popular because to the conventional researcher they offer an absolute result—either the panelist prefers one product or the panelist prefers the other. (In some cases the researcher also allows for a judgment of "no preference," when the panelist truly has no preference between the two products, or "no difference" when the products have the same sensory level of an attribute.)

Paired comparisons are easy for the researcher to administer and easy to analyze. The researcher simply obtains the consumer's response to two stimuli and determines which of the stimuli has more of a characteristic. The statistical analyses are easy to do because the data lends itself naturally to percentage statistics.

However, paired comparison data does not measure intensity of an attribute. The panelist assigns an "all or none" rating. That is, the panelist never rates the degree of intensity (from which the researcher could deduce which stimulus is stronger). The panelist only chooses the stronger of the two products (or the more preferred). It is assumed (albeit incorrectly) that, when many more panelists choose one product over another (in terms of an attribute), the chosen product has substantially more of the attribute than the product not chosen. Indeed the analysis of paired comparison data rests upon the assumption that consumers are inherently error-prone, and that they can rarely make discriminations perfectly. The more one product is preferred to another, the more the two products differ along an underlying subjective continuum (or at least that is what is assumed). There is no accepting that the panelist can act as a direct measuring instrument.

The Three Benefits and Five Problems of Paired Comparison Research vs. Direct Scaling

Benefit #1—Simplicity and Apparent Directness of the Question. Paired comparison procedures are simple to execute. Panelists have little trouble comprehending the task, at least intellectually. They simply select which of two stimuli possesses more of a characteristic. In many paired comparison procedures the panelists must choose either one of the stimuli on a series of attributes, respectively. For some attributes Product A will have more of the attribute. For other attributes Product B will have more of the attribute. This deceptive simplicity both intrigues and seduces researchers.

Benefit #2—No Need for the Panelist to Act as a Sophisticated Measuring Instrument. Paired comparisons simply require a choice. Paired comparison does not demand that the panelist accurately rate the characteristics of the product. Scaling procedures require that the panelist have some well-defined concept of "magnitude" or intensity in order to properly create a scale.

Benefit #3—Ease of Administration and Data Collection. Because paired comparison is so simple, it is easy to administer in the field. The panelist need not be trained, nor even "oriented" in scaling. To provide valid data the panelist must merely comprehend the idea of "more than/less than." In contrast, scaling procedures require that the panelist understand how to evaluate intensity in order to perform as a valid measuring instrument.

Problem #1—Only All-or-None Responses Allowed. The panelist cannot assign a graded response. The panelist simply makes a generalized response—choose one sample or choose the other. There is no room for equivocation. For instance, even when two products seem subjectively close in intensity, the panelist must choose one or the other when choosing "stronger." In contrast, scaling procedures allow the panelist to assign these two stimuli "similar" (but not identical) ratings. If the products seem quite different from each other in intensity then the panelist can assign these two stimuli quite different ratings. The subjective scale mirrors the panelist's perception. It does not force an arbitrary all-or-none response.

Problem #2—A Convoluted Approach to Creating a Subjective Scale. In order to create a subjective scale of magnitude and thus advance beyond the simple paired comparison data to an underlying scale which can be better interpreted, the researcher must transform the paired comparison results to scalar data. L. L. Thurstone (1927) developed many of these transformation methods. Transformation requires a theory relating paired comparison data (percentage preferences) to an underlying scale assumed to be used by the panelists. Rather than the results being based squarely on the responses (assuming that the panelist can act as a measuring instrument), the higher order transformation and analysis of paired comparison data requires a complicated theory. There is no "free lunch." In truth, the researcher does not simplify matters with paired comparisons. The panelist's task may be easier, but the subsequent higher level data analyses are far more complicated than the simple averaging applied to subjective ratings of perceived intensity.

Problem #3—Difficulty Working with Multiple Products. Many researchers like to work with many more than two products in their studies. However, paired comparison only allows for the assessment of two products at a time. In contrast, scaling allows for the assessment of an unlimited number of products at any one time, because the panelist evaluates each product by itself. Researchers have developed round robin procedures to create an underlying scale from the evaluation of many products, using paired comparisons of just two products at a time. However, this round robin analysis again requires a theoretical leap of faith, rather than simple averaging.

Problem #4—Paired Comparison Data Requires Much More "Fieldwork" for the Assessment of Many Products. For two products, paired comparison procedures require one evaluation of both products. For three products, paired comparison procedures require that each product be compared to each other. This requires three pairs, each of which comprises two components (A-B, A-C, B-C). The consumer must evaluate six samples (not three as would be the case for direct scaling). The situation becomes far worse as the number of samples grows. For instance, with six samples, direct scaling requires the evaluation of six products, whereas paired comparison requires evaluation of 15 pairs (6 × 5/2 = 15), and thus the evaluation of 30 products. Often when a researcher or marketer is faced with the need to evaluate so many products, the task seems so daunting that the researcher drops some of the products because of the inherent difficulty in executing the project. In contrast, the researcher who uses scaling techniques can incorporate many products into the test, including samples that would ordinarily not be tested. In a paired comparison test every additional sample costs much more money. Not so with direct scaling.

Problem #5—The Panelist Does Not Truly Act as a Measuring Instrument When Choosing One Product over the Other. Paired comparison procedures assume very little in the way of a panelist's ability. They simply assume that the panelist can choose which of two products possesses "more" of a characteristic. Consequently, paired comparisons do not truly measure the consumer's response. For instance, in terms of overall liking, knowing that a person likes one product more than another does not tell the researcher whether the person likes the preferred product much more, or only slightly more.

Why Scaling?

Given the researcher's need to measure the characteristics of a product from the consumer's viewpoint, it is vital to have the consumer act as a measuring instrument by scaling the perceptions. Properly applied, scaling links the consumer, the researcher, and the product. Through the numbers assigned by the panelist the researcher can begin to better understand the consumer's response to products. For product development and for marketing purposes, numbers are one key by which the subjective pattern of responses can be uncovered. It is purposely "shooting oneself in the foot" when researchers avoid having the panelist scale his responses, and rely only on the statistical analysis of paired comparison data. The subjective patterns underlying the stimuli will be significantly harder to uncover, and the ultimate utility of the panelist as an intrinsic link in the development process will be substantially weakened.

6. Avoiding Biases: Rotating the Order of Attributes

Introduction

Most researchers are familiar with order bias in product evaluation. When a consumer evaluates many products, ratings for the product "tried first" in the series often differ from those for the same product rated later on. There is an equally prevalent but less explored bias—the bias of order of attributes. This bias manifests itself as a systematic distortion of the data when the questions are always asked in a fixed order.

Unlike order bias in products, order bias in attributes is more subtle and difficult to discover, but may be unavoidable. The researcher can vary the order of products with little trouble. However, in a questionnaire it is sometimes awkward to vary the order of the attributes (to remove order bias), without at the same time introducing another bias (e.g., asking some attributes in the wrong order, relative to the order of appearance of those attributes in the actual product).

Recommended Practices for Attribute Order

In order to obtain the least biased data, many researchers recommend that the panelist rate the product attributes in the temporal order in which those attributes appear. For instance, appearance generally precedes fragrance, and fragrance generally precedes product application and resulting skinfeel. The conventional wisdom is for the panelist to rate the attributes in the order of appearance, smell, and feel/touch. This order insures that the panelist can attend to the attributes without jumping back and forth and changing focus.

By having the panelist rate the order of attributes in which they appear, however, the researcher may bias the data. It is not known whether the appearance ratings influence the fragrance ratings or the texture/feel ratings. A skeptic could argue that the panelist is already biased by the initial ratings, and that a more valid procedure would randomize the order of attributes so that for some panelists texture/feel attributes come first and appearance attributes come last.

An Alternative—Full Product Use Followed by Attribute Ratings

If the researcher feels that there is a strong order bias in attributes, then one way to overcome the bias instructs the panelist to use the product, pay attention to all the attributes (with which the panelist has already become familiar—e.g., through an orientation session), but then rate the attributes in randomized order. The panelist does not rate the attributes in the order of their appearance during the actual use of the product. Rather, the panelist uses the product as he normally would, and then just afterward rates the product. This sequence differs from the approach of using the product and rating the product during the course of use, as the attributes appear and are sensed.

This alternative approach enables the researcher to systematically vary the order of the attributes, and thus to discover whether or not there is an order effect for attributes. The key drawback of this alternative approach is that it does not capture the attributes in the order of appearance. The panelist loses the ability to record any rapidly changing sensations (e.g., change in perceived thickness as the product is rubbed in), because all of the ratings are obtained after the evaluation, not during the evaluation.

When Should the "Overall Rating" Be Assigned?

In most research the overall rating of product acceptance is the key attribute used for a research decision. Most attribute ratings are used to support the overall rating, or to diagnose reasons for a good rating or a poor rating.

If the attributes are evaluated in either a fixed or a randomized order, when should the researcher obtain a rating of overall product performance? (The overall attribute could be acceptability, purchase intent, or any other general rating of product quality which integrates all of the consumer's impressions.) Should the overall rating be the first attribute rated or the last attribute rated? Should the overall rating be assigned to some random position in the questionnaire?

There are two key issues to resolve here:

Information Available. At what point does the panelist have enough information to assign a valid overall rating? The overall rating requires the panelist to integrate all of the information available into a single number. The panelist must have the relevant information and therefore should have full experience with the product. Full experience requires the overall rating to be assigned at the end of the product ratings (if all of the attributes are rated in a fixed order). The overall rating should not be assigned before the panelist has had a chance to experience the product fully. The researcher should refrain from an immediate rating of overall acceptability upon first exposure, especially if the product is complex and requires multiple sensory inputs which appear in a natural temporal sequence (e.g., products like shampoo, lotion, and anti-perspirants which excite multiple sensory impressions that are experienced in a clear temporal order).

Biases (Preference Justification). Panelists usually try to be consistent when rating products. Consumers do not want to appear as if they are responding randomly and as a consequence they attempt to justify their ratings. If the product is rated high overall, then subsequently the panelist may try to "justify" that high overall rating by looking for high ratings on "attribute liking." If the overall rating (assigned first) is low, then the consumers may try to "justify" that low overall rating by rating the attribute likings as "low." The reverse also holds— consumers may uprate all of the attribute likings, and consequently uprate overall liking in order to appear consistent. Preference justification is an intellectual bias that can be overcome by randomization of the attributes. By randomizing, the researcher reduces the effort due to justification. If the preference justification bias is sufficiently severe (viz., with obsessive panelists), then the best research strategy exposes the panelist to the product, has the panelist use the product, and only afterward allows the panelist to rate the attributes.

An Overview

Attribute order bias and its corollary, preference justification bias, are potential problems in product testing. Many researchers either forget about these biases completely because of the natural order of attributes, or remove the bias by randomizing attributes. This section presents one possible compromise. The panelist should fully experience the product. Then the panelist should immediately rate the product on the attributes, with the order of attributes randomized in order to reduce bias. The location of the overall rating should be randomly placed within the set of questions in order to remove the preference justification bias.

7. Appropriate Base Size For a Study: Criteria of Representativeness, Data Stability

Introduction

The number of panelists in a study directly affects cost. The larger the base size of consumers, the higher the study cost. Whether at R&D or in marketing, it is vital for any researcher to design the study with the optimal number of panelists. Too few panelists make the data unstable. Too many panelists force the researcher to pay for data that does not add as much as it costs to obtain.

What are the criteria for base size? Is correct base size simply a matter of a "comfort feeling?" Are there any additional criteria that the researcher can impose on the study? This section deals with base size from the point of view of study objectives, panel representativeness, and data stability.

Traditional Approaches to Base Size

The traditional approach to selecting the optimum number of panelists mixes intuition with statistical considerations. The data becomes more stable as the panel size increases. Furthermore, the larger the panel size the more likely it is that the researcher will have appropriately sampled the ultimate consumer population.

In conventional research the number "100" has been the magic base size. In the absence of other criteria the number "100" "feels good" to many researchers and marketers. One hundred consumers seems a sufficiently large number to insure a proper consumer sample. Furthermore, the same 100 consumers can be divided into different subgroups (e.g., males/females, users of product A versus users of product B), thus allowing for more insight. Finally, a base of 100 panelists suffices to stabilize the mean.

Base Size, Expertise, and Replicate Samples

Expertise. Some researchers believe that the base size issue can be resolved by training a consumer to become an expert. Some practitioners feel that expert panels of highly trained consumers can obviate the problem of base size because as few as 5 or 6 panelists will suffice. It is not clear, however, how realistic is this point of view. Can 5 or 6 expert panelists generate truly stable data? There is little published literature on this topic.

Replicate Ratings from Panelists. There are other researchers who understand the concept of data stability, but believe that they can achieve the same level of stability by using few panelists but obtaining more ratings per panelist. Most consumer research secures one rating per product. Replicate ratings are based on the belief that stability can be achieved by having every panelist rate each product several times. Researchers feel that some instability occurs because with one rating per product per panelist there is a significant amount of variability due to individual differences and due to the "position effect" of the first product tested. (The first product tested often provides the least stable data.) Researchers who hold this view also feel that additional stability can be introduced by inserting a training product ahead of the test products to diminish the unduly high variability of the "tried first product." It is not clear whether an initial training product can effectively reduce the necessary base size but still maintain stability.

Can a Base of 100 Consumers Represent the Population? The first question we should ask is whether or not the 100 consumers actually represent the population. How many consumers are really needed to represent the ultimate population, given the substantial person-to-person differences? This variability leads to certain qualitative guidelines:

1. Individual ratings of sensory attributes often show lower variability than do individual ratings of overall liking. Given this interindividual variability, the base size might have to be larger for liking attributes and smaller for sensory attributes.
2. Are there significant market-to-market variations in consumer preference? If there are many markets of interest, then the base size of 100 panelists may be too small. Similarly with regard to product usage. If there are many products in the category, each with a distinct sensory profile, and if consumers are loyal to the different products, then a base size of 100 may not capture the richness of the market. The researcher may require more than 100 consumers.

3. On the other hand, the consumer market may be significantly more homogeneous than believed. In this case the base size of 100 consumers may be too large. The data may become stable after a base of 30–50 is reached.

At What Base Size Does the Data Become Stable?

Can we develop some rules of base size based simply upon data? Is there something in a database that allows us to identify the point at which data becomes stable?

Fragrance Test. Let us consider the results of an experiment in which the panelists evaluated four fragrances in a central location (see Table 5). Each of 100 panelists rated every one of the four fragrances on a series of attribute scales. The panelists rated each fragrance using a 0–100 point scale, anchored at both ends.

We can analyze the data for sensory and liking attributes. The analysis is straightforward:

1. Compute the mean rating on attributes for a specific fragrance, randomly selecting 10, 20, 30 . . . 100 panelists. That is, begin by selecting ten panelists at random. Compute the mean of their ratings. Then add an additional ten new panelists, and compute the mean of the 20 panelists. Continue adding groups of ten panelists from the set, and compute the mean. At some point the mean should stabilize so that the addition of ten new panelists scarcely affects the mean at all.
2. Compare the patterns across sensory and liking attributes to determine whether there are different patterns. Even though panelists differ from

Table 5 How the Mean for Fragrance Ratings Stabilizes with Increasing Base Size (Examples with Sensory, Liking, and Image Attributes)

Base size	Liking	Intensity	Sweetness	Feminine	Unique
10	45	34	67	78	34
20	51	37	63	72	37
30	48	40	60	73	41
40	46	38	59	74	40
50	45	37	59	72	38
60	46	39	61	74	39
70	47	37	60	73	38
80	46	37	62	73	37
90	47	37	62	73	37
100	47	37	62	73	37

Note: Each additional base of ten respondents is added to the previous base (total panel), and the mean recalculated.

Practical Considerations

each other more on liking than on sensory attribute ratings, does this reflect itself in the base size needed for stability? That is, do sensory attributes stabilize more rapidly than do liking attributes?

Results from the Fragrance Evaluations. Table 5 shows the results of the fragrance study. The results are shown as a function of increasing base size. The empirical data suggests the following:

1. Overall the mean stabilizes at a base size of approximately 40–50. Beyond 50 the mean does not budge.
2. The sensory ratings (e.g., sweetness) converge at the same rate as the liking ratings. Although there are usually greater interindividual differences in liking as compared to sensory attributes, this does not show itself in the rate at which the mean converges.

Random Data Test. Do the same analysis, but with random data, and assess the degree to which additional samples change the mean. At some point even the random data should stabilize. (It stabilizes at 50; see Table 6.)

Table 6 How Random Data (True Mean = 50, Varying Standard Deviations) Converges to the True Mean, with Increasing Numbers of Observations

Cases	SD=5	SD=20	SD=15	SD=20	SD=25
5	46	47	46	56	37
10	51	50	57	44	51
15	49	49	55	46	52
20	51	49	51	46	56
25	50	49	50	48	55
30	50	48	51	47	55
35	50	49	51	47	53
40	50	49	51	47	53
45	50	49	51	49	52
50	50	50	51	48	52
55	50	50	52	47	52
60	50	49	52	47	50
65	50	50	51	46	50
70	50	49	51	47	51
75	50	50	51	47	52
80	50	49	51	47	51
85	50	49	50	49	52
90	50	49	50	48	52
95	50	49	50	49	50
100	50	49	51	49	51

SD = Standard deviation

An Overview. Base size is a controversial topic in research. There are two schools of researchers. One school comprises the "maximalists" who want as large a base size as possible to insure that they sample the individuals who truly represent the population. Maximalists are influenced by the sociological model which looks at large numbers of consumers. The other school are the "minimalists," who want as few panelists as necessary to generate stable data. Minimalists are more influenced by the physical sciences, where the base size is chosen to produce stability.

If strength of opinion itself were the sole governing factor and one had to select base size to accord with a "view" of the world, then there would be no single answer. Both large and small base sizes would be relevant because the researcher could appeal to an underlying "theory" or world view. From a numerical viewpoint, however, a base size of 50 generates a stable mean, at least for products. This base size is not engraved in stone, however. It may be that a base size ranging from 40–60 is the most appropriate. Clearly a base size of 20 is too little, and a base size of 200 is too large, if one's primary concern is the stability of the mean.

Does an Orientation Product Increase Stability and Decrease Base Size?

The issue here is whether an orientation product tested by panelists at the start of the evaluations generates "better" data (in terms of data being more stable). The specific study evaluated lipsticks. The panelists rated eight lipsticks on a set of liking and sensory attributes. The lipsticks varied in color level. Each of 160 panelists evaluated all eight lipsticks on the lips, in a sequential 8-day home use test. Every product was tested in every position (1–8) equally often (20 ×). The original purpose of the design was to screen various red shades for a product line. However, the results can illustrate the effect of an orientation or "first" product.

The study had an external criterion of "validity"—the ability of the ratings to track the known levels of "redness." Table 7 shows the results, on a position-by-position basis. The products tried in the first position are assigned ratings of "redness" that clearly do not track the physical degree of redness. However, the products tried in the second and further positions do track physical redness. The same set of 20 ratings/product shows greater "validity" in the second and subsequent positions than in the first position.

We can conclude from this study that an orientation product (or a training product in the first position) can reduce the base size for a study by producing more valid, more stable data. The ratings become more stable when the products are tested in the second and subsequent positions.

Table 7 How the Presence of a First Product (e.g., Training Product) Enhances the Ability of a Panelist to Show Differences (Example - Eight Lipsticks, Varying in Shades of Red)

Actual red dye	Mean	Position							
		P1	P2	P3	P4	P5	P6	P7	P8
Level 1	35	41	34	31	35	36	35	29	34
Level 2	36	34	36	36	36	38	35	32	40
Level 3	39	26	33	46	38	47	43	33	43
Level 4	44	42	45	49	40	43	46	36	47
Level 5	43	31	42	48	46	58	42	32	43
Level 6	50	36	57	53	49	55	56	47	49
Level 7	59	46	56	55	48	59	55	48	58
Level 8	61	65	61	65	57	64	59	54	61

Attribute rating = Perceived redness of lipstick

8. Statistical Differences vs. Meaningful Differences: On Interpreting Inferential Statistics

Introduction

With the ever growing availability and popularity of statistical packages, and with the increasing appreciation by researchers of the use and power of statistical techniques, interest has focused on the concept of "statistical difference."

What is statistical significance? In common parlance (but in incorrect usage) statistical significance is thought to be some magic dividing line which differentiates two products. If the products do not statistically differ from each other by some recognized test, then they are assumed to be the same. In contrast, if the products statistically differ from each other, then they are assumed to be different.

The situation becomes substantially worse when the test involves how much consumers like two products. The researcher and the brand manager hold their breaths until the test shows that the new product is "significantly" better than either the competitor or the current product that it has been designed to replace.

Matters become even worse when the notion of probability of difference is invoked. Few researchers in the testing business have escaped hearing the boast (or plaint) that "two products differ from each other at the 95% confidence level." The claim sounds so enticing. But what exactly does it mean?

What Is Inferential Statistics About?

Inferential statistics, the branch of statistics underlying the tests of difference, considers the ratings assigned to two or more products and attempts to understand whether these two products could have possibly arisen from the same basic distribution. Here are the three assumptions:

1. When a researcher measures a property of the product many times, inevitably there is variation. No measurement is perfectly reliable. For instance, if the attribute being rated is overall liking on a 0–100 point scale, then a product may be rated differently by panelists (or even differently by the same panelist at different times). The results of the measurement generate a "distribution." Some typical distributions appear in Figures 1A–1D.

2. The distribution can be characterized by two major indices. These are the mean (average, measure of central tendency, centroid) and the variability or scatter (indexed by the standard deviation). Figure 1 shows four

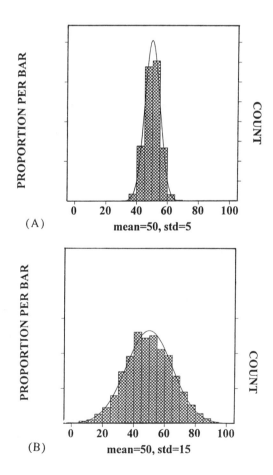

Practical Considerations

distributions with low and high means, and low and high standard deviations. Typically the mean and the standard deviation are reasonably independent of each other (except if the mean lies at the very bottom area or top of the scale, in which case the standard deviation will be quite small).

3. Suppose we know the mean for the current product is 65 on the 0–100 point scale. Now, let's select a prototype product to test. The prototype achieves a 70 on the scale. Can we conclude that the prototype differs

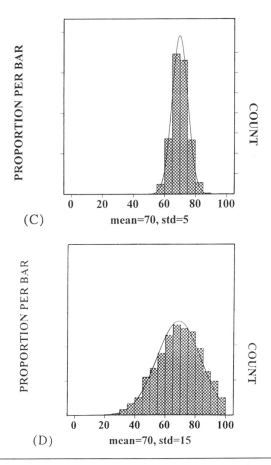

Figure 1 Distribution of four products with different means and standard deviations: A. High mean (50) and low standard deviation (5) B. Low mean (50) and high standard deviation (15) C. Low mean (70) and low standard deviation (5) D. High mean (70) and high standard deviation (15).

from 65? What is the probability that the 70 we just observed is simply one observation from the distribution, with a true mean of 65? What if the prototype we tested achieves a value of 80, rather than 70? We probably would conclude that this 80 means that the prototype does not originate from a distribution with a true mean of 65. What if the prototype achieves a mean of 90? The odds are that most people, ourselves included, would conclude that the prototype does not originate from the distribution with a true mean of 65. Inferential statistics is a way to quantify the probability that the value we measure for the prototype comes from a reference distribution with a known mean and a known standard deviation.

The T Test—Quantifying the Odds of the Null Hypothesis

Let us run a simple T test between the mean of a distribution and a fixed value. Suppose we manufacture a mouthwash. Our objective is to come up with a prototype that performs "well" against norms. Our norms suggest that on a scale of 0–100 (0 = does not taste at all like it will clean my breath → 100 = tastes like it will clean my breath), a good performing product must score 65 (on average). We know that the prototype must score 65 because over the past 10 years we have done many mouthwash studies. Products around 65 or higher tend to be successful. Products lower than 65 tend to be unsuccessful.

Of course the value 65 is not engraved in stone. We know that the prototype may score 63 or 64 and still have a chance of success because of many other factors, such as price. We also know that there are products that were introduced which, in our previous tests, scored 67, 70 and 73 and still failed. We believe that the failure was not due to a poor performance on this key attribute, but rather due to faulty distribution in the marketplace, poor advertising, and in one unfortunate instance an advertising campaign that "broke" two weeks before the products were on the shelves in the supermarkets and drug stores. However, over the years the manufacturer has come to realize that a score of 65 represents a level which promises to be adequate.

We now have a new prototype to test. The prototype has passed many hurdles in development. The prototype tastes fairly good, can be manufactured at a reasonable cost to be profitable, and differs in sensory character from the other competitors in the mouthwash category.

How well does the new prototype perform on this critical aspect of perceived efficacy? We run the typical test, using 100 consumers. [100 is not necessary, but often makes both researchers and marketers feel that they have done their homework by selecting an adequate sample]. The population of 100 consumers evaluates the product. They take the product home, rinse for 10 seconds in the morning and rate the liking (among other attributes). Here are four possible scenarios:

Scenario A—High Mean Score, Low Variability Around the Mean. The product receives a mean rating of 70 and a standard deviation of 20, with a base size of 100. Table 8 shows the calculations to determine the odds that the mean rating, with this variability, comes from a distribution whose true mean is 65. The odds are very low that this product comes from a distribution whose true mean is 65. We conclude that the prototype scoring 70 does not come from the distribution with a mean of 65. We also conclude that the prototype comes from another distribution with a mean greater than 65. [Our best guess about the mean of that "true" distribution is that it has a value of 70].

Scenario B—High Mean, High Variability Around the Mean. The product receives a mean rating of 70, and a standard deviation of 60. We conclude from Table 8 that the odds are very high that the prototype comes from a distribution whose mean is 65. The high variability associated with the observations (S.D. = 60) means that a score of 70 could have been achieved simply from random variation. Our major conclusion is that the product still passes muster on the key liking attribute. We feel comfortable that the product is acceptable to consumers, but not quite as comfortable as we felt with scenario A.

Scenario C—Low Mean, Low Variability Around the Mean. The product receives a mean rating of 60 and a low standard deviation of 20. We conclude from Table 8 that the odds are very low that the prototype comes from a distribution where the mean is 65. The true mean is probably lower than 65.

Table 8 Four Scenarios of Results from Mouthwash Product Test, Calculation of Statistical Differences, and Interpretation

	Category = Mouthwash	True mean = 65		
	A	B	C	D
Base Size	100	100	100	100
True mean of product	65	65	65	65
Observed mean from test	70	70	60	60
Difference (observed - true)	5	5	-5	-5
Standard deviation	20	60	20	60
Standard error of the mean (Standard deviation)/(Base $^{0.5}$)	2	6	2	6
T value for the difference (Difference)/Standard error	2.50	0.83	-2.50	-0.83
Conclusion	Higher	Same	Lower	Same

Scenario D—Low Mean, High Variability Around the Mean. The product receives a mean rating of 60 and a standard deviation of 60. We conclude from Table 8 that the odds are very high that the prototype comes from a distribution where the mean is 65. Our major conclusion is that the product still passes muster on the key attribute. We don't feel very comfortable that the product is less acceptable to consumers, but the statistical analysis tells us that the odds are high that the observed mean of 60 could have come from a sampling distribution with a true mean of 65.

The T statistic quantifies the odds that the mean empirically obtained from a study comes from a sampling distribution with a specified mean value. The T test needs four numbers:

1. the mean we obtain in the study
2. target mean for comparison
3. the variability of the distribution (standard deviation), and
4. the base size or number of observations which are used to compute the mean.

As the base size increases, the T statistic increases for any fixed difference and any fixed standard deviation. Although the mean stays the same, the odds of the observed mean coming from a specific sampling distribution decrease. In fact, with a base size of 1,000,000 virtually any difference higher than one point will be significant (see Table 9).

What Is a Meaningful Difference?

We see from Table 9 that the statistical significance of a fixed difference becomes greater as the base size increases. The mean remains the same. All that increases is the statistical probability or odds that the observed mean differs from a pre-defined mean. The same difference remains.

If the statistical significance of a difference can be affected by base size, then perhaps it is time to use another measure of "difference." This alternative measure might simply be the numerical value of the difference. The best guess for performance of a test product is the mean that we measure. If the mean differs by some value from the target value (e.g., norm), then we conclude that the product may truly differ from a target goal. The notion of significant difference applies to our confidence in making that statement. It is not that "product A is significantly lower than a target," but product A is 5 points lower than a target norm, and the probability of this occurring by chance is less than 2%.

Although the change in language seems like a minor thing, it is not. The change in language focuses our attention on the key issue—how well the test product scores relative to norms. We are no longer interested in "difference" but rather in level of performance. Significance is simply an additional fact to be added to the conclusion. The real interest is the level of performance.

Practical Considerations

Table 9 How the T Value for Difference Changes as a Function of Base Size: T Values Corresponding to Various Differences, with a Standard Deviation of 30

Base size	\multicolumn{8}{c}{Point difference between two products on a scale}
	1	2	3	4	5	6	8	10
10	0.10	0.20	0.30	0.40	0.50	0.60	0.80	1.00
20	0.15	0.29	0.44	0.58	0.73	0.87	1.16	1.45
30	0.18	0.36	0.54	0.72	0.90	1.08	1.44	1.80
40	0.21	0.42	0.62	0.83	1.04	1.25	1.67	2.08
50	0.23	0.47	0.70	0.93	1.17	1.40	1.87	2.33
60	0.26	0.51	0.77	1.02	1.28	1.54	2.05	2.56
70	0.28	0.55	0.83	1.11	1.38	1.66	2.22	2.77
80	0.30	0.59	0.89	1.19	1.48	1.78	2.37	2.96
90	0.31	0.63	0.94	1.26	1.57	1.89	2.52	3.14
100	0.33	0.66	0.99	1.33	1.66	1.99	2.65	3.32
110	0.35	0.70	1.04	1.39	1.74	2.09	2.78	3.48
120	0.36	0.73	1.09	1.45	1.82	2.18	2.91	3.64
130	0.38	0.76	1.14	1.51	1.89	2.27	3.03	3.79
140	0.39	0.79	1.18	1.57	1.96	2.36	3.14	3.93
150	0.41	0.81	1.22	1.63	2.03	2.44	3.26	4.07
160	0.42	0.84	1.26	1.68	2.10	2.52	3.36	4.20
170	0.43	0.87	1.30	1.73	2.17	2.60	3.47	4.33
180	0.45	0.89	1.34	1.78	2.23	2.68	3.57	4.46
190	0.46	0.92	1.37	1.83	2.29	2.75	3.67	4.58
200	0.47	0.94	1.41	1.88	2.35	2.82	3.76	4.70
250	0.53	1.05	1.58	2.10	2.63	3.16	4.21	5.26
300	0.58	1.15	1.73	2.31	2.88	3.46	4.61	5.76
350	0.62	1.25	1.87	2.49	3.11	3.74	4.98	6.23
400	0.67	1.33	2.00	2.66	3.33	3.99	5.33	6.66
450	0.71	1.41	2.12	2.83	3.53	4.24	5.65	7.06
500	0.74	1.49	2.23	2.98	3.72	4.47	5.96	7.45
600	0.82	1.63	2.45	3.26	4.08	4.89	6.53	8.16
700	0.88	1.76	2.64	3.53	4.41	5.29	7.05	8.81
800	0.94	1.88	2.83	3.77	4.71	5.65	7.54	9.42
900	1.00	2.00	3.00	4.00	5.00	6.00	8.00	9.99
1,000	1.05	2.11	3.16	4.21	5.27	6.32	8.43	10.54
10,000	3.33	6.67	10.00	13.33	16.67	20.00	26.67	33.33
100,000	10.54	21.08	31.62	42.16	52.70	63.25	84.33	105.41
1,000,000	33.33	66.67	100.00	133.33	166.67	200.00	266.67	333.33

Numbers in the body of the table are T values corresponding to the difference and the standard deviation for that specific base size. T values of 1.96 or higher are considered statistically significant.

9. On the Use of Norms in Product Testing

Introduction

As a scientific discipline matures it accretes standard methods and data. In product evaluation many researchers compare their ratings to "norms" for the category. The norm may comprise a set of benchmark measurements which indicate that the product or concept is acceptable or not acceptable. Other norms may be the scores of competitive products in the category. Whether the researcher uses fixed benchmarks or competitive scores as norms, research conventionally interprets scores against benchmarks and norms. Indeed, numerical scores by themselves are just that—numbers. It is only with the aid of benchmarks that these numerical values take on substantive meaning above and beyond the numbers themselves.

Fixed Benchmarks

In some product evaluation studies the benchmarks are fixed. For instance, in product tests using the 0–100-point scale for liking, blind products are often benchmarked against the norms shown in Table 10. A product is considered "poor" if its score is 50 or lower on the 100-point scale, and "very poor" if its score is lower than 40.

Most manufacturers who engage in repeated product testing, with whatever scale, develop a battery of test scores and correlate performance of the product with other tests or with in-market performance. This correlation of scores obtained on a test with external measures of product performance generates the norms. More often than not, the benchmark scores on a test emerge after dozens of studies in which the researcher compares performance scores on the test with external performance.

Fixed benchmarks are intuitively attractive. At the most primitive level they provide the user with a sense of where the product stands relative to an independent measure of "good or poor." One should never underestimate the power and appeal of benchmarks and descriptor words. Research users are not usually statisticians familiar with scales and the "ins and outs" of testing and test

Table 10 Interpretative Norms for the Moskowitz Jacobs Inc. 0–100 Liking Scale

0–40 = very poor, needs significant amount of repair
41–50 = poor, needs work
51–60 = fair, but needs a little work
61–70 = good
71–100 = excellent

Practical Considerations

interpretation. Most research users are simply interested in the test as a means to learn about their product or concept. Fixed benchmarks yield a qualitative impression or an idea about the product or concept. Benchmarks transform the numerical information into a simple, easily understood and easily communicated result.

What Happens to Norms When the Category Truly Changes?

From the research supplier's perspective it is ever-tempting to develop a set of norms and use those norms ad infinitum. As long as the test is conducted in the same rigorous fashion, year after year, with the same type of population and with relatively similar types of panelists, it seems unnecessary to recalibrate the norms. Or is it unnecessary? Do norms remain invariant when the scale remains the same, and most other features of the project remain unchanged (except perhaps for year-to-year variation in the number and types of products in the category)?

Product categories change over time. In most categories, manufacturers upgrade their entries in order to maintain a competitive advantage. This quality improvement effort is critical because new products enter the market all the time. In many cases these new products are simply line extensions which have no impact on a category (e.g., new fragrances, new forms). In other cases, the product might combine two previous products into one new product, or the product might contain a revolutionary ingredient or provide a dramatically new method of action. For these truly new products the old norms may not apply. These new products change the consumer's point of view. Historical data may no longer apply. For instance, if the entire category scored between 40 and 60 in overall liking on a 0–100 point scale, then what can the researcher conclude if the new product scores 70 or higher? No product ever scored that high, and there is nothing against which to compare these scores. Even more confusing is the potential restructuring of the ratings of the category. Consumers might score the truly new product 60, and downrate all of the other products which currently exist.

Keeping Up with the Times—Recalibrating on a Regular Basis

Norms traditionally have been thought of as engraved in stone. Although many practitioners give verbal acquiescence to the fact that norms only "suggest," in actuality researchers and research employers like to use data that they feel to be "rock-solid." When a research supplier discovers that the category has changed, it causes discomfort to both the supplier and research user when they must discard the norms previously developed.

What is needed is a different way to look at norms. Categories change dynamically. The pace of business is more rapid than ever before. Norms need not be absolutely invariant numbers which have an existence in some platonic realm of perfection. Rather, the researcher must develop relative norms at the time of research. For instance, in a product test, the researcher may obtain ratings of many products in the competitive frame. The researcher can locate the new prototypes (or in-market products) within the spectrum of existing products measured at the same time and in the same way. As a result, there are no fixed norms. There is only relative performance on an ongoing basis. This relativistic approach need not replace all fixed norms (e.g., it need not replace norms for expected market share, predicted on the basis of test scores). However, for measurement of relative product acceptance, a combination of measurements of all major competitors (the dynamic norms) and the absolute scores of acceptance (the fixed norms) should allow the researcher and research user a more profound sense of how well the product actually performs.

The dual approach outlined above requires that the researcher measure the response to many products. This may be more expensive, but it avoids the dependency on potentially outdated norms. Furthermore, the approach tracks category performance over time. Rather than creating the norms at the start of the process and using those norms for years, the researcher obtains consumer ratings for products in a category and continually updates these consumer ratings as a time series. From an inspection of the time series (e.g., acceptance of given competitors over time), the researcher learns about true performance in a category. There is significantly more data available on each product, which in turn allows for a meaningful revision of the norms as times change.

10. Predicting Wearout Through Repeated Use

Introduction

Products have natural life cycles. Even if competitive activity does not intervene, consumers try a product, become loyal users, and eventually discard the product in favor of another product in the same class. What makes a consumer lose interest in the product? Is there such a phenomenon as "wearout" or "boredom," and if so, can this phenomenon be measured?

Short Term Satiety and Long Term Wearout

This section deals with the effect of time and repeated exposure on acceptance. It deals with a technology developed for foods that may have potential application to personal products.

The underlying idea is that preferences are not static, but change. People like variety. One need only go to the store to find the many different products that are available to consumers, even within the same general category. Fragrances, forms, colors differ. Furthermore, the same package may become boring over time. The package designer has to refresh the package, perhaps with

Practical Considerations 205

a major change, perhaps with a set of minor changes. Advertising loses its effectiveness as consumers become first aware of the message, then intrigued, then inured, then indifferent. Built-in psychological mechanisms lead to wearout of stimuli, so that over time the same stimulus loses its ability to attract the consumer.

Can the researcher, product developer and marketer investigate the nature of this "wearout" behavior, in order to understand it and eventually to create products and concepts that are less affected? Furthermore, given the inevitable wearout of product acceptance over time and with repeated use, can the researcher learn the dynamics of such wearout in order to optimize the development process?

The Role of Hedonic Adaptation in Product Boredom and Switching

Consumers get tired of products and switch. They especially get tired of products with different fragrances. This is an example of hedonic habituation. What started out pleasant becomes neutral. What started out unpleasant may either become increasingly unpleasant or neutral.

Consumers are novelty-seeking organisms. They like pleasant novelty. Consequently, often consumers report that they are bored with the product that they are currently using, and switch to another product in the same category, with a somewhat different sensory profile. This is hedonic habituation. The panelists perceive the stimuli just as intensely as before, but their hedonic response to the stimulus has changed as a result of repeated exposure to the stimulus.

Hedonic habituation occurs most frequently with the chemical senses—taste and smell—and in categories which feature many flavors or fragrances. Functional fragrances, ice cream flavors, and the like represent the types of stimuli that are most likely to undergo hedonic habituation. To a far lesser extent, appearance characteristics are also subject to habituation. Whereas a person might say that he or she no longer "likes the smell" of a particular perfume after repeated exposure (a true indication of a hedonic shift), the same person might say that he or she "has gotten tired of the look of a certain lipstick." The chemical stimulus smells different, and acceptability has truly changed. The visual stimulus does not look any different. Psychologically the consumer does not truly like it "less" as desires an alternative because the visual stimulus is less interesting. There is far less hedonic habituation in the sense of touch than in vision, and far less in vision than in the chemical senses.

Long Term Sensory Adaptation with Repeated Use

From time to time consumers complain that they can no longer "smell" the fragrance of a soap or a deodorant. They complain that, over time, the product has lost its intensity and has become substantially weaker than it used to be. This is an unusual, but not altogether rare occurrence.

There is little research on the long term sensory effects of repeated exposure to products. We know very little about the potential shifts in the salience or intensity of stimuli that are used on a repeated basis. It may well turn out that some significant loss in perceived intensity occurs with repeated exposure, not so much due to physiological adaptation (a momentary phenomenon), but rather due to a psychological or cognitive change in the responsiveness to a particular stimulus. (This would be similar to the observation that residents living near train tracks simply "do not hear" the train after living in the area for a while.)

Three Types of Satiety

When a person says that he cannot "smell" a product or gets tired of a fragrance, three things could have occurred:

1. *Sensory Fatigue.* The person may actually have "adapted" to the fragrance, so that physiologically he cannot smell it. The olfactory system is exceptionally sensitive to repeated presentation of a stimulus. If a person stays in a room filled with an odor, eventually the odor will "disappear." The person will report that he cannot smell anything. This sensory fatigue is a well-established phenomenon for most fragrances. Sensory fatigue occurs when the person remains in a confined space with an odor. Eventually the sensory system stops responding to the odor, and fatigues. However, fatigue is not permanent. Remove the odorous stimulus and fatigue dissipates rapidly. Let the person leave the room, even for a moment, and the smell returns, sometimes in full force, sometimes slightly weakened.

2. *Sensory Satiety.* Sensory satiety is characterized by a decrease in acceptance of the product with increasing exposure. However, sensory satiety is not sensory fatigue. The stimulus seems just as strong, except that the consumer feels differently about it. Ask the consumer to describe it, and the consumer may state that "I just don't like the product any more." Sensory satiety results from a long-term exposure to the product.

3. *Habituation—Long-Term Inability to Perceive a Product in the Normal Course of Events.* If a person lives near a railroad and the train passes several times a day, eventually the person stops hearing the train, even if the train is loud and rumbling. The person "tunes out" the sound of the train, despite its clarity. Unlike sensory satiety (a change in liking), habituation is characterized by the person actually reporting that he doesn't hear anything. If a person lives in a neighborhood permeated by a noxious smell that appears at random, often the person stops smelling the odor entirely. The sensory system appears to shut out the specific smell on a particular basis. Call the panelist's attention

Practical Considerations

to the smell, however, and the smell reappears in its full intensity and noxiousness. There is no loss in sensory intensity (as occurs with sensory fatigue), nor a change in the hedonic character (which occurs with sensory satiety).

The Case of Real Products

Despite the apparent clinical nature of satiety, it occurs in fact quite often with personal products. Here are a few examples:

1. A fragrance smells different on the skin, after repeated wearing, than it does in the bottle. The consumer cannot smell the fragrance quite as strongly. A fragrance evaporates over time, surrounding the panelist with a cloud of perfume vapor, and thus desensitizing the consumer. These are examples of sensory fatigue.
2. With repeated exposure a consumer often reports that the product smells weaker than it once did. This is an example of sensory habituation. The sensory intensity drops, so that after using the product for a few months the product smells different.
3. With repeated exposure a consumer soon begins to like a fragrance that was originally disliked, or dislike a fragrance that was originally liked. Or, in some cases the consumer loses interest or aversion, so that the fragrance hedonics (likes, dislikes) head towards indifference. The consumer ceases caring about the fragrance, one way or the other. This an example of hedonic habituation.

If Boredom Exists, How Can We Measure It and Use It?

Can we quantify boredom? Can we predict which products will be most subject to hedonic habituation so that the manufacturer can concentrate on those stimuli which will maintain their acceptance over a long time span and pay less attention to those stimuli that will lose their attractiveness once they are widely used? Why are some fragrances popular year after year, while other fragrances (or other "looks" in cosmetics) are popular for only a limited time?

Part of the answer to the problem of prediction comes from food science rather than from cosmetics and toiletries. Food researchers know that consumers like some foods more than they like others, and that some items (e.g., hamburger) never seem to lose their attractiveness. Other foods that score well on food attitude surveys (e.g., steak) quickly lose their attractiveness if the consumers have to eat the same product day after day. (In contrast, hamburger maintains much of its intrinsic acceptability if the consumer eats hamburger day after day.) Why? Are there any general rules or trends that could generalize across categories, and help the manufacturer of personal products become smarter in the selection of sensory characteristics?

Observation 1—Some Products Become Boring Faster than Do Others. In food product evaluations we can ask consumers to rate their interest in consuming a food, given a variety of occasions during which they previously consumed the same food. (This is an attitudinal question, not a behavioral question. We ask the consumers to estimate how interested they would be in the product, assuming a previous length of time since last consumption.) Some simple foods like bread never achieve high levels of acceptance, but also never really become boring. Other foods, such as steak, becoming boring rather quickly, even though these items intrinsically score very highly. Figure 2 shows the results of some of these "time-preference curves."

Could it be that fragrances follow the same patterns? That is, some of the more complex yet identifiable fragrances can become very boring after repeated use, whereas other fragrances which are less identifiable, less immediately acceptable but also less "salient" maintain their modest level of acceptance.

Observation 2—The Degree to Which a Product Is "Boring" Is Inversely Related to the Last Time the Product Was Selected and Consumed. The time preference curve suggests that interest in a product increases as a function of the time since the product was last consumed. This means that even the most boring product can increase its acceptance (however little) if the consumer stays away from the product. Does this hold for a fragrance? Does time since the fragrance was last used affect boredom or wearout, or is boredom with fragrance

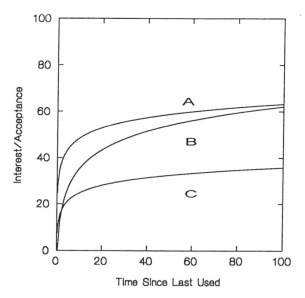

Figure 2 Example of time-preference curves—Foods.

an "all-or-none" phenomenon (i.e., once the consumer becomes bored with the fragrance the fragrance never again regains its acceptance)? This is an area worth pursuing for basic and applied research. It may well turn out that fragrances follow one time preference curve (all-or-none—once boring then never again accepted), whereas products designed for more visual or tactile use show a more graded time preference curve (continuous—acceptance increases up to a maximum asymptotic level if the consumer does not use the product for a long period of time).

Asking the Proper Question for Time-Preference

Time factor in product acceptance is a new area of research for which the scientific literature is not well-developed. Consequently there are no "tried and true" procedures whereby the researcher can construct a questionnaire to accurately probe the time effect of repeated usage and boredom. One could ask simplistic questions such as, "How likely are you to become bored with this product if you were to use it every day?" One could ask a more complicated question such as, "How interested are you in using the product if the last time you used this type of product, with these same sensory characteristics, was x weeks ago?" The former question asks the consumer to assign an overall rating of estimated potential for boredom. The latter question asks the consumer to hypothesize about interest in the product, given various rates of usage. By asking the panelist to rate liking as a function of time since last usage rather than as a function of actual rate of usage, the researcher "concretizes" the problem of usage frequency. Panelists have an easier time responding to concrete questions rather than responding to more theoretical questions. Time since last usage is a concrete question. Frequency is a more theoretical, more abstract question.

The Concept of Time Preference

In the late 1960's, Joseph Balintfy and his colleagues began to investigate the phenomenon of changing food preferences. Balintfy noticed that food acceptance could not be characterized by a single number. Rather, there seemed to be two components—one a basic level of acceptance, and the other an effect of time and repeated exposure to the food. According to Balintfy's theory (and subsequent confirmatory observations), the acceptance of a food product is not a static phenomenon, but rather a dynamic one. Foods are optimally acceptable when they have not been consumed for a very long time. However, just after consuming a food, an individual might not want to consume the food again. Acceptance drops as a function of recent eating. Over time the food regains its initial level of acceptance. Eventually, after a sufficiently long wait, the food becomes just as acceptable as it had been originally. Balintfy also noted that some foods are more "time sensitive" than others. For instance, meats and complex dishes were more sensitive to "time since last consumed" than were other foods (e.g., breads). Although steak is highly acceptable, when the consumer has just eaten it, its acceptance plummets. It takes a while for the consumer to desire steak again. On the other hand, some products like milk and bread are not highly acceptable to begin with, but also never lose their acceptance, even on repeated consumptions. These items seem to be time independent.

Applying the Theory of Time Preference to Personal Products

Can we take the same approach to products? Is there anything that we can learn from the theory of time preference for foods? The answer may be yes, but we have to rework the theory:

1. The theory applies primarily to items that are judged visually or by the chemical senses (taste/smell).
2. The theory applies to items where there is potential boredom or wearout. These items include fragrances, lipsticks, etc.
3. The theory applies to items which have a large potential substitution set, so that the consumer can switch from one product to another. Again, fragrances, lipsticks, etc. are good examples. These products are judged on their sensory characteristics. There are many equally good alternatives in the marketplace from which the consumer can choose. Switching is easy to do.
4. The wearout factor is not as rapid for personal products as it is for foods. The effects so easily noticed for foods (almost immediately after a person eats the food) are much more subtle and long-lasting for personal products. After wearing a fragrance, an individual doesn't immediately conceive of a distaste for wearing the same fragrance the next day. It takes a while for the time preference to establish itself.

Implications of the Time Preference Method for Product Design and Selection

If the time preference approach describes the changes in product acceptance, then there may be a major opportunity for researchers to identify those specific products and even features that lead to consumer satiety. The approach follows these six steps:

1. Develop a method to quantify the effect of repeated exposure on product acceptance. The method may either be behavioral (measurement of liking after product use) or attitudinal (estimation of degree of liking after one exposure, by asking the respondent to rate the degree to which he would be interested in the product given different frequencies of usage).
2. Create an index of "sensory satiety" or "habituation," based upon the results in Step 1.

Practical Considerations

3. Systematically vary the physical characteristics of a product, have consumers use these products in the normal fashion, and quantify the degree of potential "sensory satiety" (from Step 2).
4. Relate the physical characteristics (Step 3) to the index of "sensory satiety," by means of an equation. The equation uses linear terms, square terms, and cross terms for the independent physical variables. The equation enables the researcher to assess the effect of each independent physical characteristic.
5. Identify the physical characteristics which drive sensory satiety by means of a sensitivity analysis. The sensitivity analysis uses the equation developed in Step 4. The researcher sets all of the physical characteristics to a specific, fixed level, and then varies one characteristic in small steps over a wide range. For each level of that characteristic the researcher estimates the degree of satiety by means of the index.
6. For optimization, the researcher identifies the minimum possible level of sensory satiety, and determines the combination of independent characteristics that generates that minimum level of satiety.

A Worked Example

Table 11 shows one example. The study comprised an evaluation of 15 different fragrances, systematically varied in character by means of changing the proportions of the keys. There were four keys in the experiment (A-D), which in different combinations provided a variety of different qualitative notes. The panelists rated various characteristics, two of which are of particular interest: overall interest (Like), and degree to which the fragrance would be acceptable after continuous wearing for 1 month (Live). The panelists did not actually "live with the fragrance," but only estimated how they would feel. Table 11 also shows a third column "Retain," which is the ratio of Live to Like (Retain = Live/Like). The higher the ratio, the more likely it is (in the consumer's mind) that the fragrance would maintain its acceptance with repeated use.

The experimental design shown in Table 11A is a mixture model. The four fragrance keys add to 100%. The design enables the researcher to optimize the formulation to maximize liking, or any other attribute measured by consumers. Table 11B shows the optimized fragrance. Furthermore, the results would be the same, whether the attribute being optimized were "Like" or "Live" (although those two attributes need not generate the same results).

Finally, by doing a sensitivity analysis, the researcher can identify those specific keys that lead to acceptance and maintain interest in the face of repeated use. Table 11C shows the expected change in attribute ratings when each of the four keys is increased from a low of 20% in the fragrance blend to a high of 40%. As the key increases to take up more of the blend, the remaining three

Table 11A Experimental Design for Fragrances

Product	Formula				Rating		Expected
	A(%)	B(%)	C(%)	D(%)	Like	Live	% Retained liking
1	40	20	20	20	75	68	91
2	30	30	20	20	83	71	85
3	20	40	20	20	40	31	77
4	30	20	30	20	53	33	62
5	20	30	30	20	30	15	50
6	20	20	40	20	38	34	89
7	30	20	20	30	61	34	55
8	20	30	20	30	77	60	77
9	20	20	30	30	77	49	63
10	20	20	20	40	46	26	56
11	20	20	33	27	41	30	73
12	27	20	27	27	59	50	83
13	20	33	20	27	28	27	95
14	27	20	20	33	41	30	73
15	27	27	20	27	69	51	73

keys must decrease to keep the total at 100%. These three keys decrease in such a way as to maintain the proportion between the three of them.

A sensitivity analysis rapidly shows which fragrance keys lead to potential boredom and which do not. Furthermore, as Table 11C shows, the dynamics of each fragrance key differ. A fragrance key that increases (or decreases) in concentration may increase liking, but leave unchanged the consumer's estimate of whether the fragrance would be boring.

Table 11B Optimal Formulation to Maximize Fragrance Acceptance

Formula	(%)
A	34
B	26
C	20
D	20
Expected rating	
Like	87
Live	74
Retain	85%

Table 11C Sensitivity of Consumer Ratings to Changes in Fragrance

A	20%	23%	27%	30%	33%	37%	40%
B	27%	26%	24%	23%	22%	21%	20%
C	27%	26%	24%	23%	22%	21%	20%
D	27%	26%	24%	23%	22%	21%	20%
Like	43	51	58	64	68	72	75
Live	28	34	41	47	54	60	67
Retain	62	63	66	70	75	81	89
A	27%	26%	24%	23%	22%	21%	20%
B	20%	23%	27%	30%	33%	37%	40%
C	27%	26%	24%	23%	22%	21%	20%
D	27%	26%	24%	23%	22%	21%	20%
Like	53	54	54	53	51	46	41
Live	35	37	38	38	38	37	36
Retain	67	65	65	67	72	79	89
A	27%	26%	24%	23%	22%	21%	20%
B	27%	26%	24%	23%	22%	21%	20%
C	20%	23%	27%	30%	33%	37%	40%
D	27%	26%	24%	23%	22%	21%	20%
Like	65	58	52	48	45	44	45
Live	49	41	35	31	30	32	37
Retain	75	67	63	63	66	73	84
A	27%	26%	24%	23%	22%	21%	20%
B	27%	26%	24%	23%	22%	21%	20%
C	27%	26%	24%	23%	22%	21%	20%
D	20%	23%	27%	30%	33%	37%	40%
Like	58	55	54	54	55	57	61
Live	41	38	37	36	36	37	38
Retain	61	64	65	65	65	63	60

Using the Results to Engineer Products with a Long Life Cycle

By thinking about products in terms of two dimensions—formulation and time—the marketer and developer can begin to identify those variables under development control which can lead to long term product acceptance and thus long term profitability. For too many years researchers have looked at acceptability as a static variable, rather than as a dynamic one influenced by time. Incorporating time preference theory into product development allows the developer to "engineer" long term acceptance into a product while at the same time paying attention to short term acceptance and (where relevant) cost of goods, and other ancillary factors.

11. Higher Order Analyses of Products and Concepts: Tapping into the Consumer's Freely Emitted Language

Introduction

Consumer research on products and consumers traditionally has been divided into qualitative and quantitative research. Qualitative researchers focus on the "in-depth" response to a product. Through group discussions and extended "one-on-one's" (following a "moderator guide" or set of topics to cover in the interview), the qualitative researcher attempts to understand how the panelist responds to the product. What are the positives and negatives, and what are some of the underlying decision criteria?

Qualitative research is not inherently quantitative in nature. There are no hard and fast numbers, and the population of panelists is relatively small (at least compared to the many dozens or hundreds of panelists who participate in quantitative research). Qualitative research does not use attribute questionnaires per se. (Those qualitative researchers who poll the panelists in a group session and use attribute lists, are actually doing a very small scale quantitative test in the midst of the qualitative research.) The moderator guide is a type of questionnaire, but it is designed to guide the discussion across a variety of topics. The actual information elicited in a qualitative interview is much less structured, more personal, much more idiosyncratic to the panelist, and not typically amenable to quantitative analysis.

A qualitative research report will look quite different from a quantitative research report on the same product. The qualitative researcher will pore through the "verbatim discussion" of the product or concept, and try to piece together a coherent whole from the different statements of panelists. Because the qualitative research is so focused on the individual panelist's experience, and because the researcher cannot easily classify the responses, the typical qualitative research report will comprise a set of major topics, followed by some conclusions, supported by certain statements. To a great extent the researchers attempt to uncover the pattern underlying the consumer's responses to the product, basing the pattern on the collection of freely emitted descriptions and discussions.

Bringing Quantitative Methods to Qualitative Research

Experienced researchers with a quantitative or numerical bent know that there is often structure in a free discussion. For instance, a researcher can tape-record a one-on-one interview with a panelist who is discussing her response to a specific lotion, or to an entire set of lotions currently in the market. The researcher may find that there are certain recurrent patterns. As an example, many times when a specific lotion is mentioned, the descriptor terms used at the same time refer to "functionality." For other lotions, the panelist may use descriptor terms pertaining more to cosmetics and beauty.

Properly done, an analysis of the frequency with which various terms appear together in free discussion of a category can identify a person's individual "reality" about the product. Does the person verbally associate some products with certain benefits and not with others? Table 12 shows some examples of this type of correlation analysis, where all the researcher does is sample discussions about a specific product category and count the frequency with which various key terms appear together in close temporal proximity.

An Overview

Qualitative methods can be quantified. Just as the researcher can do a content analysis of products in order to identify what features exist and occur together, so can the researcher do a content analysis of language and product (or concept) description to identify what terms recur in temporal proximity. This type of analysis has tremendous possibilities for unlocking even more profound insights into a consumer's response to products, without forcing the consumer to be intellectual about the process.

Table 12 Analysis of the Patterns of Language Used to Describe Anti-Perspirants

	Anti-perspirants–Brand Q		
	Brand name of product	Protection	Product quality
Protection	20		
Product quality	12	7	
Price value	6	1	9
	Anti-perspirants–Brand R		
	Brand name of product	Protection	Product quality
Protection	3		
Product quality	18	2	
Price value	20	2	24

Number in heading of table = Number of times the row and column appeared together in a discussion
Brand Q = more "protection and performance"
Brand R = more "price"

References

Allison, R.I. and Uhl, K.P. 1964. Influence of beer brand identification on taste perception. Journal of Marketing Research, *1*, 36-39.

Balintfy, J., Duffy, W.J. and Sinha, P. 1974. Modeling food preferences over time. Operations Research, *22*, 711-727.

Basker, D. 1980. Polygonal and polyhedral taste testing. Journal of Food Quality, *3*, 1-10.

Berglund, U. 1974. Dynamic properties of the olfactory system. Annals of the New York Academy of Sciences, *237*, 17-27.

Blom, G. 1955. How many taste testers? Wallerstein Laboratories Communications, *18*, 173-178.

Cardello, A.V. and Maller, O. 1982. Relationships between food preferences and food acceptance ratings. Journal of Food Science, *47*, 1553-1557.

Dethmers, A.E. 1968. Experienced versus inexperienced judges for preference testing. Food Product Development, *2*, 22-23.

Guilford, J.P. 1954. Psychometric Methods. McGraw-Hill, New York.

Hayes, T.J. 1989. The flexible focus group: designing and implementing effective and creative research. *In:* Product Testing with Consumers for Research Guidance, L. Wu (ed.), STP 1035, pp. 77-84. American Society for Testing and Materials, Philadelphia.

Hubbard, M.R. 1990. Statistical Quality Control for the Food Industry. Van Nostrand Reinhold/AVI, New York.

Indow, T. and Stevens, S.S. 1966. Scaling of saturation and hue. Perception & Psychophysics, *1*, 253-272.

Jacoby, J., Olson, J.C. and Haddock, R.A. 1971. Price, brand name and product composition characteristics as determinants of perceived quality. Journal of Applied Psychology, *55*, 570-590.

Johnson, J. and Vickers, Z. 1987. Avoid the centering bias or range effect when determining an optimum level of sweetness in lemonade. Journal of Sensory Studies, *2*, 283-292.

Kramer, H.C. and Thiemann, S. 1987. How Many Subjects: Statistical Power Analysis in Research. Sage, Newbury Park.

Makens, J.C. 1965. Effect of brand preference upon consumers' perceived taste of turkey. Journal of Applied Psychology, *49*, 261-263.

Marks, L.E. 1968. Stimulus range, number of categories, and the form of the category scale. American Journal of Psychology, *81*, 467-479.

Meiselman, H.L. 1978. Scales for measuring food preference. *In:* Encyclopedia of Food Science, M.S. Petersen and A.H. Johnson (eds.), pp. 675-678. Van Nostrand Reinhold/AVI, New York.

Moskowitz, H.R. 1977. Magnitude estimation: Notes on what, how, when and why to use it. Journal of Food Quality, *1*, 195-228.

Moskowitz, H.R. 1981. Relative importance of perceptual factors to consumer acceptance: linear versus quadratic analysis. Journal of Food Science, *46*, 244-248.

Moskowitz, H.R. 1983. Product Testing and Sensory Evaluation of Foods: Marketing and R&D Approaches. Food and Nutrition Press, Trumbull, CT.

Moskowitz, H.R. 1984. Cosmetic Product Testing: A Modern Psychophysical Approach. Marcel Dekker, New York.
Moskowitz, H.R. 1985. New Directions for Product Testing and Sensory Analysis of Foods. Food and Nutrition Press, Trumbull, CT.
Moskowitz, H.R. and Fishken, D. 1979. What effects do repeated evaluations play in fragrance perceptions? Perfumer & Flavorist, May, 45-52.
Moskowitz, H.R., Jacobs, B. and Firtle, N. 1980. Discrimination testing and product decisions. Journal of Marketing Research, *17*, 84-90.
Moskowitz, H.R., Jacobs, B.E. and Lazar, N. 1985. Product response segmentation and the analysis of individual differences in liking. Journal of Food Quality, *8*, 168-191.
Pangborn, R.M. 1970. Individual variations in affective responses to taste stimuli. Psychonomic Science, *21*, 125-128.
Peryam, D.R. and Pilgrim, F.J. 1957. Hedonic scale method of measuring food preferences. Food Technology, *11*, 9-14.
Riskey, D.R., Parducci, A. and Beauchamp, G.K. 1979. Effects of context in the judgments of sweetness and pleasantness. Perception & Psychophysics, *26*, 171-176.
Rothman, L. 1990. Just right scales: making the best of a bad situation. Newsletter of the Sensory Evaluation Division, Institute of Food Technologists (U.S.), October, 4.
Schutz, H.G. 1965. A food action rating scale for measuring food acceptance. Journal of Food Science, *30*, 365-374.
Schutz, H.G. 1989. Beyond preference: appropriateness as a measure of contextual acceptance of food. *In:* Food Acceptance, D.M.H. Thomson (ed.), pp. 115-134. Elsevier Applied Science, Barking, Essex, U.K.
Sheen, M.R. and Drayton, J.L. 1989. Influence of brand label on sensory perception. *In:* Food Acceptability, D.M.H. Thomson (ed.), pp. 89-99. Elsevier Applied Science, Barking, Essex, U.K.
Shepherd, R., Farleigh, C.A. and Land, D.G. 1984. Effects of stimulus context on preference judgments for salt. Perception, *13*, 739-742.
Shepherd, R., Farleigh, C.A., Land, D.G. and Franklin, J.G. 1985. Validity of a relative-to-ideal rating procedure compared with hedonic rating. *In:* Progress In Flavour Research, J. Adda (ed.), pp. 103-110. Elsevier Science Publishers, Amsterdam.
Shepherd, R., Smith, K. and Farleigh, C.A. 1989. The relation between intensity, hedonic and relative-to-ideal ratings. Food Quality & Preference, *1*, 75-80.
Smithies, R.A. and Buchanan, B.S. 1991. Substantiating a Taste Claim. Transcript Proceedings, NAD Workshop, New York, Council of Better Business Bureaus, and the Advertising Research Foundation.
Stevens, S.S. 1953. On the brightness of lights and the loudness of sounds. Science, *118*, 576.
Stevens, S.S. 1975. Psychophysics: An Introduction to Its Perceptual, Neural and Social Prospects. John Wiley & Sons, New York.
Stone, H., Sidel, J.L. and Bloomquist, J. 1980. Quantitative descriptive analysis. Cereal Foods World, *25*, 642-644.
Szczesniak, A.S. 1987. Correlating sensory with instrumental texture measurements—an overview of recent developments. Journal of Texture Studies, *18*, 1-15.

Szczesniak, A.S., Brandt, M.A. and Friedman, H.H. 1963. Development of standard rating scales for mechanical parameters of texture and correlation between the objective and sensory methods of texture evaluation. Journal of Food Science, *28*, 397–403.

Thurstone, L.L. 1927. A law of comparative judgment. Psychological Review, *34*, 273–286.

Vickers, Z. 1988. Sensory specific satiety in lemonade using a just right scale for sweetness. Journal of Sensory Studies, *3*, 1–8.

Weitz, B. and Wright, P. 1979. Retrospective self insight about the factors considered in product evaluations. Journal of Consumer Research, *6*, 280–294.

4
Experimental Design and Product Optimization

1. Introduction

Product optimization comprises the disciplined creation and testing of product prototypes in order to achieve the following two goals:

1. Understand how the factors under the developer's control affect consumer responses or instrumental measures
2. Discover the specific combination of factors which generates a desired goal, whether this goal be maximum product acceptance, maximum acceptance under constraints (e.g., within specified cost of goods), or a product which generates a desired profile (goal-fitting).

Product optimization has been a part of the scientific tool box for several decades. In the 1940's, researchers became aware that they could understand the dynamics of products in a more profound way when they systematically varied the physical components of the products. Statisticians discovered that researchers could estimate the part-worth contribution of product components by arraying these according to an "orthogonal design" (where the components vary independently of each other).

The key word is "vary" the components. This means that the product developer must *create* prototypes. Unlike a category appraisal (which tests in-market products, or prototypes not varied in a systematic fashion), experimental design requires that the product developer actually create products of known composition. Only in this way does it become possible to assess what every component in the product contributes to acceptance, sensory aesthetics, and performance, respectively.

As one might surmise, experimental design was not met with unadulterated joy and universal acceptance. It is one thing to promote a technique that helps the developer do his job better. It is quite another thing to require that the developer expend more effort. Today, most developers are strapped for time and resources. To those developers uninitiated in the value of experimental design, the investment of time and money seems overwhelming, and the promise of payout insufficient relative to the perceived effort. However, developers experienced with experimental design understand the effort involved and the payout. To them the choice to use experimental design is far less problematic. These experienced developers recognize that by using designed experiments they reach the right answer more quickly, efficiently, and inexpensively. Designed studies create a database of information which answers many more questions "down the road," without requiring the researcher to return to the development bench again and again.

2. Steps in Experimental Design and Product Optimization

Researchers who want to develop products by experimental design follow a series of ten steps described below. These ten steps are the same whether the product being optimized is a fragrance (at the micro level) or the combination of ingredients and processing conditions (at the macro level).

Step 1 - Identify the Formula or Process Variables Under Technical Control

Are these variables under study continuous or discrete? How many levels or options should each variable comprise? The more levels or options investigated, the better the resulting data. This early stage requires discussion and a compromise between completeness and feasibility. Product developers want the correct answer. No one else wants to create so many prototypes that the study becomes unduly long and expensive.

All too often Step 1 immediately aborts the development process. Since an optimization project costs money and entails significant corporate effort, there is a tendency for developers and marketers to "want and expect everything" from an optimization study. On the other hand, there is the tendency for the participants in the planning stage to reject the entire process when they discover that they cannot assess all possible options to answer their stored-up questions. Consequently, correctly setting up the study goals and reining in expectations is critical. The number of variables to be studied should be reasonable, not overly large. The study should neither promise nor even imply more than can be easily delivered by any product test.

There is no easy way to expedite this first stage, nor to identify precisely what variables to control, other than through the discussions and inputs from

Experimental Design and Product Optimization 221

R&D and marketing. There is no magic formula which, for a given product, lists the variables to be studied. Despite the attempts of some researchers to transform the process to a "cookbook," in actuality the planning step is awash with dangers. It takes the sincere, open interaction between R&D product developers and marketing to truly and efficiently identify the variables to study, and to accept that this is a scientific experiment for learning rather than the ultimate magical answer.

Step 2 - Develop the Experimental Design (Systematic Product Array)

The key to experimental design and optimization is the proper array or combination of variables. Over the years researchers have struggled with the problem of developing efficient combinations of ingredients in order to accomplish two objectives:

1. Maximize the amount of information to be obtained from the study.
2. Minimize the number of prototypes in the design.

These two objectives pull in opposite directions. The more prototypes one creates the more information and learning will be obtained, but at the same time the longer and more expensive the process. In the most extreme case when the need is to learn as much as possible from the study, so many combinations of ingredients (viz., prototypes) are required that the project collapses from the weight of its own ponderousness.

Wheel and Spoke Designs. Experimental designs come in many variations. At the very simplest level are the wheel and spoke designs. Here, all factors are held constant but one. The one factor to be investigated is varied up and down in small increments. Sometimes that factor is only investigated at one additional level. Sometimes the factor is investigated at two levels. In some cases the changes are quantitative (e.g., amount of emollient versus current). In other cases the changes are qualitative (e.g., new color versus current color). Table 1A shows an example of the wheel and spoke design for six factors (A-F).

Wheel and spoke designs derive from classical psychophysics, where interest is focused on the effect of one factor at a time. At first glance the design is attractive because it appears to isolate the effect of a single factor quite clearly, holding all other factors constant. The wheel and spoke is useful for a limited analysis of a few variables. It shows the effect of changing from one level of a factor to another, estimating the effect quite clearly. However, the wheel and spoke design cannot predict what would happen when two variables change simultaneously (e.g., from current to "new," for instance). More sophisticated experimental designs are required to deal with multiple variables simultaneously.

Screening Designs. Screening designs are useful when the product developer wants to identify which factors in a product really make a difference (e.g., either

Table 1A Example of a Wheel and Spoke Design for Six Variables in Seven Runs

Product	Variable					
	A	B	C	D	E	F
1	0	0	0	0	0	0
2	1	0	0	0	0	0
3	0	1	0	0	0	0
4	0	0	1	0	0	0
5	0	0	0	1	0	0
6	0	0	0	0	1	0
7	0	0	0	0	0	1

Note: 1 = High/new/present level.
2 = Low/old/absent level.

to the consumer or in a measurable, physical way). The key difference between a screening design and a wheel and spoke design is that the screening design allows the investigator to assess each factor at different levels in combination with different levels of every other factor. Consequently, the researcher can be assured that the combination of two factors at the "new" state (e.g., higher level than current) will be empirically measured, rather than strictly deduced (as would have been the case from the wheel and spoke).

Table 1B shows two screening designs for six factors and Table 1C for 12 factors, respectively. These two designs are members of a family of experimental designs known as the Plackett Burman screening designs. These designs come in multiples of four factors. There are designs available for 4, 8, 12 ... 100 factors (each factor at two levels). If one goes to the trouble of analyzing the design, one quickly sees that every factor is statistically independent of every other factor (an important point for creating the model), and that every factor appears equally often with every other factor.

Screening designs are powerful tools to identify the key factors that affect responses. However, a major shortcoming is that they are "unsaturated" designs. For instance, with six variables under analysis and an experimental design that calls for 12 prototypes, the design really addresses only 12 of the possible $2^6 = 64$, or approximately 20% of the possible combinations. The unsaturation becomes even worse when the screening design investigates more variables, such as 12 factors. This design requires, at the bare minimum, 16 products, out of the possible $2^{16} = 65536$. With this level of unsaturation the screening design identifies only the "key variables" in the formulation, but nothing else.

Issues to Consider—Saturation, Non-linearities, Interactions. Most product developers are not so rigid that they must test all of the combinations. Yet, product developers need experimental designs that are sufficiently robust and

Table 1B Examples of Two-Level Plackett Burman Screening Designs for Six Variables in 12 Runs

	Variable					
Product	A	B	C	D	E	F
1	0	0	0	0	0	0
2	1	1	0	1	1	1
3	0	1	1	0	1	1
4	1	0	1	1	0	1
5	0	1	0	1	1	0
6	0	0	1	0	1	1
7	0	0	0	1	0	1
8	1	0	0	0	1	0
9	1	1	0	0	0	1
10	1	1	1	0	0	0
11	0	1	1	1	0	0
12	1	0	1	1	1	0

Note: 1 = High/new/present level.
0 = Low/old/absent level.

Table 1C Example of Two-Level Plackett Burman Screening Design for 12 Variables in 20 Runs

	Variable											
Product	A	B	C	D	E	F	G	H	I	J	K	L
1	0	0	0	0	0	0	0	0	0	0	0	0
2	1	1	0	0	1	1	1	1	0	1	0	1
3	0	1	1	0	0	1	1	1	1	0	1	0
4	1	0	1	1	0	0	1	1	1	1	0	1
5	1	1	0	1	1	0	0	1	1	1	1	0
6	0	1	1	0	1	1	0	0	1	1	1	1
7	0	0	1	1	0	1	1	0	0	1	1	1
8	0	0	0	1	1	0	1	1	0	0	1	1
9	0	0	0	0	1	1	0	1	1	0	0	1
10	1	0	0	0	0	1	1	0	1	1	0	0
11	0	1	0	0	0	0	1	1	0	1	1	0
12	1	0	1	0	0	0	0	1	1	0	1	1
13	0	1	0	1	0	0	0	0	1	1	0	1
14	1	0	1	0	1	0	0	0	0	1	1	0
15	1	1	0	1	0	1	0	0	0	0	1	1
16	1	1	1	0	1	0	1	0	0	0	0	1
17	1	1	1	1	0	1	0	1	0	0	0	0
18	0	1	1	1	1	0	1	0	1	0	0	0
19	0	0	1	1	1	1	0	1	0	1	0	0
20	1	0	0	1	1	1	1	0	1	0	1	0

Note: 1 = High/new/present level.
0 = Low/old/absent level.

complete in order to test many of the combinations in an efficient manner. These researchers and product developers recognize that there are other issues that are important besides the number of variables. For example, the relation between an independent variable and a consumer response may not be linear, but rather non-linear. (Add an increasing amount of an ingredient to a product and consumer acceptance will change. The acceptance rating does not plot as a straight line against the physical stimulus level. The relation is generally a curve, often looking like an inverted U.) The non-linearity requires the researcher to investigate at least three levels of each of the variables when those variables are continuous (e.g., fragrance level or oil level). Furthermore, there are pairwise interactions between the independent variables. The variables do not act separately. Often the level of one variable determines how another variable will affect consumer ratings. The interaction can be synergistic (so that the presence of one variable substantially enhances the consumer response), or suppressive (so that the presence of one variable substantially reduces the consumer response).

Given these considerations, statisticians have resorted to other experimental designs. The most popular of these for product developers is known as the "central composite design." Two examples of this design appear in Table 2A, one for three variables and one for six variables. The central composite design allows the investigator to assess three (or five) levels of a specific factor. Furthermore, the design can identify significant pairwise interactions between ingredients (but not higher order interactions). For the most part the central composite designs have proven very popular among researchers because they are robust. They enable the researcher to identify both curvilinearity (the optimum may be in the middle) as well as significant interactions between pairs of variables.

Working with Constraints. When creating combinations of independent variables, researchers often discover that some combinations are simply impossible to create. They are physically not realizable, either in the product development itself or subsequently (e.g., the product will not flow out of the bottle). Statisticians recognize this problem and create experimental designs with "constraints." Constraints define combinations of independent variables that the experimental design must avoid. The constraints will "unbalance" the previously balanced design, for the sake of avoiding impossible combinations.

As an example, consider the design shown in Table 3A for a lotion. The lotion comprises three key variables. For this example, let us assume that each variable ranges between a low level of 1 to a high level of 3. (These are relative values that can be converted back into the actual feature levels later on.) The initial design with 15 prototypes appears in Table 3A. However, upon investigation or based upon previous knowledge, the product developer "knows" (or learns by experience) that the combination of the three variables, A, B and

Table 2A Example of Central Composite Designs for Three Variables

	3 Level				5 Level		
Product	A	B	C	Product	A	B	C
1	1	1	1	1	1	1	1
2	1	1	-1	2	1	1	-1
3	1	-1	1	3	1	-1	1
4	1	-1	-1	4	1	-1	-1
5	-1	1	1	5	-1	1	1
6	-1	1	-1	6	-1	1	-1
7	-1	-1	1	7	-1	-1	1
8	-1	-1	-1	8	-1	-1	-1
9	1	0	0	9	2	0	0
10	-1	0	0	10	-2	0	0
11	0	1	0	11	0	2	0
12	0	-1	0	12	0	-2	0
13	0	0	1	13	0	0	2
14	0	0	-1	14	0	0	-2
15	0	0	0	15	0	0	0

Note: -2 = Lowest
 -1 = Low
 0 = Medium
 1 = High
 2 = Highest

C must lie within the range of 4.5 and 5.5. That is, no matter what values A, B and C take on, the sum A + B + C must lie between 4.5 and 5.5. Thus, the combinations 1,1,1 and 3,3,3 would not be allowed because the sums are lower than 4.5 and higher than 5.5, respectively. Table 3B shows a revised experimental design with these constraints. There are various texts and computer programs that deal with, and solve the problem of creating an experimental design with constraints versus without.

Step 3—Identifying the Competitive Frame and Prototypes to Test

In most optimization studies the prototypes are developed according to the experimental design, and then tested. The only remaining issues are to select the competitive frame against which to test the product, and to decide how the products will be tested (blind versus branded).

In years gone by the traditional way to test products was "blind" and "monadic." That is, the products were not identified as to manufacturer or ingredients (except in the most general way). Each consumer tested only one

Table 2B Example of Central Composite Design for Five Variables at Three Levels

Product	Q	R	S	T	U
1	1	1	1	1	1
2	1	1	1	1	-1
3	1	1	1	-1	1
4	1	1	1	-1	-1
5	1	1	-1	1	1
6	1	1	-1	1	-1
7	1	1	-1	-1	1
8	1	1	-1	-1	-1
9	1	-1	1	1	1
10	1	-1	1	1	-1
11	1	-1	1	-1	1
12	1	-1	1	-1	-1
13	1	-1	-1	1	1
14	1	-1	-1	1	-1
15	1	-1	-1	-1	1
16	1	-1	-1	-1	-1
17	-1	1	1	1	1
18	-1	1	1	1	-1
19	-1	1	1	-1	1
20	-1	1	1	-1	-1
21	-1	1	-1	1	1
22	-1	1	-1	1	-1
23	-1	1	-1	-1	1
24	-1	1	-1	-1	-1
25	-1	-1	1	1	1
26	-1	-1	1	1	-1
27	-1	-1	1	-1	1
28	-1	-1	1	-1	-1
29	-1	-1	-1	1	1
30	-1	-1	-1	1	-1
31	-1	-1	-1	-1	1
32	-1	-1	-1	-1	-1
33	1	0	0	0	0
34	-1	0	0	0	0
35	0	1	0	0	0
36	0	-1	0	0	0
37	0	0	1	0	0
38	0	0	-1	0	0
39	0	0	0	1	0
40	0	0	0	-1	0
41	0	0	0	0	1
42	0	0	0	0	-1
43	0	0	0	0	0

1 = High
0 = Medium
-1 = Low

Table 3 Examples of Central Composite Design for a Lotion—No Constraints

Product	A	B	C
1	2.00	2.00	2.00
2	1.00	2.00	2.00
3	3.00	2.00	2.00
4	2.00	1.00	2.00
5	2.00	3.00	2.00
6	2.00	2.00	1.00
7	2.00	2.00	3.00
8	1.00	3.00	3.00
9	3.00	1.00	3.00
10	3.00	3.00	1.00
11	1.00	1.00	1.00
12	3.00	3.00	3.00
13	1.00	1.00	3.00
14	1.00	3.00	1.00
15	3.00	1.00	1.00

Table 3B Central Composite Design for Lotion with Constraints (A + B + C Must Lie Between 4.5 and 5.5)

Product	A	B	C
1	1.00	1.00	3.00
2	3.00	1.00	1.00
3	1.00	3.00	1.00
4	1.00	1.75	1.75
5	2.50	1.00	1.00
6	1.50	1.00	3.00
7	1.00	1.00	2.50
8	2.25	1.00	2.25
9	3.00	1.00	1.50
10	1.00	3.00	1.50
11	1.71	1.71	1.71
12	1.00	1.50	3.00
13	3.00	1.50	1.00
14	1.00	2.50	1.00
15	1.00	2.25	2.25
16	2.25	2.25	1.00

product, or on some rare occasions the consumer tested two products and stated which of the two products was preferred and why. Blind, monadic testing was not conducive to measuring the performance of the products against the full competitive frame. The costs were prohibitive. Each additional product required another several hundred consumers. As a consequence the researcher relied more on norms than on the response to the competitive products at that particular time and place.

With the current state of the art, along with budget considerations, many researchers have come to realize that they learn more when the same panelist evaluates a number of different products during an extended session. The extended sessions may last several hours for early stage fragrance evaluation, or several weeks or months for analysis of other products such as shampoo and anti-perspirant, where home use is involved. Furthermore, many researchers recognize that, whereas norms are adequate in general, things change in the marketplace over time. The norms that were valid a year ago in a previous study may already be obsolete because competitors in the category have shifted the category around. Consequently, it is always prudent to assess the performance of a competitive product at the time of evaluating the prototypes, in order to get an "honest read" of the viability of the prototype against today's competitors.

Step 4—The Questionnaire

The questionnaire is similar to questionnaires used in conventional product evaluation, where the product is not systematically varied. The panelist rates acceptance, and sensory and image characteristics of all relevant attributes.

Researchers often believe that with experimental designs and systematically formulated products the panelist need not assess anything other than the degree of liking. In reality, however, the panelist can provide a great deal of information that will be useful for later analysis, modeling and optimization. If the panelist simply rates overall liking then it is possible to optimize the product so as to maximize overall liking. However, the researcher and product developer cannot add constraints to the product in order to conform to a specific marketing platform. One such constraint might be to make a lotion maximally acceptable, but also make it perceived by consumers as extremely thick in order to support a medicinal image. A thin lotion might be the most acceptable product, but the thin product does not convey to the consumer the impression of "medicinal." A thicker, less optimal product does. This finding would be revealed through the attributes.

The questionnaire should probe as many attributes as possible without overloading the panelist. There is no fixed number of attributes that is best to use. Some practitioners prefer to use relatively few attributes, others prefer to use many. The researcher should use questions which can link consumer perceptions to conclusions to be drawn from the data.

Step 5—Test Execution

Evaluating systematically varied products is no different from evaluating multiple products in a category appraisal or benchmarking study. Panelists evaluate either all or a randomized subset of products in a balanced order. The nature of the session will vary depending upon the product being tested. Fragrance evaluations can be done quickly (in several hours, or over short periods across several days if the panelist evaluates multiple fragrances on the skin). Other products requiring longer-term usage (e.g., shampoos) may take several weeks to evaluate. The panelist may require a short or a long exposure to the product, depending upon the type of information that is needed. (For fragrance evaluation, even of soaps and lotions, the panelist may be able to evaluate all of the products in a few hours. However, for lotions evaluated on skin it may take 2 weeks or more for the panelists to rate all of the products, even if the panelist evaluates one product per day. For other products, such as antiperspirant, the panelist may need at least three exposures to the product in actual use before being able to rate the attributes accurately.)

Step 6—Early Stage Data Analysis

The basic data for product optimization comprises a product × attribute matrix. The columns are the products, the rows are the attributes. Table 4 shows an example of this fundamental matrix. The data in the body of the matrix are mean values from the panelists who evaluated the particular product.

From time to time researchers also use percent statistics (e.g., the percent of consumers who, having evaluated a product on an attribute, choose a specific response). Purchase intent is one of the key attributes of this type. Although the purchase intent scale ranges from 1 (definitely not buy) to 5 (definitely buy), many marketing researchers simply tally the proportion of consumers who select the top two boxes (definitely would buy, probably would buy), and use that percent statistic as the measure they optimize. This strategy is perfectly acceptable as long as the researcher understands that the dependent variable is the *proportion of the consumers who give a specific response*, rather than the intensity of that response. (Percent top two box purchase intent, which is the proportion of consumers who say they would probably or definitely purchase a product, is not the same as the average degree of purchase intent. The former is an incidence statistic, whereas the latter is a true average of the intensity of feeling towards a product.)

More often than not the researcher is interested in the response of subgroups of consumers, not just in the response of the total panel. In a marketing-driven environment different subgroups of consumers exhibit different likes and dislikes. Therefore, in product optimization studies, just as in other product evalu-

Table 4 Example of Basic Data for an Optimization Study: Experimentally Varied Products, Levels of Independent Variables, Mean Ratings Assigned by Consumers

Product	101	102
Independent variables		
A	1	1
B	3	2
C	1	2
Liking	23	43
Overall	32	39
Appearance	58	68
Fragrance	51	41
Way it applies	42	37
Way it feels	72	60
Sensory		
Color depth	75	60
Thickness	68	53
Shininess	63	24
Fragrance strength	39	23
Greasiness	39	74
Thickness	37	59
Speed to absorb	50	42
Amount of residue	22	77
Appropriate for		
Younger vs. Older	41	38
Cosmetic vs. Medicinal	20	65

ations, it is conventional research practice to divide the population into key subgroups, and then calculate the relevant statistics (e.g., mean, % top two boxes) on a subgroup basis, as well as on a total panel basis. By the very nature of subgroups, the data are less robust than they would be for the total panel because the subgroups always comprise fewer panelists than the total panel. The reduced base size is of lesser concern when the products are systematically varied than when the product set comprises unrelated samples (e.g., in a category appraisal study). In experimental design studies the objective is to establish the relation between physical stimulus level and consumer response. Even with reduced base size, that relation can often clearly be discerned. In contrast, with unrelated stimuli there is no simple pattern to uncover. Therefore, there must be substantially more data underlying each point in order to yield a stable and robust mean.

Experimental Design and Product Optimization

Early stage data analysis is primarily descriptive. It does not incorporate any other data sets (e.g., levels of the independent variables). Interest focuses on the actual data itself. Besides calculating means or distributions of ratings by product and attribute for total panel as well as subgroups, early stage data analysis assesses the following two things:

1. Intercorrelations between variables, and factor analysis to identify the existence and nature of any underlying "basic dimensions" of perception.
2. Sensory segmentation between consumers to identify the existence and nature of groups of consumers who exhibit similar sensory preferences.

These two topics are dealt with extensively in other areas of this book, and will be referred to in passing when we come to look at different product categories.

Step 7—Correlation Between Independent and Dependent Variables

The correlation statistic quantifies strength of a linear relation between two variables. For experimental design it is often instructive to calculate the correlations between the independent variables under the developer's control and the consumer ratings. Only the sensory attributes should really be correlated with the R&D independent variables. There are four reasons for this restriction:

1. We expect there to be a linear relation between the sensory attribute rating and a single independent variable.
2. The correlation measures the strength of a linear relation between two variables.
3. The experimental design presents these independent variables in a systematic fashion which insures that the independent variables minimally correlate with each other. Therefore, it is possible to estimate the linear relation between one independent and one dependent variable.
4. The relation between rated liking and an independent variable may be linear, but it just as easily may be quadratic, with an optimum level of the independent variable lying in the middle range, rather than at either the upper or the lower extreme. Curvilinearity means that the straight line or linear relation is not appropriate to fit the data. Statistics which deal only with a linear relation may miss the true underlying relation when there is significant curvature. For example, the correlation coefficient would suggest a minor, insignificant correlation between the independent variable and liking. It would fail to discover that the relation is a curve with an optimum point. It would treat that curvature as failure to conform to linearity.

The correlation analysis between an independent variable (e.g., ingredients) and sensory attributes (e.g., darkness, thickness, or fragrance intensity) reveals whether or not the panelist accurately detects the "signal" in the product. The signal corresponds to the ingredient variable, and the detection or response corresponds to the fact that the panelist can register the signal by means of the rating of a sensory attribute. For instance, if the product developer increases the concentration of fragrance, then does the panelist rate the fragrance intensity higher? If yes, then the panelist detects the signal sent by the product developer and responds appropriately. If no, then the panelist does not detect the signal.

Correlation analysis is also instructive in other ways. It shows which sensory attributes covary with physical changes in ingredients. For instance, if the product developer increases the lotion thickener and the panelist rates the product as smelling weaker, then we know that the physical viscosity affects the perception of fragrance intensity, not just the perception of thickness.

Finally, correlation analysis is a method by which to establish the validity of the test procedure. It has been averred that panelists are incapable of rating anything other than degree of liking, and that a panelist cannot accurately rate the sensory attributes of products. Correlation analysis can immediately disprove this claim. If the panelist's sensory ratings highly correlate with physical variables, then we must conclude that panelists correctly track the sensory attributes of products. Table 5 shows an example of this correlation analysis.

Step 8—Model Creation

A key benefit of systematic variation of ingredients is the ability to create a "product model." The product model comprises a set of equations. Each attribute rated by the panelist generates its own equation.

Table 5 Sample Correlation Table Between Five Independent Variables (A–E) and Three Sensory Attributes Rated by Consumers

	Independent variables				
	A	B	C	D	E
Darkness	0.40	|0.96|	0.24	0.31	0.22
Fragrance	0.51	0.31	|0.62|	0.36	0.04
Greasy	|0.62|	−0.35	−0.24	0.05	0.81

A = Emollient level
B = Coloring level
C = Fragrance concentration
D = Thickener level
E = Active ingredient level

Experimental Design and Product Optimization

The equation summarizes the relation between physical variables under the developer's control and subjective ratings assigned by the panelist. The equation has the following three properties:

1. The parameters are estimated by the method of least squares. There are well-defined statistical procedures for fitting an equation to the data. The equation is fit by a set of criteria to make the equation match the actual data as closely as possible.
2. The equation can accommodate the fact that the relation between the attribute rating and the physical variable often is not a straight line. There can be curvature, allowing the researcher to identify an intermediate level of physical ingredients as the optimum.
3. The equation can accommodate the fact that the relation between attribute rating and physical variables results from interactions between the physical variables, above and beyond the separate contribution of the physical variables. That is, the effect of one physical variable (e.g., fragrance concentration) on a consumer rating (e.g., perceived fragrance intensity) may be affected by the level of another physical variable (e.g., emollient level). In most cases the interaction of physical variables is limited to pairs of physical variables.

Most research data can be modeled by equations that are relatively simple. The equations are written as follows for two variables (A, B):

Attribute Rating = Constant + Linear Portion + Quadratic Portion
+ Interaction Portion

Constant = Adjustment factor, added to the remaining terms = k_0
Linear Portion = $k_1 A + k_2 B$
Quadratic Portion = $k_3 A^2 + k_4 B^2$
Interaction Portion = $k_5 A \times B$

Every equation generates its own unique equation with coefficients. For more variables the equation has more terms. For example, if the experimental design comprises five variables (A,B,C,D,E respectively), rather than the two variables above, then the full equation would comprise an additive constant, five linear terms, five quadratic terms and (at most) ten pairwise interaction terms.

Underlying the creation of the descriptive equation are several points of view. Some researchers believe that the equation should comprise all of the terms, even though some of the terms may be statistically insignificant and add little predictive power. Other researchers feel that only those terms which are statistically significant should be incorporated into the equation. This author believes that the equation for a consumer-rated attribute should comprise the additive constant, the linear terms, and the quadratic terms. Once these linear and quadratic terms are incorporated into the equation, then only those pairwise interaction terms which add substantially more predictability beyond that currently achieved by the equation should be added. This approach combines the best parts of both viewpoints—the equation comprises a relatively limited number of terms (parsimony), captures curvature, and allows for synergism between pairs of independent variables.

Conventional statistical procedures, such as least squares, create the equations according to specific objectives (e.g., minimize the sum of the squares of the deviations between predicted ratings and actual ratings). With the computing power available on today's personal computer, regression analysis can be done easily and quickly.

In the standard product evaluation study the consumer rates dozens of attributes. In addition, R&D can provide physical measures (e.g., of stability) as well as cost measures. The subjective ratings are modeled by equations containing square and cross terms (e.g., to allow for curvilinearity and for pairwise interactions). The measures for cost of goods and physical values are often modeled by simple linear equations (without the square and cross terms). The specific terms used in the model are determined on an attribute-by-attribute basis. In some cases (e.g., consumer perception of product color), the researcher may wish to include in the equation only those predictor terms having to do with appearance. Other terms, e.g., dealing with fragrance, may in fact be statistically "significant" and add predictive power to the ratings of "appearance," but those fragrance-related terms are not interpretable and make no sense.

Creating equations that relate physical variables to consumer ratings is both an art and a science. The actual estimation procedures are well-defined. What specific terms to include (linear, square, cross term) and what complexity of the model to develop (quadratic equation, even cubic or other non-linear equation) is left up to the researcher.

Sometimes the equation does not adequately fit the data. From time to time there are outlier products (products whose ratings simply cannot be predicted on the basis of the equation). For many of these outliers it is not clear whether the single point or observation results from normal statistical variation (simply a product that cannot be predicted because of unforeseen random variation) or whether this outlier point (or set of points) signals that the model is inadequate to capture the data. Researchers perennially face the problem of accounting for data points that cannot be predicted. Is the departure from prediction statistical error, or is it a "signal" pointing to a more complex situation (and a possible breakthrough)?

When creating a model the researcher faces the dual problem of keeping the model simple with the fewest terms possible, and yet making the model predict all of the points as closely as possible by using many terms. Fewer terms in the model make the model parsimonious and robust. It is unlikely that the model will produce a nonsensical prediction because it is "very flat." In contrast, more terms will make the model fit the data better, but could produce a nonsensical prediction for new combinations of the ingredients in regions of the "design space" where there are no empirical data.

There are various schools of thought about creating robust vs. better-fitting models. It is always tempting to create a model that fits the data. On the other hand, if new predictions have to be accurate, then the more complex model could provide a misleading prediction. Prediction error for new data may arise because the more complex model uses interaction terms (A × B). In regions of the design space where there is no data, the possibility exists that the interaction term (A × B) might actually be very large. When there is no data, the interaction terms would be unconstrained by "reality." The researcher would have no idea whether the interaction term is a "Trojan Horse," sitting in the equation ready to add (or subtract) a considerable amount to the estimated value for a specific, previously untested combination of independent variables A and B.

Step 9—Using the Models to Visualize Relations

One of the key uses of the models is to understand the relation between variables in the data set. By visual inspection of raw means (the array of averages from the study), the researcher can get a sense of how the independent variables drive consumer reactions. Nonetheless, this is an exceedingly difficult task. Discerning a pattern solely by visual inspection of raw means is easiest when only one variable changes, with all other variables held constant. The task becomes much more difficult when multiple variables change in a systematic design, with no single variable changing and all other variables held constant. As an example, consider the patterns that one can discern in Tables 6A and 6B. Table 6A shows the effect of changing one variable (Q) on three attributes (dependent variables). The pattern is easy to discover. The analyst need only line up the independent variable in descending order and inspect the empirical changes in the consumer ratings. In Table 6B the task is far more difficult because there is no simple way to line up the different products to identify a pattern. The three independent variables change in different ways, so that one cannot be sure how changes in a specific independent variable really drive the consumer rating.

In order to understand the relation between independent and dependent variables, the researcher often performs a "sensitivity analysis," using the equation. The sensitivity analysis looks at the expected ratings of attributes given small, equally spaced variations in one independent variable, with all other variables held constant. Sensitivity analysis is straightforward and simple because the investigator has already created the mathematical equations. Sensitivity

Table 6 Example of Data Patterns—Independent Variables Versus Dependent Variables

A—One Independent Variable (Q) vs. Three Dependent Variables (A-C)

Independent	Dependent		
Q	A	B	C
10	20	69	46
20	25	49	42
30	28	40	78
40	31	36	45
50	36	32	36
60	43	31	37
70	58	30	58
80	43	30	41
90	41	30	39

B—Three Independent Variables (X-Z) vs. Three Dependent Variables (A-C)

Independent			Dependent		
X	Y	Z	A	B	C
10	10	10	20	69	46
10	10	50	25	49	42
10	50	90	28	40	78
50	50	10	31	36	45
50	90	50	36	32	36
50	90	90	43	31	37
90	10	10	58	30	58
90	50	50	43	30	41
90	90	90	41	30	39

analysis simply puts numbers onto that equation to show the magnitude of the changes that one can expect. With sensitivity analysis the researcher begins to see the effects of adding interaction terms to the equation. An equation comprising interaction terms often shows stronger effects of one variable than an equation without interaction terms.

In order to demonstrate sensitivity analysis, consider the six equations in Table 7A, and the table of expected ratings in Table 7B. The equations are used as examples only. In Table 7A we see six equations predicting a response attribute. Some equations use only linear terms, other equations use linear, square and cross terms. Table 7B shows the comparable sensitivity analyses. Note the difference in expected ratings between the two equations. The simpler, more parsimonious equation fails to show the general trend that we might expect (if the attribute were liking), whereas the more complicated (but less parsimonious) equation does show the trend.

Table 7A Examples of Equations in Two Variables

Equation 1—Linear in X only:
Dependent = 20 + 5X

Equation 2—Linear in X and Y:
Dependent = 20 + 5X + 5Y

Equation 3—Linear in X and Y, with pairwise interaction:
Dependent = 50 + 5X + 5Y + 3XY

Equation 4—Quadratic in X only:
Dependent = 50 − ((3 − X)^2)

Equation 5—Quadratic in X and Y, but no cross terms (interaction):
Dependent = 100 − (10)[(3 − X)^2 − (3 − Y)^2]

Equation 6—Quadratic in X and Y, with a cross term:
Dependent = 100 − (0.051)[(3 − X) − (3 − Y)]^2

Table 7B Sensitivity Analysis—Expected Values of the Six Equations at Various Levels of Independent Variables X, Y

X	Y	Eq. 1 Linear in X	Eq. 2 Linear in X and Y	Eq. 3 Linear with interaction	Eq. 4 Quadratic in X	Eq. 5 Quadratic in X and Y	Eq. 6 Quadratic with interaction
1	1	25	30	63	30	56	71
2	1	30	35	71	45	86	57
3	1	35	40	79	50	96	41
4	1	40	45	87	45	86	22
5	1	45	50	95	30	56	1
1	2	25	35	71	30	59	73
2	2	30	40	82	45	89	60
3	2	35	45	93	50	99	44
4	2	40	50	104	45	89	26
5	2	45	55	115	30	59	6
1	3	25	40	79	30	60	75
2	3	30	45	93	45	90	63
3	3	35	50	107	50	100	48
4	3	40	55	121	45	90	30
5	3	45	60	135	30	60	10
1	4	25	45	87	30	59	78
2	4	30	50	104	45	89	66
3	4	35	55	121	50	99	51
4	4	40	60	138	45	89	34
5	4	45	65	155	30	59	14
1	5	25	50	95	30	56	80
2	5	30	55	115	45	86	68
3	5	35	60	135	50	96	54
4	5	40	65	155	45	86	38
5	5	45	70	175	30	56	18

Over the past 10 years there have been two different ways to do sensitivity analysis:

1. Examination of single variables, and
2. Plotting of iso-hyphs.

Dose—Response Plots. This approach looks only at one variable at a time, as described above. The approach is based upon the premise that the greatest learning emerges from the pattern of a response attribute versus a single independent variable, with all other independent variables held equal.

This single variable approach arises from the psychophysical way of thinking. Psychophysicists look for the relation between two variables, a single independent and a single dependent variable, respectively. Even though the equation may be complex, with many independent variables and many interaction terms, the actual plot only involves two variables. The output of the analysis is a "sense" of causality between one variable and another. When plotted on a graph the result is the conventional X-Y plot, where the abscissa is the level of the independent variable and the ordinate is the level of the consumer rating. Figures 1A and 1B present some standard X-Y plots arising from this type of sensitivity analysis. With the increasing popularity of powerful statistical programs that plot data, researchers have also been able to plot the three-dimensional dose-response or sensitivity analysis for two independent variables vs. a dependent variable as shown by Figure 1C.

Iso-Hyphs. This represents an entirely different way of thinking about the sensitivity analysis. Iso-hyphs or contour plots are plots of the levels of two variables which in concert generate equal responses or ratings. Figure 2 (A-B) shows examples of an iso-hyph or equal intensity plot. The two axes represent levels of the independent variables. The contour shows those combinations of the two independent variables which in concert generate the same value of the dependent variable. Iso-hyphs are valued by researchers because they graphically illustrate the existence of interactions between pairs of independent variables. For instance, Figure 2A shows two variables which do not interact statistically to generate an attribute rating. Both variables contribute separately, but there is no significant cross term. In contrast, Figure 2B shows what happens when there is a significant interaction. The curves depart from linearity.

Step 10—Identifying Optimal Products

By themselves the equations and the sensitivity analyses provide the researcher with a more profound understanding of the relation in the data, but do not identify solutions to problems. Optimization provides a concrete solution to the problem stated as follows: "What combination of variables, under the

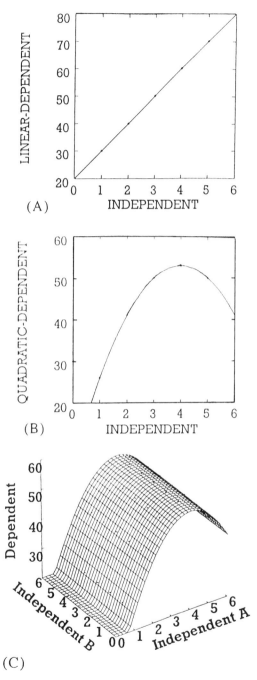

Figure 1 (A) Linear relation between two variables (one independent, one dependent), (B) Non-linear relation between two variables (one independent, one dependent), and (C) Relation between three variables (two independent, one dependent).

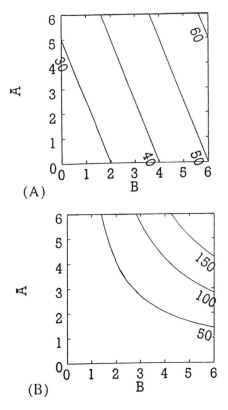

Figure 2 (A) Iso-hyphs for two variables that do not interact and (B) Iso-hyphs for two variables that interact.

experimenter's control, generates a product that possesses a specific profile of values for the dependent variable?"

The specific profile of values might be one of the following three cases:

1. The maximum level attainable for a specified consumer attribute rating (e.g., acceptance), within the range of independent variables (ingredients) tested in the study.
2. The maximum level attainable for a consumer attribute rating, subject to constraints imposed by levels of another variable (e.g., cost of goods).
3. A profile of consumer attribute ratings designated as the "target" or "goal" profile. In this case the objective is to identify a combination of independent variables which, in concert, generate a pre-designated profile of consumer ratings. This is "goal fitting," a variant of optimization.

2. Case History—Sanitary Napkins

In 1985 Lynn Johnson purchased the American Personals Company. American Personals had been in operation for over 65 years, starting out as a purveyor of "paper napkins and other items for ladies." During much of its history, American Personals was a low-profile corporation, content to ride on the coattails of an increasing demand for specialty paper goods and other personal products. Under Johnson, the new management at American Personals recognized an opportunity to branch out into many other areas of personal care. Sanitary napkins seemed the most likely candidate, given American Personals' success in paper-based products.

At first, American Personals' napkin introductions met with market success, easily fulfilling and exceeding the marketing plan. It was 5 years later, however, in 1991 that the need for a new marketing plan became apparent. As competition stiffened in the category, volume sales for sanitary napkins flattened. Management at American Personals recognized that they could no longer ride on the wave of increasing sales that had propelled them thus far.

Management had two options. The easier option was to purchase another company in order to have new products to sell. This was an attractive strategy because it meant that the company did not need to invest in technology or research. The company could simply transform old profits to new acquisitions, and maintain its share of the market. The harder option was to invest in product development by hiring a top-notch product development team. This second option was far less certain, more risky and novel to American Personals which, until Lynn Johnson took over, had never had a reputation as a strong product developer.

Lynn Johnson, the far-sighted, almost visionary president of American Personals, pushed to develop a strong R&D team. The rationale was simple. Johnson recognized that as competition stiffened, American Personals would need to push products into the marketplace at an accelerating rate. Although purchasing a company seemed to be less risky, in fact, according to Johnson, the purchase would simply mask an inherent potential weakness of American Personals. American Personals had to develop the technological muscle to compete in the marketplace over the years to come. Sooner or later top management at American Personals would have to "bite the bullet" and invest in technology. Why not now?

Putting the New Team to Work—A New Sanitary Napkin

Napkins in the early 1990's were proliferating at an unbelievable rate. Manufacturers searched for new physical formats, new benefits, new packages etc. as each manufacturer in turn tried to achieve a competitive advantage in the marketplace. The store shelves became cluttered with products as retailers were inundated with different offerings.

Dr. William Riha, the newly minted R&D director at American Personals, held a point of view that the products which American Personals offered had to differ substantially from the competition, and had to be demonstrably better. This meant that development efforts would have to be targeted to producing a product substantially better than the competitive frame in terms of both aesthetic characteristics and performance. Riha wanted the new products to undergo intense consumer evaluation to insure that they accomplished the marketing objectives.

Identifying the Proper Niche for the New Napkin

The first question was the definition of what the napkin should be. The product could be based upon new technology or simply based on a highly acceptable product with improved aesthetic characteristics.

After careful deliberation and consideration (accompanied by the inevitable shouting and intense discussion which goes with these types of serious endeavors), marketing and R&D opted for a new absorbent technology. The reasoning was straightforward. The current sanitary napkin market in that period was heating up, with many new line extensions offered by well-known manufacturers. However, there was a growing, expressed need by consumers for a thin napkin product with increased absorbent protection.

At this juncture marketing reached an impasse. One faction of the marketing group wanted to test and optimize the product using a population of consumers who were currently users of extra absorbency products. Another faction of the marketing department felt that for the product to reach its greatest success it would have to be targeted toward the general user population. That is, this second group wanted to create and market an all-purpose sanitary napkin that was more absorbent than the typical napkin currently marketed, but did not want to have their product positioned as one primarily for heavy flow days. This second group, the generalists, won out.

Identifying Variables at the Very Early Stage

At the very earliest stage, even before any exploratory work has been done, the problem often arises as to what exactly should be varied. This seems like a naïve question. Certainly most competent product developers know (or should know) the compositions of the products that they are developing.

Although the foregoing paragraph uses the word "should," in actuality, most product developers world-wide in most consumer categories do not know the "ins and outs" of their product in profound detail. Ingredients and absorbent technology change, and product developers go from project to project, and

Experimental Design and Product Optimization 243

from company to company. In today's hectic world it is sufficient for the product developer simply to be able to keep up with the demands of the business. There is precious little time for exploratory work with a product when, for the past two decades, the development timelines have shrunk drastically. Consequently, when faced with the task of determining what ingredient or feature levels should be explored, most competent product developers can create a list of feasible ingredients (or process conditions), and provide some idea of the appropriate levels of those ingredients (levels that will "work" in a product). Few product developers have the luxury to explore alternative, perhaps even optimal levels, and simply "play" with the product in a relaxed fashion in order to learn about it.

Thus, the task of choosing the appropriate product features and their levels is far more complex than one might imagine. A productive way to choose the ingredients and levels is to follow these four steps:

1. On a piece of paper lay out three columns.
2. Column 1 lists what can be varied "operationally" by the product developer (either in ingredients, structure, process conditions, etc.). These are what actually appear in the experimental design.
3. Column 2 lists how the ingredient, structural or process condition being varied is expected to affect the physical characteristics of the product.
4. Column 3 lists the expected effect on perception of that physical characteristic.

The rationale is as follows:

Independent Variables → Physical Changes in the Product
Physical Changes → Consumer Sensory Impressions
Consumer Sensory Impressions → Acceptance, Image Impressions

By laying out these variables in a list, the product developer can assess the different factors under her control. The objective is to modify the consumer acceptance or image reactions. The approach identifies what physical factors are believed to cause those acceptance reactions. The action is to select a limited (feasible) number of independent variables that will generate the requisite variation in physical changes, and thus yield the desired sensory reactions and liking.

The Experimental Design

The R&D group screened various alternative features and initially selected 15 product features. These 15 features were simply too many to work with, even though the desire was expressed several times in meetings that the research project would need to yield the ultimate product. After following the foregoing 4-step exercise, the R&D group recommended changing five variables:

length, width, thickness, amount of the structural layering and roundness of the edges. These independent variables were felt to create the greatest effects on both a perceptual and performance basis. These five variables were all continuous variables.

The experimental design comprised 19 prototypes from a full set of 43. Table 8 shows the two experimental designs—both full factorial design (Table 8A) and the 1/4 replicate finally selected (Table 8B). In many tests, R&D and marketing research have no problem creating and testing all of the requisite prototypes (which in this case would mean the full set of 43). However, in this study each panelist could only test a limited number of products. Consequently it was necessary to pare down the number of prototypes to a reasonable number. There were three considerations:

1. The greater the number of prototypes to be tested the more likely it would be that the test would yield winning products.
2. The greater the number of prototypes the more likely it would be that the data would be useful for years to come. The pattern relating consumer acceptance and product formulations would emerge more clearly with the greater number of panelists.
3. However, cost would increase with increasing number of prototypes.

In situations such as these, there is always a trade-off between cost versus knowledge as an investment. The trade-off is not always resolved in favor of lower cost, however. Often, management opts for the more expensive alternative, recognizing that the "up-side potential" of the additional information is substantially greater than the extra cost incurred. This consideration holds even more forcefully when the research can be used to create a line of products, not simply one product, or when the financial expectations for a product when "done right" are so high that management is reluctant to cut corners. For this napkin study, however, ongoing discussions between marketing researchers and R&D led to the consensus that the budget had to be maintained. With the need to test competitive products along with the experimentally designed prototypes, the consensus was that a 1/4 replicate design (19 prototypes instead of 43) was the more fiscally conservative option.

Setting Variable Levels for the Design

Experimental designs are templates for the product developer to use. The experimental design specifies neither which variables are to be used nor the levels of those variables. How then does the product developer flesh out an experimental design? There is no magical answer. It requires effort and insight to identify variables and to determine the appropriate levels. For example, if the experimental design calls for three levels, then these three levels should be perceptibly different from each other. To guarantee perceptual differences the

Table 8A Experimental Design for Sanitary Napkin

	Full central composite design				
Product	Length	Width	Depth	Internal structure	Round
1	1	1	1	1	1
2	1	1	1	1	-1
3	1	1	1	-1	1
4	1	1	1	-1	-1
5	1	1	-1	1	1
6	1	1	-1	1	-1
7	1	1	-1	-1	1
8	1	1	-1	-1	-1
9	1	-1	1	1	1
10	1	-1	1	1	-1
11	1	-1	1	-1	1
12	1	-1	1	-1	-1
13	1	-1	-1	1	1
14	1	-1	-1	1	-1
15	1	-1	-1	-1	1
16	1	-1	-1	-1	-1
17	-1	1	1	1	1
18	-1	1	1	1	-1
19	-1	1	1	-1	1
20	-1	1	1	-1	-1
21	-1	1	-1	1	1
22	-1	1	-1	1	-1
23	-1	1	-1	-1	1
24	-1	1	-1	-1	-1
25	-1	-1	1	1	1
26	-1	-1	1	1	-1
27	-1	-1	1	-1	1
28	-1	-1	1	-1	-1
29	-1	-1	-1	1	1
30	-1	-1	-1	1	-1
31	-1	-1	-1	-1	1
32	-1	-1	-1	-1	-1
33	1	0	0	0	0
34	-1	0	0	0	0
35	0	1	0	0	0
36	0	-1	0	0	0
37	0	0	1	0	0
38	0	0	-1	0	0
39	0	0	0	1	0
40	0	0	0	-1	0
41	0	0	0	0	1
42	0	0	0	0	-1
43	0	0	0	0	0

−1 = Low 0 = Medium 1 = High

Table 8B One-Quarter Replicate of Five Variables

Product	Length	Width	Depth	Internal structure	Round
1	1	1	1	1	1
2	1	1	-1	1	-1
3	1	-1	1	-1	-1
4	1	-1	-1	-1	1
5	-1	1	1	-1	1
6	-1	1	-1	-1	-1
7	-1	-1	1	1	-1
8	-1	-1	-1	1	1
9	1	0	0	0	0
10	-1	0	0	0	0
11	0	1	0	0	0
12	0	-1	0	0	0
13	0	0	1	0	0
14	0	0	-1	0	0
15	0	0	0	1	0
16	0	0	0	-1	0
17	0	0	0	0	1
18	0	0	0	0	-1
19	0	0	0	0	0

developer should create a base product, and separately vary each single ingredient, physical feature, or processing condition from low through medium, through high. This procedure often must be repeated as the developer learns about the behavior of the product from these trial and error runs. Product developers who follow disciplined product development using experimental design often report that they learn more about their product from these initial development trials than they do from previous experiences.

No one can be absolutely sure before creating prototypes that these prototypes will differ dramatically from each other. However, if the product developer adheres to the exploratory guidelines, learns about the performance of the ingredients, and goes through the very early stage exploration of levels in a systematic way (and perhaps several times), then the experimentally varied prototypes will probably generate the requisite sensory range to make optimization a worthwhile approach.

Screening the Panelists

In view of the dual nature of the study (a design for general use but also for heavy flow), marketing recommended a "loose screening." Current napkin users

Table 9 Screening Concept

Concept 1—Version 1 (More Factual)
Introducing American Personals' "New Days"

An all-new napkin created with cotton-soft. Cotton-soft has been proven in clinical tests to be 50% more absorbent than the leading sanitary napkins. New Days is available at your local store, at a competitive price.

Concept 1—Version 2 (Factual + Sales Oriented)
Introducing American Personals' "New Days"

An all-new napkin created with cotton-soft. This is a powerful, revolutionary new product for your feminine needs. Cotton-soft has been proven in clinical tests to be 50% more absorbent than the leading sanitary napkins. New Days is available at your local store, at a price that will be equal to or less than the brands on the market.

---

were eligible to participate, as long as they were non-rejecters of the screening concept (see Table 9). The screening concept was a short paragraph describing the product in "non-selling terms" (Concept 1, version 1 in Table 9). That is, the concept was purely descriptive. (Compare this with Concept 1, version 2 which has more of a sales orientation.) Non-rejecters were defined as consumers who rated the screening concept as either "definitely," "probably" or "might/might not" purchase on a 5-point purchase intent scale. The remaining two categories, "definitely not" purchase and "probably not" purchase, were defined as rejecter categories. (Anyone who assigned the concept to either of the two rejecter categories during the telephone screening was rejected as a potential participant.)

There is a continuing controversy about who to accept as a potential participant on a product evaluation panel. Should consumers be "top two box" accepters (those who say that they would definitely or probably buy the product)? Or, should the panel comprise consumers who do not reject the product? Arguments can be adduced for both positions. Those in favor of working only with concept accepters feel that these individuals are the most likely candidates to purchase the product, and thus should be the prime targets. (An even more focused group would be consumers who rate the concept as "definitely would buy." However, in most cases this selection criterion unduly restricts the potential population of panelists.) Those in favor of working with non-rejecters espouse a different point of view. They feel that as long as the consumer does not reject the product at the "concept stage" there is the ever-present possibility that the consumer will eventually convert to become a product accepter. Concept rejecters will never purchase or use the product. Those who are indifferent, however, can be enticed to use the product based upon strong consumer trade promotion or sampling (free distribution of small, trial-sized containers).

In this study, the marketing group felt that because the product straddled two consumer segments (regular users, users demanding strong protection) it was best to leave the screening criteria as loose as possible.

The Actual Product Evaluation—Product Setup

The study comprised 19 experimentally varied products (see Table 8B) and five commercially available products to serve as benchmarks. All products were presented to consumers in the same format (white box, no manufacturer brand). In this way the R&D group attempted to standardize the presentation. All products were labeled with a three-digit identification number. Thus, the products were tested "double blind" so neither the interviewer nor the panelist would know what prototype or commercial product corresponded to each package of product. It was impossible to totally disguise the competitors, but R&D did the best job they could.

The Questionnaire

In order to maximize the amount of information but not fatigue the panelists, the researchers agreed to limit the questionnaire to 25 attributes. The panelists were instructed to fill out the questionnaire at the end of using each product (one product per day).

The questionnaire comprises questions about the product, the person–product interaction, and the end benefits of the product. Table 10 shows the questionnaire attributes. All product attributes were rated on an anchored 0–100 point scale, except purchase intent (which was rated on a 5-point scale).

Parenthetically, it is always useful to pretest any questionnaire. In many research studies investigators feel that panelists can rate a seemingly unlimited number of attributes, especially when the panelists fill out a rating sheet at home. However, in many cases panelists are frustrated by the undue number of attributes and by the repetitiveness of the questionnaire. Boredom often occurs when the questionnaire comprises a "laundry list" of attributes, with many of these attributes highly correlated with each other. A pretest with a small group of 4–8 consumers (not corporate employees!) will quickly reveal whether the questionnaire is interesting or whether it is repetitive and boring. Furthermore, the pretest can identify ambiguous questions, or incomplete instructions.

Setting Up the Field

The napkin study was deemed to be sufficiently important so that the data would be obtained from several markets, not just the local market near the R&D laboratory. R&D was in the habit of commissioning a local field service to do inexpensive tests. The results of those small-scale tests were used to guide development. Here, however, R&D recognized a need for a wider representation of

Table 10 Attributes Used for the Sanitary Napkin Study

Abbreviation	Question
Absorbency	How well does this napkin absorb? (0 = poorly→100 = extremely well)
Initial	How much do you like this napkin when you first use it? (0 = hate→100 = love)
After 2 Hours	After 2 hours how much do you like this napkin? (0 = hate→100 = love)
After 4 Hours	After 4 hours how much do you like this napkin? (0 = hate→100 = love)
Final	When you remove the napkin how much do you like it? (0 = hate→100 = love)
Purchase	How interested are you in purchasing this napkin? 5 = definitely buy 4 = probably buy 3 = might/might not buy 2 = probably not buy 1 = definitely not buy
Length	How long is this napkin? (0 = very short→100 = very long)
Thickness	How thick is this napkin? (0 = very thin→100 = very thick)
Width	How wide is this napkin? (0 = very narrow→100 = very wide)
Like length	How much do you like the length? (0 = hate→100 = love)
Like width	How much do you like the width? (0 = hate→100 = love)
Like thickness	How much do you like the thickness? (0 = hate→100 = love)
Fit concept 1	How well does it fit this concept? (#1) (0 = poorly→100 = extremely well)
Fit concept 2	How well does it fit this concept? (#2) (0 = poorly→100 = extremely well)
Fit concept 3	How well does it fit this concept? (#3) (0 = poorly→100 = extremely well)
Like removal	How well do you like the way the napkin removes? (0 = hate→100 = love)
Like application	How well do you like the way the napkin applies? (0 = hate→100 = love)
Unique	How unique is the napkin? (0 = similar to others→100 = very different)
Absorbs	How well does the napkin absorb? (0 = poorly→100 = extremely well)
For heavy days	How appropriate is the napkin for heavy days? (0 = inappropriate→100 = very appropriate)
For light days	How appropriate is the napkin for light days? (0 = inappropriate→100 = very appropriate)
For me	How appropriate is the napkin for a person like you? (0 = inappropriate→100 = very appropriate)
Secure	How secure do you feel wearing the napkin? (0 = insecure→100 = very secure)
Discreet	How discreet is the napkin? (0 = not discreet at all→100 = very discreet)
Fits well	How well does the napkin fit? (0 = fits poorly→100 = fits well)

consumers. The marketing research group recommended using five markets to insure geographical representation.

The selection of base size was made by taking into account the following six considerations:

1. There were 24 products. Each product had to be evaluated by 50 women in order to obtain readable samples. The 50 panelist base size was deemed

to be sufficient for the researcher to analyze subgroups (for "directional insights") as well as being cost effective.
2. With the 50 panelists per product and 24 products a total of 1200 product ratings was required.
3. Each panelist was recruited to participate for 2 months during which time each panelist would test 8 of the 24 products. Thus, the 1200 product ratings (usage occasions) had to be allocated among the panelists so that each product would appear equally as often (50 ×), and each panelist would evaluate eight products. This necessitated 150 panelists for the total study (1200/8 = 150).
4. Researchers recognized that there would be the inevitable dropouts in the study, and thus felt that it would be more prudent to recruit 180 panelists rather than 150 panelists.
5. Each of the five markets had its own field service—marketing research professionals trained to set up and implement studies. The five field services were given screening questionnaires and a screening concept, and instructed to recruit 36 consumers to participate for a study. The consumer panelists were to be screened by telephone and invited to participate. The participants were told that they would be compensated for completing the study.
6. Participants came into the first session for orientation. In order to make the study more efficient the panelists showed up in groups of 18 for the 1-hour orientation. The orientation followed these six steps:
   a. Explained the purpose of the study
   b. Explained the scales to be used at home
   c. Oriented in scaling by means of a practice exercise
   d. Delivered the first four products to the panelist
   e. Instructed about product use and questionnaire
   f. Instructed when to return to the test site.

Some Observations on Field Execution

When one reads accounts of scientific research, one is often struck by the apparently smooth and effortless execution of the study. In consumer research this is rarely the case. The two events described below illustrate problems that the researchers and the field services encountered:

1. The questionnaires for the markets were coded with different panelist identification numbers. In two markets, and due to human error, some of the numbers overlapped. This is an easily correctable problem, but exemplifies some of the foibles of research. Not all research is conducted in a pristine, pure, unhurried environment. Mistakes happen, even in the simplest of situations where any observant person would quickly notice the error and correct it.

Experimental Design and Product Optimization 251

2. Some of the product was damaged. It was not clear whether the damage was due to faulty packaging or product mishandling on the shipper's part. Product damage often occurs, and indeed far more frequently than one is willing to admit. After the first report of damage, R&D decided to send one back-up box to each market. The back-up box contained three samples of each of the 24 test products. It is a good idea to insure product safety and security by shipping multiple boxes of product packed so that each box contains one to three samples of each of the products. Thus, each package is similar to every other package, and each package contains all products, not just one product. This method of packing the product is more difficult and unusual, since manufacturers are accustomed to sending all prototypes of one kind in one box. By having each box contain an assortment of products, the shipper insures that loss of, or damage to one box does not ruin one product prototype completely. Rather, the damage affects each of the products, but only slightly. Furthermore, by dividing up the products to create boxes of mixed products the shipper insures against loss. Even if one box is lost during transit, the researcher still has virtually all of the product left for the test, save one to three units of each prototype.

Topline Results—Product × Attribute Matrix

The basic analysis for all of the multiproduct studies uses a rectangular data matrix, with the columns constituting the products and the rows constituting the attributes. If only one physical variable is changed, then it is easiest to portray the data in the order of that physical variable. Quite a lot of insight can then be gained by looking at the changing values of the ratings when the independent variable increases from low to high.

In the current study there is no single obvious way to lay out the rows of products. The experimental design with five variables is significantly more complex. One could arrange the rows of products in descending order of one variable. That is unproductive, however, because the remaining four variables also change. Table 11(A-C) shows three potential orders in which one could lay out the experimental design.

If there are no compelling reasons to lay out the products in a given order, then there are two recommended orders:

Order of Product Identification Number. The order is the same from table to table, and follows the order of product numbers. The lowest identification numbers lie at the top of the table, and products with the highest identification numbers lie at the bottom of the table. This layout is best when one wishes to compare two or more products on a continuing basis. Laying out the products in order of identification number allows the user to quickly locate products for visual comparison of data.

Table 11 Layout of Data for the Napkin Products: Three Different Formats

A—Rank Order by Product Number

Product	Length	Width	Depth	Internal structure	Round	Liking
1	3	3	3	3	3	60
2	3	3	1	3	1	61
3	3	1	3	1	1	57
4	3	1	1	1	3	33
5	1	3	3	1	3	36
6	1	3	1	1	1	57
7	1	1	3	3	1	48

B—Rank Order by Liking

Product	Length	Width	Depth	Internal structure	Round	Liking
15	2	2	2	3	2	62
2	3	3	1	3	1	61
16	2	2	2	1	2	60
1	3	3	3	3	3	60
8	1	1	1	3	3	58
3	3	1	3	1	1	57
6	1	3	1	1	1	57

C—Rank Order by Length and Width Levels

Product	Length	Width	Depth	Internal structure	Round	Liking
1	3	3	3	3	3	60
2	3	3	1	3	1	61
9	3	2	2	2	2	50
3	3	1	3	1	1	57
4	3	1	1	1	3	33
11	2	3	2	2	2	34
13	2	2	3	2	2	47

Order of Liking (or Other Criterion Variable), Descending from High to Low. Order of liking is quite popular in research circles. It enables the user to concentrate on the winning products (lying at the top of the table) and to discard the losing products (lying at the bottom of the table). Order of liking also lets the researcher "sweep" down the table, at a glance get an idea of the range of liking, and learn where every product lies against the totality of all other products. There is a tremendous visual impact and potential insight which accompanies a column of decreasing numbers.

Experimental Design and Product Optimization

Table 12 Performance of the 24 Sanitary Napkin Products

Product	Physical variable					Initial liking
	Length	Width	Depth	Internal structure	Round	
15	2	2	2	3	2	62
2	3	3	1	3	1	61
16	2	2	2	1	2	60
1	3	3	3	3	3	60
8	1	1	1	3	3	58
3	3	1	3	1	1	57
6	1	3	1	1	1	57
22	Competitor C					56
20	Competitor A					53
17	2	2	2	2	3	52
9	3	2	2	2	2	50
10	1	2	2	2	2	49
7	1	1	3	3	1	48
13	2	2	3	2	2	47
19	2	2	2	2	2	44
23	Competitor D					43
21	Competitor B					42
12	2	1	2	2	2	40
18	2	2	2	2	1	40
24	Competitor E					36
5	1	3	3	1	3	36
11	2	3	2	2	2	34
4	3	1	1	1	3	33
14	2	2	1	2	2	31

The data in this study were arranged in order of descending liking. Table 12 shows the liking ratings (initial liking/when applied) assigned to the 24 products (means on a 0–100 point scale). The left-most portion of Table 12 shows the physical variables of the product from the experimental design or (where appropriate) of the competitive product. The right-most portion of Table 12 shows the average liking rating.

The remaining data can also be presented in this rectangular fashion. It is probably most instructive and easiest for the user to divide the attributes in a logical fashion:

1. Liking ratings first (in order of appearance, so that appearance precedes fragrance which precedes texture/feel)
2. Sensory ratings second
3. Image and performance ratings third.

Although a detailed discussion of order of attributes might seem a bit pedantic and although every practiced researcher intuitively "knows" the right order, data actually "comes alive" when presented properly. There is nothing more disconcerting than hunting through a table in order to make relevant comparisons. Presenting attributes in order of their appearance on the questionnaire often hinders the analyst's understanding. It is better to group the attributes in logical order, anticipating the user's needs, than to force the user to cope with a wall of numbers.

Establishing Validity—Do Ratings Track Physical Variables?

In experimental designs the researcher can easily demonstrate that the panelists track or do not track the physical changes produced by the changes in the features. The researcher correlates the sensory attribute ratings with the physical variables. The higher the correlation the better the ratings track the physical variations. The correlation statistic varies from a low of 0 (no linear relation between ingredient level and consumer attribute rating) to a high of 1.0 (perfect linear relation) or -1 (perfect inverse linear relation).

If the attribute ratings accurately track the physical changes then the researcher can feel confident about the validity of the data. Table 13 shows the correlation table, and reveals that the panelists can discern the physical differences, albeit to varying degrees.

Creating a Product Model (Equation)

Correlation analysis can only hint at a linear relation between two variables. Correlation analysis neither shows the quantitative nature of that relation, nor accounts for non-linearities, nor allows multiple independent variables to jointly determine the sensory attribute rating. Finally, correlation analysis does not account for synergistic or suppressive interactions between variables. All of these are considerations which require equations that relate independent variables

Table 13 Simple Linear Correlations Between Physical Features and Consumer Sensory Ratings

Sensory	Physical variables				
	Length	Width	Depth	Internal structure	Round
Long	0.62	0.16	0.13	−0.18	−0.01
Thick	0.00	0.05	0.32	0.00	0.02
Wide	0.27	0.49	0.16	−0.01	−0.14

under the developer's control (features, ingredients, process conditions) to dependent variables (consumer sensory, liking, or image ratings, respectively).

Fitting equations to data is a straightforward procedure today. With the widespread use of statistical packages on personal computers, virtually anyone can create equations relating one or more independent variables to a dependent variable. The procedures are known as "regression analysis."

Regression analysis involves the following four steps:

1. *Identify the dependent variables.* Select a specific consumer attribute rating as the dependent variable. The attribute may be a sensory, liking or image attribute. For this case let us use the attribute of initial liking (when first using the napkin).
2. *Select the "cases" or products to be used for the regression analysis.* In our napkin study we will use the 19 physically varied prototypes as the cases. These are the 19 cases or observations.
3. *Select the independent variables.* These will be the five physical features varied by R&D. We will develop several models.

    a. Use only one feature (e.g., Length, L). In this case the equation is:

    $$\text{Initial Liking} = k_0 + k_1(L)$$

    b. Use 2 features, e.g., length (L) and width (W). In this case the equation is:

    $$\text{Initial liking} = k_0 + k_1(L) + k_2(W).$$

    c. Continue developing the equation until all independent physical variables are incorporated in their linear terms. In this case the final linear equation is:

    $$\text{Initial Liking} = k_0 + k_1(L) + k_2(W) + k_3(D) + k_4(S) + k_5(R)$$

4. *Estimate the coefficients $(k_1 \cdots k_5)$ and the additive constant (k_0).* There are standard statistical procedures to do this. Table 14 presents the coefficients of the equation and measures of "goodness-of-fit." Two considerations to keep in mind are:

    a. How well does the equation fit the data? This is shown by the multiple r^2 values for the equation. When multiplied by 100 these multiple r^2 values show the percentage of variation in the consumer attribute ratings accounted for by changes in the predictors (physical variables). As the number of independent variables increases, the multiple r^2 also increases. This is a standard outcome—increasing the number of predictors (even nonsense predictors) increases the proportion of the variability accounted for by the equation. However, with five linear predictors, the equation never predicts more than 16% of the variability ($r^2 = 0.16$, $r = 0.40$).

b. How close is the expected rating to the actual rating? The standard error of estimate is the variability of the actual data around the estimated data, corrected for the number of predictors. Low standard errors of regression mean that the model fits the data accurately. High standard errors of regression mean that the model does not really fit the data well, so that there are discrepancies.

Are Linear Equations Enough?—Interactions, Non-linearities

Quite often the relation between a sensory attribute rating and an ingredient level is not a straight line, but rather a curve. For instance, as the physical feature increases, consumers first perceive the increase in intensity. However, beyond a certain level the same increase in the ingredient fails to increase perceived intensity. The panelist's ratings stop increasing and flatten out. The asymptotic relation may begin anywhere at any feature level.

Simple linear equations (Table 14) cannot account for this "curvilinear" behavior. Furthermore, by using only linear terms but introducing many predictor variables to improve the goodness-of-fit, the researcher misses this asymptotic relation. The equation is far less valuable and not really valid. Non-linearity will be covered up. The loss of validity (e.g., missing a curve where the curve exists) will not be obvious immediately when the researcher fits the data by an equation. However, the loss will become evident when the researcher uses the equation to predict and to develop. The linear equation predicts the attribute to increase or decrease in a linear fashion (at an unchanging, constant rate) with the increasing level of a product feature. The prediction will be wrong. However, by holding all features constant but one, the researcher would soon discover that as he increased the level of one feature, the attribute would initially increase. The increase would then cease after attribute intensity reached a certain level. The equation would predict one pattern, while observation would reveal another pattern.

How can the researcher guard against this error when creating a model? Unfortunately, with the widespread use of statistical packages and the virtually instantaneous estimation of equations (often without much thought given to the mechanism which underlies consumer perceptions), it is difficult to resist the temptation to produce a "better-fitting equation" simply by adding more uncorrelated predictor terms. Over the past 40 years statisticians have been taught that the best equation is the equation which has the least number of terms, and yet accounts for the greatest amount of the variability in the data. The linear equation comprising uncorrelated predictors or independent variables often wins out, because it is parsimonious and simple to understand.

There is, however, a remedy. The remedy requires us to stand back from the data for a moment and hypothesize about the mechanism underlying the perception. For the current problem (creating a model to predict or at least to describe an attribute's intensity) there are these three possibilities:

Table 14 Sequence of Linear Equations Relating "Initial Overall Liking" to the Independently Varied Napkin Features

Multiple r: 0.10		Multiple r^2: 0.01		
		Standard error of estimate: 10.57		
Variable	Coefficient	Std error	T	P(2-Tail)
Constant	45.52	7.11	6.40	0.000
Length	1.40	3.34	0.42	0.680

Multiple r: 0.13		Multiple r^2: 0.02		
		Standard error of estimate: 10.86		
Variable	Coefficient	Std error	T	P(2-Tail)
Constant	43.26	10.03	4.31	0.000
Length	1.40	3.43	0.40	0.680
Width	1.13	3.43	0.33	0.740

Multiple r: 0.14		Multiple r^2: 0.02		
		Standard error of estimate: 11.20		
Variable	Coefficient	Std error	T	P(2-Tail)
Constant	41.98	12.54	3.347	0.004
Length	1.40	3.54	0.396	0.697
Width	1.13	3.54	0.319	0.754
Depth	0.64	3.54	0.181	0.859

Multiple r: 0.35		Multiple r^2: 0.13		
		Standard error of estimate: 10.95		
Variable	Coefficient	Std error	T	P(2-Tail)
Constant	32.94	14.08	2.347	0.035
Length	1.40	3.46	0.401	0.691
Width	1.13	3.46	0.181	0.749
Depth	0.64	3.46	0.182	0.856
Internal Structure	4.52	3.46	1.306	0.213

Multiple r: 0.40		Multiple r^2: 0.16		
		Standard error of estimate: 11.15		
Variable	Coefficient	Std error	T	P(2-Tail)
Constant	37.92	15.98	2.373	0.034
Length	1.40	3.52	0.398	0.697
Width	1.13	3.52	0.321	0.753
Depth	0.64	3.52	0.181	0.859
Internal Structure	4.52	3.52	1.281	0.222
Round	2.49	3.52	0.706	0.493

Pure Linearity. The attribute is really described entirely by a linear equation. That is, each of the five independent variables acts independently to drive perceived intensity. Furthermore, each of the five independent variables acts in a linear fashion, so that the effect of changing ingredient level by a constant amount is the same, no matter what level of the ingredient is being studied. This is the thinking which underlies the linear equations in Table 14.

The Sensory Attribute Perception Follows an Asymptotic Curve. As the ingredient or feature level increases, the attribute rating first increases but then flattens out. In order to accommodate this pattern, the researcher must introduce an additional term into the predictor equation. This term is the square term. The equation is then written as a quadratic function:

Perceived Initial Liking = $k_0 + k_1(L) + k_2(L^2)$ ....

The equation contains linear term (L) and quadratic or square terms (L^2). With five variables with which to contend, there are a variety of alternative equations that could fit the data. All five variables could enter the equation in linear terms. However, there are various combinations of square terms that can enter the equation. Table 15 shows a sample of six of the 1023 different equations that could be developed. [The most complicated equation contains all five linear and all five square terms; the least complicated equation contains only one linear term].

Fortunately for researchers who want to create models, the problem is not as difficult as it might seem. The researcher can develop a linear model incorporating all five linear terms. Then, the researcher can instruct the computer program to identify any square terms which add substantial additional predictability to the data beyond that already achieved. It may turn out that once all of the linear terms are incorporated into the equation, no square term adds any additional predictability beyond that already achieved by using only linear terms.

Table 15 How Adding Additional Quadratic Terms to a Linear Model Increases the Goodness-of-Fit of the Model to the Empirical Data

Term in the equation	Dependent variable = Initial liking of the napkin	
	Multiple r^2	Standard error
L + W + D + I + R	0.16	11.6
Adding L^2	0.22	11.1
Adding W^2	0.26	11.3
Adding D^2	0.27	11.8
Adding I^2	0.60	9.3
Adding R^2	0.60	9.8

Sometimes the researcher has an underlying theoretical reason for using square terms in the equation (even if the square term does not add all that much additional predictability). It is perfectly possible to force in the linear terms and specific square terms (based in theory), and then search for other square terms which increase predictability. The bottom line, however, is that fitting the equation is a straightforward procedure. Choosing the terms that belong in the equation is more of an art, more subjective, and more based on one's experience without the product category and intuition rather than a pure, scientific activity.

Interactions Between Pairs of Variables. Quite often variables interact with each other, beyond the additive effects. The linear equation assumes no interactive effect. Interaction occurs when the effect of one variable is influenced by the level of another variable. The equations below contrast models with and with pairwise interactions between two variables (A,B):

No Interaction: Rating = $k_0 + k_1(A) + k_2(B)$

Interaction: Rating = $k_0 + k_1(A) + k_2(B) + k_3(A \times B)$

The interaction term ($A \times B$) implies that the effect of ingredient A depends both upon the level of A (denoted by the term $k_1 A$) and the level of B which multiplies the effect of A.

Interaction terms ought not be used unless the researcher is confident that there are interactions between variables. This point of view contrasts to that of many researchers who try to fit the equation with any available terms, as long as they are statistically significant. This author believes that the model should comprise the least number of terms possible, not the most. Furthermore, the model need not fit the data "perfectly." There is always a trade-off between the number of predictors used in the equation and the goodness-of-fit. It is also more prudent to sacrifice some degree of predictability in order to insure that the equation is "parsimonious." The more parsimonious the equation the more conservative the prediction.

Table 16(A–D) shows equations for liking and length, respectively, using linear terms, quadratic terms, and a variety of pairwise interaction terms. For the most part the interaction terms add some predictability, but not always and not to every attribute.

Bringing Life to Equations

For most practitioners the equations (such as those in Table 16) have little or no "life" attached to them. Equations come to life when they are used to estimate the consumer ratings that would be assigned to a set of products.

To bring life to an equation, consider the "sensitivity analysis." Sensitivity analysis holds all of the independent variables constant but one, and then varies that one independent variable from low to high, in small steps. For each

Table 16 Comparison of Models for Two Attributes (Initial Liking, Perceived Length)—Models Created Using Linear, Square and Cross Terms

A—Pure Linear Equation

Dependent variable—Initial liking
Multiple r^2	=	0.16
Adjusted multiple r^2	=	−0.17
Standard error of estimate	=	1.16

Regression equation:
Additive constant	38.00
Length	1.41
Width	1.13
Depth	0.64
Internal structure	4.52
Roundedness of edge	−2.49

Dependent variable—Perceived length
Multiple r^2	=	0.44
Adjusted multiple r^2	=	0.22
Standard error of estimate	=	7.57

Regression equation:
Additive constant	12.34
Length	7.08
Width	1.11
Depth	1.49
Internal structure	−2.05
Roundedness of edge	−0.12

B—Quadratic Function: Using Linear and Square Terms

Dependent variable—Initial liking
Multiple r^2	=	0.60
Adjusted multiple r^2	=	0.09
Standard error of estimate	=	9.85

Regression equation:
Additive constant	59.41
Length	−15.66
Width	34.45
Depth	24.74
Internal structure	−58.00
Roundedness of edge	−5.06
Length * Length	4.27
Width * Width	−8.33
Depth * Depth	−6.03
Internal structure * Internal structure	15.63
Roundedness of edge * Roundedness of edge	0.64

Table 16 Continued

Dependent variable—Perceived length
Multiple r^2 = 0.52
Adjusted multiple r^2 = -0.08
Standard error of estimate = 8.89

B—Quadratic Function: Using Linear and Square Terms

Regression equation:
Additive constant	8.77
Length	3.61
Width	7.09
Depth	-5.76
Internal structure	-16.77
Roundedness of edge	23.46
Length * Length	0.87
Width * Width	-1.49
Depth * Depth	1.81
Internal structure * Internal structure	3.68
Roundedness of edge * Roundedness of edge	-5.90

C—Equations Forcing in Linear Terms, Allowing in Significant Quadratic and Cross Terms

Dependent variable—Initial liking
Multiple r^2 = 0.60
Adjusted multiple r^2 = 0.35
Standard error of estimate = 8.33
Regression equation:
Additive constant	100.26
Length	1.41
Width	1.13
Depth	0.64
Internal structure	-49.67
Roundedness of edge	-15.99
Internal structure * Internal structure	10.17
Internal structure * Roundedness of edge	6.75

Dependent variable—Perceived length
Multiple r^2 = 0.63
Adjusted multiple r^2 = 0.39
Standard error of estimate = 6.68
Regression equation:
Additive constant	-19.49
Length	7.08

(continued)

Table 16 Continued

Width	1.11
Depth	9.80
Internal structure	13.87
Roundedness of edge	7.48
Depth * Internal structure	-4.16
Internal structure * Roundedness of edge	-3.80

D—Equations Forcing in Linear and Square Terms, Allowing Significant Cross Terms to Enter

Dependent variable—Initial liking
Multiple r^2	=	0.79
Adjusted multiple r^2	=	0.45
Standard error of estimate	=	7.67

Regression equation:
Additive constant	86.41
Length	-29.16
Width	34.45
Depth	11.24
Internal structure	-58.00
Roundedness of edge	-5.06
Length * Length	4.27
Width * Width	-8.33
Depth * Depth	-6.03
Internal structure * Internal structure	15.63
Roundedness of edge * Roundedness of edge	0.64
Length * Depth	6.75

Dependent variable—Perceived length
Multiple r^2	=	0.71
Adjusted multiple r^2	=	0.14
Standard error of estimate	=	7.95

Regression equation:
Additive constant	-23.06
Length	11.92
Width	7.09
Depth	-5.76
Internal structure	-9.16
Roundedness of edge	39.37
Length * Length	0.87
Width * Width	-1.49
Depth * Depth	1.81
Internal structure * Internal structure	3.68
Roundedness of edge * Roundedness of edge	-5.90
Length * Roundedness of edge	-4.16
Internal structure * Roundedness of edge	-3.80

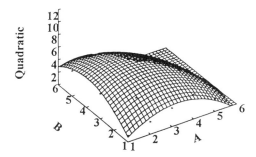

Figure 3 Example of a response surface with an optimum in the middle.

value of the physical variable, the model enables the researcher to estimate the expected profile of attribute ratings. Table 17 shows five sensitivity analyses.

Optimizing the Product for Overall Acceptance

The surface relating formula variables and consumer ratings can be visualized as a hill, similar conceptually to the hill shown in Figure 3. Optimization consists of climbing that hill in order to reach the top. The top of the hill corresponds to the maximum level of an attribute that can be reached. The only caveat is that the optimization algorithm must identify a combination of ingredients or product features that lies within the boundaries tested (interpolation) and may not go outside of these boundaries (extrapolation). It may turn out that the relation describing liking (or other consumer attributes) follows a plane as shown in Figure 4. The optimum for the plane is clearly beyond the highest level tested. Indeed, the plane suggests that were the product developer to keep increasing the physical ingredient level the consumer's rating would continue to increase.

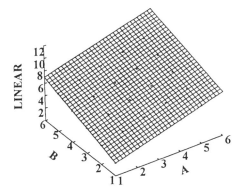

Figure 4 Example of a linear response surface with no optimum.

Table 17 Sensitivity Analysis to Five Features of a Sanitary Napkin—Estimated Rating of All Attributes, When One Variable Increases from Low (1) to High (3) While All Other Variables Are Held Constant

	A—Length						
Length	1.00	1.33	1.67	2.00	2.33	2.67	3.00
Cost of Goods	58	60	63	66	69	73	76
Physical absorbency	56	53	50	49	47	46	46
Initial liking	48	46	45	45	46	48	51
Liking after 2 hours	62	58	55	55	55	58	61
Liking after 4 hours	36	35	35	35	36	37	39
Final liking	30	29	28	28	29	30	31
Purchase	39	37	36	36	37	38	40
Length	22	24	26	28	30	33	36
Thickness	39	39	39	39	39	39	39
Width	40	41	43	45	47	49	52
Like length	40	48	52	54	54	51	45
Like width	77	64	57	55	57	65	78
Like thickness	39	44	48	50	50	48	44
Fit concept 1	30	32	33	35	36	38	39
Fit concept 2	38	40	42	43	43	43	42
Fit concept 3	48	48	47	48	49	51	53
Like removal	26	30	35	40	46	53	59
Like application	57	52	49	48	50	54	60
Unique	26	29	31	33	34	34	33
Absorbs	52	52	52	52	53	54	56
For heavy days	39	39	40	41	41	42	43
For light days	62	64	66	69	72	75	78
For me	33	34	36	37	39	42	44
Secure	59	60	61	62	62	63	64
Discreet	71	68	66	66	68	70	75
Fits well	44	43	43	44	45	46	48
	B—Width						
Width	1.00	1.33	1.67	2.00	2.33	2.67	3.00
Cost of goods	62	63	65	66	67	68	69
Physical absorbency	44	46	47	49	50	52	53
Initial liking	36	41	44	45	45	42	38
Liking after 2 hours	42	48	53	55	55	53	50
Liking after 4 hours	29	32	34	35	35	34	33
Final liking	23	26	27	28	28	27	26
Purchase	29	33	35	36	36	35	32
Length	25	27	27	28	28	28	28
Thickness	27	31	35	39	43	46	49
Width	37	41	43	45	46	46	45
Like length	66	61	57	54	52	52	52
Like width	37	47	52	55	54	49	42
Like thickness	43	46	48	50	50	51	50

Table 17 Continued

Width	B—Width						
	1.00	1.33	1.67	2.00	2.33	2.67	3.00
Fit concept 1	28	31	33	35	36	37	38
Fit concept 2	39	40	42	43	44	45	45
Fit concept 3	44	46	48	48	47	46	43
Like removal	35	38	40	40	41	40	38
Like application	40	45	47	48	47	45	40
Unique	33	33	33	33	33	32	32
Absorbs	35	43	48	52	53	53	50
For heavy days	37	38	40	41	41	42	42
For light days	63	65	67	69	69	70	69
For me	33	35	37	37	37	37	35
Secure	55	58	60	62	63	65	66
Discreet	57	61	64	66	68	69	69
Fits well	45	44	44	44	45	46	48

Depth	C—Depth						
	1.00	1.33	1.67	2.00	2.33	2.67	3.00
Cost of goods	60	63	65	66	67	68	68
Physical absorbency	45	46	47	49	50	52	55
Initial liking	38	42	44	45	45	43	40
Liking after 2 hours	49	52	54	55	54	53	50
Liking after 4 hours	31	33	34	35	35	35	34
Final liking	25	27	28	28	28	27	26
Purchase	32	34	35	36	36	35	34
Length	28	28	28	28	29	30	31
Thickness	34	34	35	36	37	38	39
Width	43	43	44	45	45	46	46
Like length	46	51	54	54	52	48	42
Like width	40	48	53	55	53	49	41
Like thickness	49	51	51	50	48	44	40
Fit concept 1	35	35	35	35	35	36	37
Fit concept 2	40	42	43	43	43	42	40
Fit concept 3	41	45	47	48	47	44	40
Like removal	39	39	40	40	41	41	42
Like application	39	44	47	48	47	44	39
Unique	31	32	33	33	32	31	30
Absorbs	53	53	52	52	52	51	51
For heavy days	40	40	40	41	41	41	41
For light days	70	69	68	69	70	71	73
For me	34	36	37	37	37	36	35
Secure	57	58	60	62	63	64	65
Discreet	60	63	65	66	67	67	66
Fits well	41	42	43	44	44	44	43

(*continued*)

Table 17 Continued

	D—Internal Structure						
Internal Structure	1.00	1.33	1.67	2.00	2.33	2.67	3.00
Cost of goods	67	66	65	66	68	71	76
Physical absorbency	46	46	47	49	51	53	56
Initial liking	56	49	45	45	48	55	65
Liking after 2 hours	67	60	55	55	57	63	73
Liking after 4 hours	39	36	35	35	37	40	45
Final liking	33	30	28	28	29	32	37
Purchase	43	38	36	36	38	42	48
Length	34	31	29	28	28	28	30
Thickness	39	39	39	39	39	39	39
Width	53	48	46	45	46	48	52
Like length	72	63	57	54	55	60	68
Like width	59	58	56	55	53	52	51
Like thickness	56	52	50	50	51	55	61
Fit concept 1	39	37	36	35	35	36	38
Fit concept 2	51	47	44	43	44	46	50
Fit concept 3	58	53	49	48	48	50	54
Like removal	47	43	41	40	41	44	48
Like application	56	52	50	48	48	47	48
Unique	38	35	33	33	33	35	38
Absorbs	50	51	52	52	52	51	50
For heavy days	43	42	41	41	41	41	42
For light days	74	72	70	69	67	66	65
For me	43	40	38	37	38	39	42
Secure	63	63	62	62	61	60	60
Discreet	66	65	65	66	68	71	75
Fits well	45	45	45	44	42	40	36
	E—Roundedness						
Round	1.00	1.33	1.67	2.00	2.33	2.67	3.00
Cost of goods	63	65	66	66	66	64	62
Physical absorbency	44	47	48	49	47	45	41
Initial liking	48	47	46	45	44	44	43
Liking after 2 hours	54	55	55	55	53	50	47
Liking after 4 hours	35	35	35	35	34	33	31
Final liking	29	29	28	28	27	27	26
Purchase	37	37	36	36	35	34	33
Length	22	25	27	28	27	25	22
Thickness	38	39	39	39	39	39	39
Width	41	44	45	45	44	42	38
Like length	57	56	55	54	54	54	54
Like width	65	60	56	55	55	58	63

Experimental Design and Product Optimization 267

Table 17 Continued

Round	E—Roundedness						
	1.00	1.33	1.67	2.00	2.33	2.67	3.00
Like thickness	66	58	53	50	49	50	53
Fit concept 1	31	33	35	35	34	33	31
Fit concept 2	42	43	43	43	43	42	41
Fit concept 3	51	49	49	48	47	47	47
Like removal	37	39	41	40	39	37	33
Like application	41	43	45	48	52	56	61
Unique	40	37	35	33	31	29	27
Absorbs	58	55	53	52	52	53	54
For heavy days	39	40	40	41	40	40	39
For light days	67	68	68	69	68	67	66
For me	34	35	37	37	38	38	38
Secure	63	62	62	62	61	61	61
Discreet	62	65	66	66	65	63	59
Fits well	44	44	44	44	44	44	45

By using one of several different "optimization" algorithms (e.g., method of steepest ascent, grid search, complex method, etc.), the researcher can discover the top of the hill. The top of the hill may not necessarily be the absolute maximum because of ingredient limits. The expected true optimum may lie somewhere beyond the range of ingredient levels tested in the study. However, the conventions of optimization allow only interpolation so that the optimum formulation must lie within the boundaries of ingredients tested.

Table 18 shows the optimum products for two liking ratings (initial, final). Each of these two products corresponds to a combination of independent variables that maximizes an attribute within the range of levels tested. Sometimes the optimal level of a feature lies at the boundary of the tested range, suggesting that were the product developer to expand the range, the feature or variable might have achieved an even higher level. In other cases one or more of the variables optimizes at a middle level, suggesting that the researcher has adequately covered (or bracketed) the range of the independent variable (at least for that attribute).

Estimating the Attribute Profile Corresponding to the Optimum

The model (set of equations) enables the researcher to estimate the likely rating profile corresponding to the optimum product. The researcher need only substitute the levels of the optimum formulation into the set of equations. Each equation will generate its own prediction of the relevant attribute level. Table 18 also shows the attribute profile corresponding to the optimum.

Table 18 Optimal Features for Two Attributes—Like on Initial Wearing, and Like After Wearing

Goal: Constraint	Maximize initial like None	Maximize final like None	Final-initial
Feature level			
Length	3.00	3.00	0.00
Width	2.10	2.16	0.06
Depth	2.63	2.14	−0.49
Internal	3.00	3.00	0.00
Round	1.00	2.71	1.71
Objective measures			
Cost	83	85	2
Physical absorbency	58	54	−4
Liking			
Initial	⬚76⬚	70	−6
After 2 hours	82	76	−6
After 4 hours	51	47	−4
Final	37	⬚40⬚	3
Purchase	55	51	−4
Sensory			
Length	41	29	−12
Thickness	39	41	2
Width	61	54	−7
Sensory liking			
Length	65	59	−6
Thickness	75	49	−26
Width	71	76	5
Fit to concept			
Concept 1	46	38	−8
Concept 2	44	51	−7
Concept 3	55	60	5
Performance			
Application	44	67	23
Removal	69	61	−8
Unique	42	36	−6
Absorbs	56	55	−1
For heavy days	45	43	−2
For light days	82	69	−13
For me	45	50	5
Secure	66	63	−3
Discreet	83	81	−2
Fits well	40	41	1

To the consumer researcher the sensory profile is often far more instructive than the optimal physical formulation. The consumer researcher can easily compare the sensory profiles to determine how the optimal formulations will differ. Table 18 compares the two optimum formulations (initial and after wearing) and shows how these two products differ from each other. The products certainly differ on sensory attributes.

Optimizing a Product Subject to Cost Constraints

Cost of goods is a perennial concern for product developers. In the typical course of events, manufacturers first attempt to create an optimal product. However, cost of goods soon enters the consideration. Sometimes the product is simply too expensive to produce. Despite high acceptance, the manufacturer often may not be able to price the product sufficiently high in order to justify the high cost of goods. In this case there may be an alternative formulation that is almost as acceptable as the optimum, but has a substantially lower cost of goods.

Optimization can incorporate "objective" constraints such as costs. We can conceive of this "constrained optimization" as climbing up a mountain (corresponding to liking) and, for each alternative product configuration on this mountain, estimate the cost of goods corresponding to the product. At some point the product acceptance can be improved, but only by exceeding a pre-set upper bound. The cost constraint is a plane which intersects the liking surface. All points under the cost plane are feasible; they have a total product cost less than the decreed upper limit on costs. All points above the plane have a total cost more than the decreed upper limit. The optimum lies at the highest point on the liking surface which also lies beneath the cost plane.

Table 19 shows a set of optimal products at reduced costs. We can see the effect of imposing a cost constraint by optimizing overall liking, subject to a series of increasingly restrictive limits on total product cost. At first we find little loss in liking. In fact, if the optimum formulation lies at a middle level of product features, then we can set the cost constraint so high that the optimal product is less costly than the cost constraint. However, as we decrease the allowable cost of goods we soon must decrease the level of the features. The optimization algorithm searches for the best product while still obeying the cost constraint. The level of each feature drops, but not necessarily in the same proportion. Eventually, with increasingly severe cost limits on the product, the product is forced to lower and lower levels of features (below the optimum), and ever lower levels of overall liking.

Imposing Sensory Constraints on the Product

Cost of goods is only one attribute that can be imposed on the product formulation. Another use of constrained optimization is to create a product with de-

Table 19 Optimizing Napkin Features (Initial Liking) Subject to Cost Constraints

Constraint	None	Cost <75	Cost <60	Cost <50
Physical features				
Length	3.00	1.00	1.00	1.00
Width	2.06	2.08	1.43	1.91
Depth	2.62	1.48	1.48	1.29
Internal	3.00	3.00	2.98	1.00
Round	1.00	1.00	1.00	1.00
Objective measures				
Cost	$\boxed{83}$	$\boxed{63}$	$\boxed{60}$	$\boxed{50}$
Physical absorbency	58	57	46	51
Liking				
Initial	$\boxed{76}$	$\boxed{73}$	$\boxed{69}$	$\boxed{63}$
After 2 hours	82	81	76	75
After 4 hours	51	46	44	40
Final	37	35	33	36
Purchase	55	52	49	46
Sensory				
Length	41	17	16	13
Thickness	39	40	32	38
Width	61	40	36	39
Sensory liking				
Length	65	45	51	70
Thickness	75	76	72	53
Width	71	73	67	75
Fit to concept				
Concept 1	45	30	26	26
Concept 2	44	47	44	47
Concept 3	55	58	57	61
Performance				
Application	44	43	40	47
Removal	69	26	23	24
Unique	42	40	42	40
Absorbs	56	58	50	52
For heavy days	45	38	36	38
For light days	82	56	53	58
For me	45	33	31	33
Secure	66	56	53	59
Discreet	83	70	65	66
Fits well	40	35	35	43

Experimental Design and Product Optimization   271

fined sensory characteristics. The sensory characteristics can themselves act as constraints. For instance, Table 20 shows three optimum products, varying in perceived thickness. Each of these three products is generated by imposing one or more sensory constraints on the product. The product optimizer then uncovers the combination of physical features which simultaneously obeys the constraints and at the same time maximizes overall liking.

Using Performance and Image Characteristics

Image and performance attributes can be either optimized or can act as constraints. If the researcher wants to optimize the product, then one of the better strategies is to maximize the attribute rating while forcing overall liking to achieve at least a pre-defined level. The approach rapidly identifies a product that achieves the desired image profile, while achieving an acceptable level of liking. Table 21 shows three different optimal products, each maximizing an image or performance attribute.

Table 22 shows the same approach, using the attribute of "fit-to-concept." By optimizing for an image but insuring that liking exceeds a specific minimum level, the product developer and marketer insures that the product meets both the technical and acceptance requirements.

Constraints—How Far Can the Developer Push the System?

It may appear from the foregoing discussion that all the product developer needs to do is to supply the objectives for the product optimizer, press a button, and wait for the formulation and the expected rating profile to emerge from the system. To some extent the approach certainly is mechanical. Once the product developer and researcher have created the products, tested them, and developed models, much of the hard work has been done. It becomes easy to identify the optimal product, subject to single imposed constraints on an attribute.

In the real world of product development, however, matters are not quite as simple. Often there are at least four types of constraints that the product must satisfy:

1. The actual range of the independent variables may be narrower in product than the range that was tested. The optimal formulation may lie in a range that one later discovers to be out of the feasible boundary for the current production technology.
2. The product must achieve a cost of goods lower than the lowest feasible combination achievable with these particular product features. Perhaps during the development phase the unit cost of goods appeared to be reasonable, whereas upon later reflection it appeared that these unit costs were too high. One can set a cost constraint so low that no combination

Table 20 Optimizing Initial Liking Subject to Sensory Constraints on Perceived Thickness

Constraint	None	Thick <32	Thick <29	Thick <26
Physical features				
Length	3.00	3.00	3.00	3.00
Width	2.06	1.46	1.17	1.01
Depth	2.61	2.66	2.61	2.62
Internal	3.00	3.00	3.00	3.00
Round	1.00	1.00	1.00	1.00
Objective measures				
Cost	83	80	79	78
Physical absorbency	57	48	43	40
Liking				
Initial	[76]	[73]	[70]	[67]
After 2 hours	82	77	72	69
After 4 hours	51	49	47	45
Final	37	35	33	32
Purchase	55	52	50	48
Sensory				
Length	41	40	39	39
Thickness	39	[32]	[28]	[26]
Width	61	58	55	53
Sensory liking				
Length	65	71	75	77
Thickness	75	72	70	69
Width	71	64	59	54
Fit to concept				
Concept 1	45	42	40	38
Concept 2	44	41	40	39
Concept 3	55	54	53	52
Performance				
Application	44	41	38	36
Removal	69	67	65	63
Unique	42	39	39	38
Absorbs	56	49	43	39
For heavy days	45	43	42	41
For light days	82	79	77	76
For me	45	43	41	40
Secure	66	63	60	59
Discreet	83	78	75	74
Fits well	40	40	40	41

Table 21 Optimizing Napkin Features on Performance (with Initial Liking >58)

Maximize liking	Initial liking	For heavy days	For light days	Feel secure
Physical features				
Length	3.00	3.00	3.00	3.00
Width	2.06	3.00	2.67	3.00
Depth	2.62	3.00	3.00	3.00
Internal	3.00	1.00	1.00	1.00
Round	1.00	1.57	1.83	1.00
Objective measures				
Cost	83	85	84	82
Physical absorbency	58	38	41	34
Liking				
Initial	|76|	|58|	|61|	|59|
After 2 hours	82	71	75	70
After 4 hours	51	44	46	44
Final	37	33	34	35
Purchase	55	46	47	46
Sensory				
Length	41	44	45	39
Thickness	39	47	45	47
Width	61	63	63	62
Sensory liking				
Length	65	32	31	34
Thickness	75	42	42	49
Width	71	44	50	53
Fit to concept				
Concept 1	45	48	48	46
Concept 2	44	48	48	45
Concept 3	55	50	53	50
Performance				
Application	44	38	45	35
Removal	69	67	68	66
Unique	42	41	39	44
Absorbs	56	53	55	58
For heavy days	45	|48|	47	47
For light days	82	89	|89|	87
For me	45	45	47	42
Secure	66	74	72	|74|
Discreet	83	71	71	67
Fits well	40	52	50	52

Table 22 Optimal Napkin Features that Maximize "Fit to a Concept" Within Constraints (Must Achieve a Minimum Level of Initial Liking)

Goal maximize	Initial liking	Fit to concept 2	Fit to concept 2	Fit to concept 3	Fit to concept 3
Constraint Minimum initial Liking > 65	None	No	Yes	No	Yes
Physical features					
Length	3.00	2.67	2.72	3.00	1.00
Width	2.10	3.00	2.64	1.88	1.90
Depth	2.63	2.04	2.05	2.01	1.96
Internal	3.00	1.00	3.00	1.00	3.00
Round	1.00	2.55	2.44	3.00	1.00
Objective measures					
Cost	83	76	84	74	63
Physical absorbency	58	33	64	37	57
Liking					
Initial	☐76☐	☐51☐	☐65☐	☐60☐	☐71☐
After 2 hours	82	62	73	66	79
After 4 hours	51	38	46	39	46
Final	37	29	38	31	36
Purchase	55	40	48	43	51
Sensory					
Length	41	37	31	35	17
Thickness	39	49	46	37	38
Width	61	54	56	49	40
Sensory liking					
Length	65	65	61	63	57
Thickness	75	59	54	62	75
Width	71	58	59	90	82
Fit to concept					
Concept 1	46	44	41	39	29
Concept 2	44	☐54☐	☐52☐	51	47
Concept 3	55	56	55	☐65☐	☐60☐
Performance					
Application	44	59	56	79	49
Removal	69	53	58	55	27
Unique	42	36	37	35	41
Absorbs	56	51	53	55	55
For heavy days	45	45	44	42	38
For light days	82	80	70	80	55
For me	45	46	47	51	34
Secure	66	69	64	65	58
Discreet	83	71	81	66	74
Fits well	40	52	41	49	36

of variables within the range tested can ever be low enough to satisfy the constraints. The optimum product is physically not realizable.
3. The product must possess a sensory attribute profile that is infeasible. A single sensory attribute may be infeasible because no product in the test study ever achieved a level even near that sensory constraint. For instance, perceived thickness may be a constraint. The researcher may want a perceived thickness below 10 to support a marketing position. However, all the products in the study were assigned ratings of perceived thickness between 30 and 65. In this case it is unlikely that a product can be found within the range tested that would achieve a rating of 10 or lower on perceived thickness. That level was simply not realized in the study. It would be wishful thinking to assume that one could achieve that level by simply varying the product features within the range tested.
4. Although one constraint may be obeyed (e.g., insure that the product cost lies within stipulated limits), two or more constraints may be mutually incompatible. One constraint may permit a bounded range of ingredients. Another constraint may permit a different, bounded range of ingredients. The two ranges do not intersect at all. Consequently, satisfying one constraint may actually *insure* that the other constraint cannot be satisfied.

All too often in the development and marketing communities the business objectives are formalized as constraints. Without experimental designs and product models, a product developer might work for several months or even years in order to create a product which satisfies consumer acceptance and satisfies the imposed constraint. However, when the developer is saddled with a second constraint on top of the first, the developer is forced back to the bench, where he may spend several weeks or months fine-tuning the already optimized product in order to satisfy the newly imposed constraint. In many cases there will be no product which satisfies both constraints simultaneously. Without the benefit of an experimental design and a model, it will not be clear why the developer failed to create a product which is acceptable and which satisfies both constraints. The developer returns to the bench again and again until it becomes patently and painfully clear that no product will fulfill the requirements. In contrast, with the product model it can be very easily demonstrated ahead of time that the two constraints are mutually incompatible, and that to break the impasse, one must relax or even abandon both constraints.

3. Optimization and Reverse Engineering—Turning the Problem Around 180 Degrees

Up to now we have treated optimization as a one-directional process. We have set an objective (e.g., maximize overall liking) and imposed constraints (e.g., insure the cost of goods is maintained at a sufficiently low level). Let us turn the problem around 180 degrees. We will specify a goal or profile of attributes to be realized and search for that combination of ingredient variables under the developer's control which reproduces that profile of attributes. This is called "goal fitting."

The four premises of "goal fitting" are straightforward and in many ways similar to the premises of basic product optimization:

1. The independent variables are statistically independent of each other.
2. There is a mathematical equation relating the independent variable to the response attribute. Each attribute is described by its own equation.
3. The algorithm assesses different combinations of independent variables, attempting to match the expected profile and the goal profile. Many different combinations of independent variables are tried. For each combination the algorithm computes a measure of "badness-of-fit" (sum of absolute differences, or sum of ratios or percentages of deviations). By varying the independent ingredients in a systematic fashion, always attempting to minimize the "badness-of-fit," the algorithm eventually arrives at a set of product features which yields a product whose profile is as close as possible to the goal profile. There may be one or several different products which satisfy this criterion. A match is defined as the minimum absolute difference (or ratio) between the goal profile that is designated as the target (by the researcher) and the estimated profile to be generated by a set of levels of the independent variables.
4. One can impose constraints on the system. The constraints correspond to sensory attribute levels (or image or liking levels) beyond which the independent variables cannot extend. For instance, one may specify a sensory profile to act as an independent profile, while at the same time insuring that the liking level corresponding to the fitted profile lies above a specified level. (This type of optimization with constraints might be the case when one wishes to copy a competitor's product within one's formulation system, while at the same time insuring that the product is adequately acceptable on an absolute basis.)

Using Reverse Engineering

Let us take the product model for napkins and discover the set of independent variables which yields a match to a desired or goal profile. This profile may come from at least three different sources.

1. At the most fundamental level the profile may belong to a competitor's product. The manufacturer may wish to create a product matching the competitor's, using the manufacturer's own features. How close can the match be?

Experimental Design and Product Optimization

2. Alternatively, the manufacturer may wish to substitute a new set of features or variables for the current set, yet insure that the sensory profile of the revised product matches that of the original product as closely as possible. (This strategy can be looked at as matching a competitor, with the competitor being the manufacturer's own original product.)
3. A third use is to create a product which delivers a desired image profile. The product model relates the ingredients or features to the subjective image characteristics. One need not profoundly understand the meaning of such characteristics in order to create products with given image profiles. As long as consumers differentiate products from each other on the basis of the image attributes, there should be some equation relating the independent variables (ingredients) to the image attributes. The researcher can specify an image profile and adjust the independent variables in order to insure that the product delivers an image profile as close as possible to that which was promised.

Table 23 presents two different sets of marketing objectives to satisfy, one for "heavy days" (Table 23A), and one for "light days" (Table 23B). There are a variety of constraints to obey.

1. Not every goal can be satisfied. Some goals can be more easily satisfied than other goals.
2. If the goal lies outside the limits achieved by the product formulation, then it is unlikely that the reverse engineering approach will discover a combination of ingredients that delivers the specified "out of bounds goal." It is more likely that the matching formulation will deliver a product whose profile on image attributes lies within the range achieved by the test products. In other words, what the researcher puts in as test products will dictate what the researcher will get out in terms of optimal products. If the test prototypes do not deliver the appropriate image perceptions, then no amount of optimization analysis will create this image perception. The goals are simply out of bounds and not physically realizable.

Introducing Constraints into the Project

Quite often the optimum product defined as being of "closest fit" does not achieve the requisite level of acceptance. This may occur for two reasons:

1. The profile to be matched is not sufficiently acceptable in and of itself. Consequently, a match to that profile will not be acceptable either.
2. The new levels of features or ingredients introduce an unexpected side effect—e.g., some newly emergent sensory attributes appear that were not

Table 23 Examples of Reverse Engineering

	A. Product 1—For Heavy Days, Must Have Physical Absorbency					
	#1		#2		#3	
Physical features						
Length	1.43		1.00		1.05	
Width	1.31		1.32		1.24	
Depth	2.34		2.77		1.97	
Internal	3.00		3.00		3.00	
Round	2.33		1.39		2.07	
Goals to satisfy	Expected	Goal	Expected	Goal	Expected	Goal
Physical absorbency	50	50	50	50	50	50
Length	24	60	40	60	22	60
Thickness	31	30	30	30	30	30
For heavy days	39	60	43	60	38	60
For light days	57	40	77	40	54	40
Constraints to obey						
Cost	None		None		<65	
Initial liking	None		>65		None	
Physical						
Cost	67		81		64	
Liking						
Initial	58		69		62	
After 2 hours	65		75		71	
After 4 hours	40		48		42	
Final	35		35		35	
Purchase	43		50		46	
Sensory						
Length	24		40		22	
Thickness	31		30		30	
Width	45		57		43	
Sensory liking						
Length	73		73		62	
Thickness	47		59		46	
Width	50		49		60	
Fit concept						
Concept 1	31		41		29	
Concept 2	44		43		42	
Concept 3	50		52		52	

Experimental Design and Product Optimization 279

Table 23 Continued

A. Product 1—For Heavy Days, Must Have Physical Absorbency			
	#1	#2	#3
Performance			
Application	50	38	51
Removal	37	65	31
Unique	31	36	31
Absorbs	39	42	39
For heavy days	39	43	38
For light days	57	77	54
For me	37	43	35
Secure	56	61	53
Discreet	71	80	72
Fits well	36	39	36

B. Product 2—For Light Days, Must Have Physical Absorbency				
	#4		#5	
Physical Features				
Length	2.29		2.46	
Width	1.00		1.00	
Depth	2.90		2.44	
Internal	1.78		1.04	
Round	1.70		1.19	
Goals to satisfy	Expected	Goal	Expected	Goal
Physical absorbency	50	50	50	50
Length	30	30	30	30
Thickness	26	20	27	20
For heavy days	38	40	40	40
For light days	70	70	71	70
Constraints to obey				
Cost	No		No	
Initial liking	No		>50	
Physical				
Cost	66		68	
Liking				
Initial	34		50	
After 2 hours	42		56	
After 4 hours	30		35	
Final	23		31	
Purchase	29		38	
Sensory				
Length	30		30	
Thickness	26		27	
Width	41		47	

(continued)

Table 23 Continued

B. Product 2—For Light Days, Must Have Physical Absorbency

Goals to satisfy	Expected	Goal	Expected	Goal
Sensory liking				
Length	54		73	
Thickness	38		51	
Width	26		48	
Fit concept				
Concept 1	31		32	
Concept 2	37		44	
Concept 3	40		54	
Performance				
Application	30		40	
Removal	42		50	
Unique	27		40	
Absorbs	35		39	
For heavy days	38		40	
For light days	70		71	
For me	32		38	
Secure	60		61	
Discreet	56		55	
Fits well	45		47	

expected. Such unexpected effects often occur in product development when one tries to match a current profile using new ingredients.

In order to insure sufficient acceptability (or conformance to another constraint, such as cost of goods) the manufacturer or developer might wish to incorporate constraints. These constraints will push the optimal formulation away from the region of closest fit with the target or goal profile. However, in doing so those constraints will also insure that the product meets with sufficient acceptance, enjoys a sufficiently low cost of goods, or has a sensory profile that can be used in advertising as a reinforcer of some unseen product benefits. Tables 23 A–B also show constrained optima using the goal-fitting procedure.

4. Optimizing Products to Satisfy Multiple Consumer Groups: A Saturated Category—Shampoos

Introduction

Up to now we have been dealing with different attributes as objectives to optimize. We can expand the score of optimization by considering the response of multiple consumer groups, whether these be consumers of different ages, consumers who use different brands, or even consumers who differ on the basis of the characteristics of products that they hold to be important (the "sensory preference segments").

Given this state of affairs, how does the manufacturer develop products which appeal to several subgroups of consumers simultaneously, rather than a product which appeals to only one subgroup? There are many cases where a single product has to be "targeted," either to one subgroup of consumers, or to several subgroups simultaneously. Here are four scenarios which might conceivably occur.

1. The marketer may wish to create a single product which appeals to the total panel of consumers, without paying attention to the acceptance ratings assigned by individuals or by defined subgroups of the population. That is, the manufacturer simply wishes to achieve as high a rating as possible for the total panel.
2. The marketer may wish to create a product which appeals only to one specific group of consumers. It may be that the market is so crowded that there is no need for a product appealing to the entire group of consumers. (Even more importantly, it may be impossible to achieve a strong breakthrough product in the face of segmented preferences.) A strategy of appealing to one segment alone may be more advantageous than a strategy of appealing to the entire population. When consumers are segmented on the basis of their preferences it is often easier to target a product to one group of individuals with homogeneous preferences.
3. The marketer may wish to create a product which appeals to the total panel, but does not offend any defined person or subgroup. At the most general level, this leads to a product which appeals to all consumers, or at least does not offend any individual consumer. That general case is virtually impossible to achieve. It is easier to create a product which appeals to the total panel, and is also acceptable to defined subgroups of consumers in the population. There may be two preference segments in the population with different likes/dislikes. The objective may be to attract the total population, while not alienating either segment.
4. The marketer may wish to create a product such as that created in step 3 above (appealing to different subgroups), while at the same time insuring that the product differs from competition in a specified sensory manner. The specified sensory difference enables the marketer to use product characteristics in advertising and positioning as a reinforcer of product quality.

Case History Specifics

Alexis Inc. manufactures shampoos. There are hundreds of different brands and line items in the market, with each month seeing new entries. Over time, consumer preferences change as well. With each additional month of market data coming in from syndicated services, top management at Alexis Inc. recognized that if they were to become and remain a major player in the field, they would have to innovate. As important, however, was innovation within the context of an organizing principle. There had to be some underlying rationale for developing new products and launching them. The increasing cost of shelf space in the marketplace, the ever increasing "slotting allowances" (fees charged by the trade to stock one's product), and the ongoing frustration of competing in a changing market led to some profound rethinking by top management. It was no longer a case of "throwing the products at the wall and seeing which products stick."

The most logical approach seemed to be sensory preference segmentation and the creation of products to fit those segments. The idea was simple— understand the true dynamics of the shampoo market in terms of sensory preferences, and use those dynamics to create products targeted to the consumer. By knowing the segments, Alexis Inc. would become the smartest company in the category (at least in terms of knowing what consumers wanted) and the smartest company in terms of product development. Management and R&D at Alexis Inc. would know what consumers wanted at a subjective level, and how to create physical formulations to satisfy those subjective preferences. Furthermore, it appeared possible to engineer different products in order to satisfy consumer needs/wants. At the same time this strategy would sidestep competition by altering the products so that the products would differ from competitor products by features that one could "point to" (the so-called point-at-ables).

Developing the Products and the Experimental Design

The initial steps in the shampoo project were to identify the physical features under R&D control. These comprised various colors, fragrances, surfactants, etc. An initial assessment of the potential number of prototypes revealed thousands of possible products that one could create. Indeed, with each new month there came reports of still other shampoo ingredients used in one or another product (usually small in terms of total volume). These special ingredients provided benefits to make the product somewhat unique.

The enlightened R&D management at Alexis Inc. recognized that they would have to screen down the many ingredients to a limited number. Furthermore, they wanted to explore experimental design and response surface analysis in a way that would be productive. These requirements led to the following five decisions.

1. There would be no more than seven variables in total. More than seven variables would yield too many combinations. By the judicious use of programs to create experimental designs, R&D developed a design requir-

Experimental Design and Product Optimization 283

ing 35 prototypes. Although management at Alexis Inc. first balked at the idea of creating so many prototypes, it soon became obvious that the learning from this study would be sufficiently valuable to provide guidance for several years to come.

2. The colors had to arise from a combination of three dyes which in combination generated a large number of different end colors. Rather than working with different colors from different dyes, R&D management opted to create the colors by mixing a limited number of dyes. Thus, R&D could re-create new colors by remixing the dyes in different combinations. The product model for shampoo would enable them to simulate many (albeit not all) different colors.

3. The fragrance notes had to arise from a combination of three keys for the same reason. Although it was tempting to solicit the help of many fragrance houses, each of which would submit a different fragrance, R&D management wanted to create a general shampoo model that they could use to develop products, and to maintain control over cost of goods. To do so required that they blend the fragrance keys, and that the end fragrance be under corporate control, not the fragrance supplier's control.

4. There was a special surfactant (Ingredient Q) of great interest to the R&D Director at Alexis Inc., who wanted to explore its performance in the product. Therefore, the product developer opted to include this special surfactant at three levels.

5. The experimental design for the shampoos was by its very nature highly fractionated. That is, within the limited budget available, R&D had to develop and test a reasonable number of prototypes to cover the range of ingredients, but not be so thorough as to make the test unaffordable. Often there are major disagreements about just how complete a design should be, with one party insisting upon perfection and a complete design, and the other party bringing to bear a more easygoing attitude and a recognition that often the perfect must be sacrificed for the sake of the good. In this study the second, more rational and real world orientation won out, so that the design was highly fractionated, but did cover the seven variables quite adequately.

Results from the Study

As is the case for many of the other studies in this book, the basic data comprises a product × attribute matrix. The key analytic subgroups are gender, frequency of using shampoo, age, and sensory preference segmentation. The sensory preference groups were developed using the approach described in Chapter 6. The two groups that emerged revealed different ways that consumers could approach the shampoo category. One segment was more "sensory impact" oriented, looking for darker, more intense products. We can label these individuals the "experiential-oriented" consumers. The other segment was a more bland, but efficacy-oriented consumer, less interested in the sensory impact. These were non-experiential consumers.

Table 24 shows the results of optimizing for the total panel regardless of the subgroup acceptance (#1), for each sensory segment (#2, #3), for total panel but insuring that both segments find the product highly acceptable (#4), and for the total panel, both segments satisfied, but with a darker color that enables the marketer to call attention to the product and use the new color as a differentiator (#5). As column #1 shows, optimizing total panel acceptance does not insure that both sensory preference segments will find the product highly acceptable. Columns #2 and #3 show that concentrating on satisfying one sensory preference segment may polarize acceptance and offend the other sensory segment. Finally, Columns #4 and #5 show that it is possible to optimize overall liking for the total panel, to insure that the two segments each like the product very much and, if necessary, to modify the color of the product (a non-critical variable) in order to create a leverageable difference.

5. An Overview

This chapter has presented a new but increasingly popular approach to product development. In the traditional back and forth approach to development the developer might spend weeks or months creating a product, only to learn that the product was either not sufficiently acceptable or was acceptable but differed from what the marketer had in mind. The traditional approach required tedious hours of development work, and could be undone in a moment by results from consumer testing. In contrast, experimental design and optimization make the best use of the developer's valuable time. The developer needs to create a limited number of prototypes, albeit systematically related to each other. From these prototypes, one quickly understands how that which is under the developer's control generates consumer responses. More important, it is possible to identify a set of prototypes within the range tested (but not themselves necessarily tested) which will achieve a desired result. That result could be maximal acceptance, maximal acceptance subject to cost constraints, or fit to a concept.

The 1990's are witnessing an acceleration of product development like no other in history. The risks of market failure are higher, the competition is increasingly severe, and the difficulties of creating truly new, breakthrough products on a consistent basis are becoming apparent. Old ways of developing products must yield to new ways. The developer must use the new tools available to him. The traditional abhorrence of computers and mathematics in favor of insight and creativity must pass. The risks are great, and the opportunities

Experimental Design and Product Optimization

Table 24 Optimal Formula Levels for Shampoo Ingredients

Objective	#1	#2	#3	#4	#5
Maximize subgroup	Total panel	Segment 1	Segment 2	Total panel	Total panel
Liking rating:					
Constraint 1:				Seg 1>69	Seg 1>69
Constraint 2:				Seg 2>69	Seg 2>69
Constraint 3:					Dark>80
Ingredient					
Dye A	1.15	2.93	2.05	2.35	2.88
Dye B	1.30	1.21	2.21	2.06	1.35
Dye C	3.00	3.00	1.88	3.00	2.60
Fragrance A	1.11	1.00	1.74	1.06	4.06
Fragrance B	1.52	1.69	1.34	1.53	1.53
Fragrance C	2.26	3.00	2.11	2.81	2.69
Ingredient Q	2.07	2.17	2.05	2.07	2.07
Liking—By subgroup					
Total panel	|71*|	61	70	|69*|	|69*|
Males	63	52	69	60	59
Females	78	69	71	77	79
Frequent shampooers	67	56	70	63	62
Infrequent shampooers	74	63	71	73	74
Younger users	71	58	68	68	67
Older users	73	62	70	71	71
Sensory segment 1	62	|75*|	54	|69*|	|69*|
Sensory segment 2	76	53	|78*|	|69*|	|69*|
Sensory likes/Total panel					
Color	56	60	51	56	53
Fragrance	67	59	65	66	66
Feel of liquid	74	59	73	71	71
Amount of lather	69	54	69	65	65
Sensory attributes					
Darkness	34	68	30	76	|80*|
Thickness	58	54	54	54	54
Fragrance intensity	48	53	49	55	55
Amount of residue	24	34	27	30	30

equally great. All of these lead naturally to the recognition that systematic work, doing one's homework, modeling the results, and using straightforward optimization algorithms are the way to enhanced market success and corporate profits.

References

Baxter, N.E. 1989. Research guidance: not giving it your best shot. *In:* Product Testing with Consumers for Research Guidance, L. Wu (ed.), STP 1035, pp. 10–22. American Society for Testing and Materials, Philadelphia.

Beausire, R.L., Norback, J.L. and Maurer, A.J. 1988. Development of an acceptability constraint for a linear programming model in food formulation. Journal of Sensory Studies, *3*, 137–149.

Best, D. 1991. Designing new products from a market perspective. *In:* Food Product Development, E. Graf & I.S. Saguy (eds.), pp. 1–27. Van Nostrand Reinhold/AVI, New York.

Box, G.E.P., Hunter, J. and Hunter, S. 1978. Statistics for Experimenters. John Wiley & Sons, New York.

Carr, B.T. 1989. An integrated system for consumer-guided product development. *In:* Product Testing with Consumers for Research Guidance, L. Wu (ed.), STP 1035, pp. 41–53. American Society for Testing and Materials, Philadelphia.

Claycomb, W.W. and Sullivan, W.G. 1976. Use of response surface methodology to select a cutting tool to maximize profit. Journal of Industrial Engineering, *98*, 64–65.

Deming, S.M. 1983. Experimental design: response surfaces. *In:* Chemometrics, Mathematics and Statistics in Chemistry, B.R. Kowalski (ed.), pp. 251–266. D. Reidel, Dordrecht, The Netherlands.

Draper, N.R. and Smith, H. 1981. Applied Regression Analysis. John Wiley & Sons, New York.

Fishken, D. 1983. Consumer-oriented product optimization. Food Technology, *37*(11), 96.

Giovanni, M. 1983. Response surface methodology and product optimization. Food Technology, *37*(11), 41–45.

Griffin, R. and Stauffer, L. 1990. Product optimization in central location testing and subsequent validation and calibration in home use testing. Journal of Sensory Studies, *5*, 231–240.

Hare, L.B. 1974. Mixture designs applied to food formulation. Food Technology, *28*(3), 50.

Hill, W.J. and Hunter, W.G. 1966. A review of response surface methodology: a literature review. Technometrics, *8*, 571–590.

Hlavacek, D. and Finn, J.P. 1989. Hitting the target more frequently: a systematical approach to research guidance tests. *In:* Product Testing with Consumers for Research Guidance, L. Wu (ed.), STP 1035, pp. 5–11. American Society for Testing and Materials, Philadelphia.

Joglekar, A.M. and May, A.T. 1991. Product excellence through experimental design. *In:* Food Product Development, E. Graf & I.S. Saguy (eds.), pp. 211–230. Van Nostrand Reinhold/AVI, New York.

Khuri, A.J. and Cornell, J.A. 1987. Response Surfaces. Marcel Dekker, New York.

Moskowitz, H.R. 1981A. Sensory intensity versus hedonic functions: classical psychophysical approaches. Journal of Food Quality, 5, 109–138.

Moskowitz, H.R. 1981B. Relative importance of perceptual factors to consumer acceptance: linear versus quadratic analyses. Journal of Food Science, 46, 244–248.

Moskowitz, H.R. 1984. Cosmetic Product Testing: A Modern Psychophysical Approach. Marcel Dekker, New York.

Moskowitz, H.R. 1991. Optimizing product acceptability and perceived sensory quality. *In:* Food Product Development, E. Graf & I.S. Saguy (eds.), pp. 157–188. Van Nostrand Reinhold/AVI, New York.

Moskowitz, H.R. 1994. Food Concepts and Products: Just-In-Time Development. Food and Nutrition Press, Trumbull, CT.

Plackett, R.L. and Burman, J.D. 1946. The design of optimum multifactorial experiments. Biometrika, *33*, 305–325.

Rabino, S. and Moskowitz, H.R. 1984. Detecting buyer preferences to guide product development and advertising. Journal of Product Innovation Management, 2, 140–150.

Williams, A.A. 1989. Procedures and problems in optimizing sensory and attitudinal characteristics in foods and beverages. *In:* Food Acceptability, D.M.H. Thomson (ed.), pp. 297–309. Elsevier Applied Science, Barking, Essex, U.K.

5
The Role of Psychophysics and Experimental Psychology

I. Introduction

Psychophysics is the branch of experimental psychology which relates physical stimuli to subjective responses. Psychophysics is the oldest branch of experimental psychology, tracing its history back almost two centuries. The very earliest research studies focused on classification (the basic sensory dimensions of perception) and discrimination (the smallest physical difference between two stimuli needed in order to insure that the two stimuli are perceived to be different). The initial work in psychophysics was done by physiologists, and later continued by experimental psychologists. This chapter deals with measures of discrimination, sensory, and hedonic magnitudes.

Psychophysics provides an extraordinarily valuable tool for researchers. Psychophysics describes the relation between physical intensity level and sensory intensity as reported by the panelist. Psychophysics can also reveal laws of sensory mixtures, including suppression and enhancement. That is, when two stimuli are mixed what is the sensory intensity (and the sensory quality) of the combination?

These topics span a broad range of phenomena of vital interest to product developers. While going through the topics, the reader should keep in mind that many of the topics have direct relevance today. Furthermore, even though a specific finding may apply to "model systems" (e.g., odorant dissolved in a diluent, such as ethyl acetate dissolved in diethyl phthalate), the same approach may be very instructive for applied problems.

Relevant Topics in Psychophysics

The four relevant topics in psychophysics are the following:

a. Discriminable differences
b. Sensory scales of magnitude
c. Psychophysical functions for sensory intensity and liking
d. Laws of mixtures

2. Discrimination Testing

Discrimination testing is the assessment of two or more samples in order to discover whether they are different or the same. Psychophysicists are interested in discrimination between stimuli because through discrimination testing they can learn how the sensory system operates. For instance, consider the perception of weight. Give the panelist two packages of equal volume, weighing X and Y grams, respectively. Ask the panelist whether these two packages differ in weight. If the physical weights differ by more than 2%, then at least 50% of the panelists will report that the two samples are not equally as heavy. The perceived weights differ. If the physical weights differ by less than that critical 2%, then more than half of the panelists will report that the two samples seem equally heavy.

Discrimination or difference testing intrigued the early psychophysics researchers because it promised an objective measurement of the sensory capacities of individuals. (The reader should keep in mind that this research was done 50 years before researchers began to accept the fact that human beings could act as measuring instruments and validly assign numbers to match the intensity of their perceptions.)

Soon after the earliest research on discrimination testing (conducted in the 1850's and earlier), researchers began extensive studies on the discrimination ability of people for a variety of stimuli. Much of this information lies buried in the archival literature of psychophysics, published a century ago or more. Researchers measure our discrimination ability for appearance attributes (e.g., line length, color saturation, color hue), tactile and kinesthetic attributes (e.g., weight, pressure), taste attributes (e.g., differences needed to discriminate between the same taste solutions at different concentrations, such as sucrose or sodium chloride), and olfactory attributes (e.g., differences needed to discriminate between two concentrations of the same odorant). Table 1 lists some of these just noticeable differences, or JND's.

In today's hectic business environment, the patient, almost laborious measurement of a just noticeable difference may seem a topic relevant only for those who have time to "fill in the holes in the scientific literature." However, there is at least one relevant aspect of just noticeable differences. They give the re-

Table 1 Just Noticeable Differences (Weber Fraction) for Sense Modalities: Proportion by Which a Stimulus Must Be Increased So the Panelist Perceives That the Stimulus Intensity Has Changed

Sense modality	Weber fraction
Deep pressure (skin, subcutaneous tissue)	1/77
Visual brightness	1/60
Lifted weights	1/52
Loudness (1000 Hz tone)	1/12
Smell (rubber)	1/11
Cutaneous pressure	1/7
Taste (salt solution)	1/5

searcher a sense of what differences are relevant to product developers. For example, if the just noticeable difference in color saturation is 5%, then the product developer must change the saturation by at least 5% in order to insure that at least 50% of the panelists will report that the two stimuli "look different." If the developer is conservative and creates only marginally small changes in the color saturation (e.g., 2% changes in physical saturation), most likely the consumer will not perceive the change. The developer might take a lesson from the psychophysical literature and make significantly larger changes in physical saturation.

Criterion- or Payoff-Related Biases In Discrimination Testing

If the researcher rewards the panelist for correctly guessing the "odd sample," or punishes the panelist for incorrectly guessing, then the panelist changes his or her discrimination behavior. For example, consider two nail polish removers, one of which has a modified fragrance from a new supplier designed to match the current fragrance. The researcher wishes to determine whether the two fragrances smell the same. The easiest approach is to provide the panelist with two samples, and instruct the panelist to report "different" if the products smell different, or report "same" if the products smell the same. On the surface the task seems to be quite easy to do. It is certainly easy to explain to the panelist.

Let us now introduce "reward" and "punishment." One scenario is that from a business perspective it is vital that the products smell exactly the same. Perhaps research has uncovered the fact that many consumers prefer the manufacturer's current product on the basis of its fragrance. The product may have enjoyed considerable success because of its unique smell. Marketing management is reluctant to change the fragrance, even though the change could significantly enhance profitability by reducing the cost of goods. As a consequence, the scenario is set for the panelist to be absolutely sure that the samples are the same before the panelist reports that they are the same. (Often these business expectations are conveyed to the panelist, albeit quite subtly.) The panelist who picks up this covert message about being absolutely certain before saying that two products are the same will look for differences. *It is less risky to say that the products differ.* That is the safe course. The cost of false discrimination (reporting that two samples smell different when they really smell the same) is quite low.

The payoff scenario can be reversed. The potential cost saving obtained by adopting the new fragrance may be so great that management wishes to ensure that only in the most certain of cases should the panelist say that two products differ. If the panelist is at all unsure of his or her judgment, it is best to say that two products are the same. Only when the panelist is absolutely sure that two products differ should the panelist state that they differ. In this case the covert reward is high for "no discrimination," and the punishment is great for a "false positive" (saying two stimuli are different when they are really identical). *It is less risky to say that the products are the same.*

Although the examples above represent the extremes, there are usually significant payoff considerations in discrimination testing. A great deal of product development work is commissioned to save money, but with the caveat of not alienating one's current consumers. The constraint of not alienating current consumers leads to a stringent criterion of what constitutes "difference." Only in the most obvious of cases should the panelist report that two products are the same. It is better to say that they are different.

Response biases in discrimination tests inevitably occur whenever the panelist must make one of two different responses (viz., say "same" or "different"). The biases change the probability of one response, either increasing or decreasing it, depending upon the "payoff" associated with each outcome (e.g., correctly reporting that there is a difference when there is a difference, incorrectly reporting a difference when there is no difference, etc.).

From Two Samples to Three Samples—The Triangle Test

In order to reduce the response bias, the researcher must change the task. Rather than letting the panelist report "same" or "different," the researcher instructs the panelist to select the "odd sample" from an array. In this test the researcher has three samples. Two of the three samples are the same, the third sample differs. The panelist inspects each of the three samples, and selects the odd sample from among the three. Good research practice dictates that in half of the trials, the odd sample should be the new prototype and the other two samples should be the current product. In the other half of the trials the arrangement is reversed.

The odd sample is the current product, and the other two samples are the new prototypes.

The panelist's job is simply to select the sample that differs from the other two. The panelist cannot be affected by criteria, because the response "same" or "different" has been replaced by a choice. One of the products is definitely different from the other two (at least according to the researcher). The panelist need only identify the odd sample.

By guessing or chance alone, the panelist will be correct 33% of the time. Thus, the observed discrimination results must be corrected for guessing. Table 2 shows the computation, and the true probability of a panelist perceiving a difference between two samples, given the observed probability of "correct."

The triangle test is a good procedure for these types of evaluations in which a consumer can assess multiple products in a short period of time. The triangle test was developed for research projects where the product (or stimulus) needed to be experienced for only a few seconds. Even with that short time period, the researcher had to remain cognizant of the fact that memory plays a part in the triangle test. The panelist has to keep a trace of the previous stimuli in memory. As the time needed to experience the product increases, the panelist's memory diminishes for the previously tested products. Performance diminishes, not so much from limits on sensory acuity as from limits on memory.

From Three Samples to Multiple Samples—Sorting Tasks

The triangle test is the simplest version of a sorting task in which the panelist has to assign multiple products to different categories. The better the panelist performs on a sorting task (in terms of properly identifying the samples as belonging to different groups) the greater the panelist's discrimination. A sorting task can be made more difficult in two ways:

1. *Increase the number of samples to be sorted.* The greater the number of samples to be sorted the lower will be the probability of correctly guessing by chance alone. Demonstrating one's discrimination ability through correctly classifying multiple samples becomes more impressive the more samples one has to classify (see Table 2, remaining columns).
2. *Do not tell the panelist how many categories there are.* Instead, simply present to the panelist the set of samples and instruct the panelist to sort these samples into groups, with those samples falling in the same group perceived to be identical and those samples in different groups differing from each other.

Discrimination vs. Product Differences

Discrimination testing assumes that the products in one category are identical to each other. The original discrimination tests were run with easy-to-manipu-

Table 2 Probability of Being "Truly Correct (P_t)" in a Discrimination Test* (Panelist Task - Choose the "Odd Sample")

Observed discrimination (P_o)	Number of stimuli in test			
	2	3	4	5
0.00	0.00	0.00	0.00	0.00
0.05	0.00	0.00	0.00	0.00
0.10	0.00	0.00	0.00	0.00
0.15	0.00	0.00	0.00	0.00
0.20	0.00	0.00	0.00	0.00
0.25	0.00	0.00	0.00	0.06
0.30	0.00	0.00	0.07	0.13
0.35	0.00	0.03	0.13	0.19
0.40	0.00	0.10	0.20	0.25
0.45	0.00	0.18	0.27	0.31
0.50	0.00	0.25	0.33	0.38
0.55	0.10	0.33	0.40	0.44
0.60	0.20	0.40	0.47	0.50
0.65	0.30	0.48	0.53	0.56
0.70	0.40	0.55	0.60	0.63
0.75	0.50	0.63	0.67	0.69
0.80	0.60	0.70	0.73	0.75
0.85	0.70	0.78	0.80	0.81
0.90	0.80	0.85	0.87	0.88
0.95	0.90	0.93	0.93	0.94
1.00	1.00	1.00	1.00	1.00

*Formula used to compute "true probability"

$P_o = P_t + (1 - P_t)(P_g)$

P_t = probability of truly discriminating
P_o = observed proportion of discrimination
P_g = probability of being correct by guessing alone
    0.50 for two stimuli in the task
    0.33 for three stimuli in the task
    0.25 for four stimuli in the task
    0.20 for five stimuli in the task

late stimuli (e.g., shapes, sounds). The stimuli could be presented in rapid and randomized order. Within a very short time the panelist assessed the entire range of stimuli, and then chose the "odd" sample. Furthermore, the stimuli in these early studies did not fatigue the sensory system. In contrast, applied research uses more complex stimuli that take a long time to evaluate. Other procedures to establish product difference may be more appropriate. We will deal with difference testing later on in this chapter.

Decisions from Discrimination Testing

Most discrimination tests are used for quality control. For instance, an ingredient supplier may change. The quality assurance group may wish to insure that the ingredient change does not affect product quality. In this situation, discrimination testing is a good way to establish whether product quality has remained unchanged. Other discrimination tests are used for process variations, again to insure ongoing product quality and sensory identity.

Discrimination Testing and Panelist Experience

With repeated product usage panelists become familiar with a product. Oftentimes the panelist perceives more characteristics in the product with repeated usage (even though the typical consumer may not articulate this depth of perception). An experienced panelist will report differences that an inexperienced panelist will not (Moskowitz & Gerbers, 1974). For instance, in fragrance evaluation the consumer who always uses the fragrance will perceive more "notes" than will a "matched" consumer (in terms of age, demographics) who does not use the fragrance, and is thus unfamiliar with it.

Unfortunately for researchers there is no way to determine the panelist's "concept" of a product prior to the discrimination test. Some panelists may be more aware of the characteristics of the products and may know what to look for. These "aware" panelists are more likely to discover differences between samples than are panelists with the same sensory acuity, but who have no concept of the product, and thus do not know what to look for. The criteria for difference (and even the characteristics to which the panelist attends) differ from person to person, in an unknown and probably dynamic fashion, substantially hidden from the researcher.

When Does Discrimination Testing Fail?

Discrimination testing does not work for all judgments of "difference." Many products naturally vary from batch to batch. The alert panelist can often perceive differences between the same product from different batches, especially when the product varies naturally during the course of production. In this case discrimination testing really becomes a test of concept formation. Simply stated, discrimination between two products as being different from each other reflects the panelist's conclusion that the perceived difference between the products is greater than the panelist's expectation of random batch-to-batch difference.

On the basis of these considerations, discrimination testing is appropriate when the product is relatively simple (e.g., a fragrance), when the panelist can assess multiple samples in a single setting (which reduces memory effects), and when the decision is to accept or to reject the new sample based upon the degree to which it substitutes for the current sample (or ingredient).

Discrimination testing should not be done when the stimulus is complex and randomly varies, when the panelist must use the product for an extended period of time and cannot easily inspect multiple product samples, and when the decision to accept or reject a sample is based on factors other than "identical" vs. "different."

Practical Applications of Discrimination Testing

Although discrimination testing is rooted in the history of experimental psychology, both the basic discrimination testing and difference testing still are widely used today to answer practical problems. For instance:

1. If the manufacturer changes the amount of liquid in a bottle, then does the consumer perceive the difference? Here, classical discrimination testing is relevant. The objective is to investigate very small changes in the stimulus (all other factors held equal). Furthermore, the business goal for the researcher is to identify the greatest decrease in the amount of liquid without having the consumer (panelist) recognize that the new product differs from the current product.
2. Often, manufacturers change ingredients (e.g., in a fragrance or soap) because new suppliers offer a lower price. Again, the objective is to replace the current, more expensive ingredient with a new, less expensive ingredient, without having the panelist recognize that there has been a switch. Classical discrimination testing is useful in this regard.

Misusing Discrimination Testing

Many practitioners feel that discrimination testing is the key test for ingredient substitution. If the "revised" or "new" product is sensorially identical to the current product, then the practitioner feels that the ingredient can be substituted. However, there are often situations where the ingredient substitution changes the sensory profile slightly, but maintains overall acceptance. A rigid use of discrimination testing here would argue against using the new ingredient. But what if the revised product differs slightly from the current product (i.e., the two are not precisely identical) but is also more acceptable than the current product? Rigid adherence to discrimination testing would argue against the change. A more global view of product difference would argue for the change.

3. From Discrimination to "Difference"

The early research on discriminable differences concentrated on the smallest difference a consumer could perceive. However, as experimental psychologists developed more procedures and researchers began to open their minds to the different aspects of products, they soon realized that discrimination testing is just a small corner of a larger area of psychology which deals with the "differences" or "dissimilarities" between stimuli.

We can view sensory differences from a different perspective than we viewed discrimination testing. A researcher can present the panelist with two stimuli and instruct the panelist to rate how different these two products seem to be. The stimuli need not vary on one dimension in a small, almost insignificant fashion (as they vary when the researcher performs discrimination tests). The stimuli can vary on a variety of dimensions. The panelist's job is simply to rate how different the stimuli seem to be.

Once we begin to look at psychophysical difference testing rather than discrimination testing, an entirely new world of research and potential opens up. Here are two of the issues that can be considered:

1. How different on a subjective scale do the products seem to be? The panelist is not asked to rate any specific attribute, but simply to compare the products and rate overall "dissimilarity." Table 3A shows the results of such a study where the panelists evaluated five different nail polish removers, varying in color and fragrance, chosen from competitors in the marketplace. Each of the five products was definitely "discriminably different" from the other four. A discrimination test would not have told the researcher anything about the products because virtually all of the panelists could discriminate between the five products. However, the products vary in the degree to which they differ from each other. The matrix in Table 3A shows the average ratings from 37 panelists, who compared each pair of products and rated dissimilarity on a 0–100 scale (0 = identical → 100 = extremely different). (Were this to have been a discrimination test, then the products would have been selected to be almost identical with each other.)

2. Can we create an integrated measure of dissimilarity between the two samples based upon the panelist's ratings of attributes? That is, if the panelist rates the products on an array of attributes (Table 3B), then can we create an overall measure of "dissimilarity" based upon the differences between pairs of profiles?

Table 3C shows two calculations of "dissimilarity" based upon profiles. The calculation assumes that every attribute is equally important. Overall dissimilarity is equal to the sum of the absolute differences between the two products on the different attributes. The summation (or the mean) of these absolute differences generates the measure of dissimilarity. This first calculation is known as the city block metric or distance, because the distance is computed in the way one computes distances in a city (i.e., by the sum of the blocks that the pedestrian has to walk to go from one point (product) to another point (product).

The second approach calculates dissimilarity by the standard formula for geometric distance. This is called the Euclidean metric or distance. The key difference is that the Euclidean distance gives more weight to the large differences.

Psychophysics and Experimental Psychology

Table 3A Direct Ratings of Dissimilarity Between Pairs of Five Nail Polish Removers

	A	B	C	D	E
A	0	—	—	—	—
B	50	0	—	—	—
C	5	50	0	—	—
D	33	61	33	0	—
E	61	9	61	44	0

(0 = Identical, 100 = Extremely Dissimilar)

Table 3B Profile of Nail Polish Removers on Five Attributes

Product	Fragrance	Color	Thick
1	2	2	4
2	3	4	3
3	2	2	4
4	6	1	4
5	4	5	3

Table 3C Computation Formulas and Measures of Dissimilarity Between Pairs of Nail Polish Removers, Based upon Attribute Ratings

City Block Distance: Distance = Average Absolute Difference Between Product X and Product Y on All Attributes

	A	B	C	D	E
A	0.0	—	—	—	—
B	1.3	0.0	—	—	—
C	0.0	1.3	0.0	—	—
D	1.6	2.3	1.6	0.0	—
E	2.0	0.6	2.0	2.3	0.0

Euclidean Distance: Distance = Square Root of the Sum of the Squares (Product X - Product Y) on All Attributes Divided by Number of Attributes

	A	B	C	D	E
A	0.0	—	—	—	—
B	1.4	0.0	—	—	—
C	0.0	1.4	0.0	—	—
D	2.4	2.5	2.4	0.0	—
E	2.2	0.8	2.2	2.6	0.0

Practical Application of Difference Testing—A Case History with Tissue

Consider the problem of facial tissues presented below. The Siragusa Napkin Company (SNA Ltd.) manufactures a variety of facial tissues. Over a 30-year period SNA enjoyed a consistent, uninterrupted growth in market share, total volume, and profitability.

With increasing growth, management opened five new factories to produce the SNA products. Within 6 months, reports came in from different sources (consumers, the trade) that the SNA products were just not up to the quality standards that SNA had set for itself. Management was of course distressed, and decided to investigate the problem and cure it.

An initial visual inspection of the products from the five factories revealed that the same product differed from factory to factory. One research analyst suggested that SNA might do well to run discrimination tests between pairs of products from different plants in order to identify which batches were out of "spec" and which products were in specification. After thinking through the issues, the marketing research director at SNA realized that conventional discrimination tests would not be appropriate. The production runs from the different plants clearly differed from each other. The data from a conventional discrimination test would simply reveal that, indeed, the consumers recognized the batch-to-batch differences which were known to exist (and which were causing the problems in the first place).

SNA's task force recognized that they needed to develop a measure of overall sensory "difference" between the various plant production runs, rather than to conduct simple discrimination tests. It seemed reasonable at the time to have panelists compare different pairs of production runs to each other and rate the degree to which the products differed from each other. The scale was to be an anchored 0–100 scale (0 = identical, cannot tell the products apart → 100 = radically different from each other, easy to tell the products apart).

Consumers had no problems doing this type of evaluation. Some consumers said that they were unclear as to how to use the scale. The 0 was unambiguous; if the panelist could not discriminate between the two products, he was to use a 0. The 100 was very ambiguous. Was 100 to represent two products of the same type that were quite different from each other because of modest but noticeable differences? (This is the more conservative approach. The panelist scales the dissimilarity or difference within the framework of the product itself.) Or was 100 to represent two facial tissues that were dramatically different from each other? (This is the more radical approach. Two products of the same type that differed perceptibly from each other would here be assigned a small number, nearer 0, to denote the fact that they differed, but did not differ as dramatically as two facial tissues of different construction.) The research team decided to use the more conservative approach. The panelist was instructed to scale difference within the framework of the specific tissue alone, and not within the framework of a whole series of qualitatively different tissues.

Panelists evaluated the overall difference of every one of the five pairs, leading to 10 different comparisons of five items taken two at a time. The researchers also recognized that panelists always search for differences, even when these differences are non-existent. In difference testing these biases show up as non-zero ratings of "difference" even when the two products being compared are identical to each other. All ratings of difference must be corrected for this bias by subtraction. The reduced bias measure of difference is defined as:

Rating of Difference − Rating Assigned to Identical Pairs

(We saw these "response biases" in discrimination testing. Panelists are eager to please the researcher, and look for differences even when they cannot find the differences. These are called "false positives" in discrimination testing.)

Table 4 shows the results for the analysis and reveals that factory C is producing products with the greatest differences, whereas factories A, B, and E are producing more similar products.

Table 4 Direct Ratings of Dissimilarity Between Facial Napkins Produced at Five Plants

	A. Direct Dissimilarity Ratings of Pairs				
	A	B	C	D	E
A	23	—	—	—	—
B	36	10	—	—	—
C	44	32	19	—	—
D	46	44	46	12	—
E	32	36	51	35	14
	B. Direct Dissimilarity Ratings—Correction Factor				
	A	B	C	D	E
A	—	—	—	—	—
B	20.4	—	—	—	—
C	28.4	16.4	—	—	—
D	30.4	28.4	30.4	—	—
E	16.4	20.4	35.4	19.4	—

Correction factor = Average of ratings assigned to same pairs.

$$\text{Correction factor} = \frac{(23 + 10 + 19 + 12 + 14)}{5} = \frac{78}{5} = 15.6.$$

4. Perceived Intensity

The Correlation Trap

In sensory analysis and marketing research, it has been customary to run correlation statistics between stimulus levels and consumer intensity ratings. The research heritage in both fields has been to look at correlations as measures of the strength of a linear relation between two data sets (physical level, consumer ratings). However, correlations do not show the parameters of the relation. It is more instructive to create equations and models than to run correlations.

Psychophysical Intensity Functions

More insight into sensory processes and significantly greater practical application can be gained by having the panelist act as a measuring instrument. (In contrast, discrimination testing requires that the panelist make one of two responses—same or different.) By having panelists assign numbers to match perceived stimulus intensities, the researcher can develop a model showing how perceived sensory intensity (or liking) varies with changes in the physical stimulus.

In order to develop an intensity function, the panelist uses a numerical (or equivalent) scale. The panelist "adjusts" the numbers on that scale to match perceived intensity. The average ratings from a group of panelists can be plotted against the physical stimulus level in order to create the psychophysical intensity scale.

It is very simple to create an intensity function. The researcher first chooses a physical continuum to vary. In the simplest of cases the physical stimulus is unidimensional (e.g., apparent physical viscosity of a lotion; area or volume of a package, etc.). The physical stimulus is systematically changed on this continuum to yield a series of stimuli differing from each other in perceptible steps. These stimuli are randomized (in order to avoid order bias), and presented to the panelist. The panelist assigns numbers from the scale to match perceived intensity. By plotting the data from several panelists (or even from one panelist) against the physical intensity of the stimulus, the researcher creates the intensity function.

Figure 1 shows two of these intensity functions for the very simplest of cases. The Y axis shows the average rating assigned by the panelist, and the X axis shows the physical intensity value. If the data from a group of panelists is combined and averaged, then the average rating tracks the physical intensity quite closely. The curve may not be perfectly straight, however. It may curve upwards, it may increase and then asymptote, or it may be a straight line. Furthermore, the slope of the line may vary from flat to steep. In most cases, however, the relation is monotonic. Monotonicity means that increases in the

Psychophysics and Experimental Psychology 301

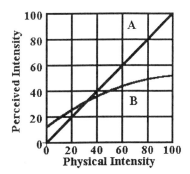

Figure 1 Example of monotonic relations between physical stimulus intensity (independent variable) and a subjective response (dependent variable).

independent physical stimulus are accompanied by increases in the consumer's rating.

Utility of the Psychophysical Intensity Function

By creating an intensity function, the researcher learns a lot about how the sensory system functions. This is the scientific merit of the approach. The applied developer also learns about the relation between ingredient or physical factors under R&D control and expected consumer responses. For instance, one might ask these three questions:

1. Do panelists accurately track the changes in the physical stimulus by means of their ratings of intensity?
2. What is the relation between physical and sensory changes? Is the relation linear? Curvilinear? Does the curve asymptote so that after a certain intensity level is reached the panelist no longer perceives any change, despite the physical increase?
3. Where does one's current product lie on the intensity curve? Where does one want one's product to lie? Can that sensory level be achieved by a simple quantitative change in the physical stimulus level?

Magnitude Estimation and the Sensory Power Law of Perceptions

Although it is easy to demonstrate that panelists can accurately perceive stimulus differences and scale them, the nature of the results depends strongly upon the scale that the researcher uses. For instance, if the panelist is permitted only 3 points (1 = weak, 3 = strong), then the sensory curve will differ from that obtained when the panelist uses a 0–100 point scale (0 = not perceptible, 1 = weak → 100 = strong).

In the psychological laboratories at Harvard University, S. S. Stevens began his extensive research on the nature of the "real" relation between perceived intensity and physical magnitude. Stevens (1953) reasoned that the scale used by the panelist should have ratio properties. That is, the scale had to be similar in mathematical properties to scales used in physics, that measured weight, length, force, etc. At that time the scales used by the researchers were arbitrary (in terms of the number of categories in the scale, the verbiage attached to the scale, etc.). Stevens erected a validated ratio scale using the method of "magnitude estimation." Panelists were instructed to match numbers to the perceived intensity of a stimulus so that the ratios of the numbers assigned matched the perceived ratios of intensity.

To the skeptic and to the novice alike, magnitude estimation seems hard to accept. Just because the panelists are instructed to match ratios of numbers to ratios of perceived intensity does not mean that they can actually do so in a reproducible fashion. Yet, again and again Stevens and his colleagues reported that panelists felt comfortable doing so, and that their numbers were reliable from study to study. Furthermore, Stevens demonstrated that he could relate the numbers assigned by the panelist to the physical intensity measurement of the stimulus. The two data sets were highly correlated with each other. Furthermore, Stevens demonstrated that time after time the relation followed a power function, written below:

Sensory Intensity = $K(\text{Physical Intensity})^m$

The power function exponent m was reproducible from study to study, and differed across physical continua. Table 5 lists power function exponents for a variety of continua. Table 5A shows the exponents for a variety of different sensory continua. Table 5B shows results for different odorous stimuli.

If Stevens' results had been limited to "model systems" (e.g., brightness of light, loudness of simple sounds, odor intensity of pure chemicals), then the results might have remained within the confines of academe. However, applied researchers soon recognized two implications of the power function. Specifically:

1. The reliability of the function demonstrated that it is possible to develop an equation relating perceived sensory intensity to physical intensity for many different types of variables.
2. The exponent, m, of the power function tells the product developer a great deal about the reaction of consumer panelists to the stimulus. When m is considerably lower than 1.0 (e.g., for the perception of viscosity or the perception of odor intensity), then it requires a very large percentage increment in the physical stimulus intensity in order to generate a noticeable subjective increase. For fragrance intensity (as an example), when the panelist reports that the fragrance intensity is "too weak," the developer must substantially increase the fragrance concentration in order to

Table 5 Power Function Exponents (m) for Different Sensory Continua—Sensory Intensity = K(Physical Intensity)^m

Continuum	Stimulus	Power function exponent	Sensory change upon doubling	Physical change to double perception
A. Various Continua				
Brightness	5 degree target/dark adapted	0.33	1.26	8
Tactile	Thickness/solution	0.50	1.41	4
Loudness	Tone/binaural	0.60	1.52	3
Vibration	Finger/250 hertz	0.60	1.52	3
Visual area	Projected square of light	0.70	1.62	3
Taste	Sweet/saccharin	0.80	1.74	2
Tactile	Hardness/rubber squeezed	0.80	1.74	2
Vibration	Finger/60 hertz	0.95	1.93	2
Temperature	Cold/arm	1.00	2.00	2
Repetition	Light, sound, touch, shock	1.00	2.00	2
Finger span	Thickness/wood blocks	1.00	2.00	2
Visual length	Projected line	1.00	2.00	2
Duration	White noise stimulus	1.10	2.14	2
Pressure	Palm/static force on skin	1.10	2.14	2
Vocal effort	Sound pressure/vocalization	1.10	2.14	2
Lightness	Reflectance/gray paper	1.20	2.31	2
Visual velocity	Moving spot of light	1.20	2.31	2
Taste	Salt/sodium chloride	1.30	2.46	2
Taste	Sweet/sucrose	1.30	2.46	2
Heaviness	Lifted weight	1.45	2.73	2
Temperature	Warm/arm	1.50	2.83	2
Tactile	Roughness/emery grits	1.50	2.83	2
Force	Hand grip/hand dynamometer	1.70	3.25	2
B. Smell				
Smell	Phenyl acetic acid/liquid	0.12	1.09	323
Smell	Amyl acetate/liquid	0.13	1.09	207
Smell	Heptanol/air	0.15	1.11	102
Smell	Anethole/liquid	0.16	1.12	76
Smell	Guaiacol/liquid	0.16	1.12	76
Smell	Citral/liquid	0.17	1.13	59
Smell	Phenylethanol/liquid	0.19	1.14	38
Smell	Geraniol/air	0.20	1.15	32
Smell	Methyl salicylate/liquid	0.20	1.15	32
Smell	Ethyl acetate/liquid	0.21	1.16	27
Smell	Iso-valeric acid/liquid	0.21	1.16	27
Smell	Butyric acid/liquid	0.22	1.16	23
Smell	D-Menthol/liquid	0.24	1.18	18
Smell	Eugenol/liquid	0.27	1.21	13
Smell	Butanol/liquid	0.31	1.24	9
Smell	Coumarin/air	0.33	1.26	8
Smell	Butyl acetate/air	0.58	1.49	3
Smell	Butanol/air	0.64	1.56	3

have the consumer perceive a noticeable increase. On the other hand, when consumers report that the fragrance is "too strong" (but also of the right quality), then the developer can substantially decrease the concentration of the fragrance (and save money). Only after the substantial decrease in concentration does the panelist begin to notice the difference in fragrance intensity.

The opposite pattern holds for other continua, such as the roughness of particulates. The formulator of a toothpaste or dermal abrasive must take care to control the "grit" of the product. Consumer panelists are exceptionally sensitive to the grit size. Small changes in the grit level produce substantial changes in perceived roughness. For quality control the formulator must force the product to remain within carefully monitored limits. Small excursions outside those limits may seem insignificant when measured on the physical stimulus scale but may be dramatic when measured on the subjective intensity scale.

Stevens' power law became the subject of numerous scientific papers published from the mid-1950's until the mid-1980's. Indeed, there is significant research still being conducted on the true nature of the psychophysical function. Not all researchers agree that the power law is as general as Stevens had conjectured, nor do they agree that the exponent is so robust as to defy the many biasing conditions in research designs. For instance, the exponent can be increased by giving the panelist a very small range of stimulus intensities. The panelist will try to exhibit differentiation by expanding the range of numbers assigned beyond the range that the stimulus levels would command in less-biased testing. Despite the issues, conjectures, arguments, etc., that have been raised with regard to the validity and universality of the results, Stevens' research has provided a powerful theoretical and practical approach for relating the physical world of stimuli to the private sensory world of experience.

Using the Power Law of Sensory Intensity to Set Ingredient Limits

The power function for sensory intensity shows how the perceptual system transforms physical information to sensory intensity. The developer can use this information in order to set limits on ingredients when developing a product. Rather than conducting a discrimination test between pairs of samples (which could run into time and money), the developer should follow these seven steps:

1. Where possible, systematically vary the physical ingredient over a wide range.
2. Have a small group of panelists (approximately 10 individuals) assign magnitude estimates to perceived intensity of the systematically varied stimuli. Randomize the stimuli for each panelist to remove the order effect. Also, if possible start the panelist with the first stimulus lying some-

where around the middle of the intensity range, rather than at either extreme.
3. Calculate the median of the magnitude estimates. The median is a better statistic than the mean, because magnitude estimates are unbounded. Thus, a panelist who uses numbers in the hundreds and thousands will outweigh a panelist who uses smaller numbers, such as 1–20.
4. Plot the data either in linear or in logarithmic coordinates.
5. Look for a sensory range that is reasonably large. By reasonable we mean a range of magnitude estimates of at least 2:1 on a subjective basis.
6. Identify the physical stimulus levels corresponding to the upper and lower sensory limits.
7. The results are limits of the ingredient levels corresponding to a reasonably large sensory range.

To illustrate this approach, consider the following two products: toothpaste with abrasive and thickened hand lotion. In both cases the developer wanted to change the product because panelists reported that the toothpaste was too rough and the lotion was too thin. The problem was to identify a reasonable range of grit to test in the toothpaste and a reasonable range of viscosity to test in the lotion. There were no prior data to help the developer.

Table 6 shows the consumer rating of roughness for seven different grits of toothpaste. From this table the developer can identify a range of grit around the current product that produces a reduced sense of "roughness" or abrasiveness. Indeed, given the table and either an equation or a graph, the formulator can set sensory limits as desired and discover the physical stimulus level corresponding to the sensory change. The key observation in Table 6 is

Table 6 How Physical Changes in Viscosity of a Lotion Generate Thickness, and Physical Changes in Grit of a Toothpaste Generate Roughness

Stimulus (relative units)	Toothpaste perceived roughness	Lotion perceived thickness
1	3	3
2	9	4
3	16	5
4	25	6
5	35	7
6	46	8
7	59	8

that the consumer's perception of roughness is far more expansive and has a wider range than one might believe, based upon the physical stimulus measure alone.

Table 6 also shows the same type of data, this time for seven different lotions. Here the research objective was to increase the viscosity. Had the developer simply looked at the measured viscosity rather than at the subjective rating of thickness, the problem of increasing the viscosity might be solved in three iterations, rather than in one iteration. The perceived thickness of the lotion grows far more slowly than we might guess on the basis of the physical measure (centipoises). Thus, the developer who considers only the physical measure would make a large change in physical viscosity. That large change would, however, yield a substantially smaller increase in subjectively perceived thickness. The process would then have to be repeated until the developer finally achieves the requisite increase in perceived thickness through a very large change in physical viscosity. Two or three small changes in physical viscosity, over different iterations, would generate the large change needed. Fortunately, with tables or curves showing the relation between physical stimulus level and subjective intensity, the developer can reduce some trial and error work. The psychophysical intensity function provides the developer with a good first guess.

5. The Psychophysics of Liking and Disliking

Hedonics vs. Intensity

Researchers have applied the same scaling approach to hedonics as they have to sensory intensity. The panelist is instructed to assign numbers to the degree of "liking" or "purchase intent." These numbers are assumed to reflect the intensity of a person's attraction toward the stimulus. The scale can be magnitude estimation or a simple 9-point category scale (ranging from dislike extremely to like extremely).

The investigator systematically varies the stimulus (e.g., concentration of a sweetener in water) in order to create the function relating liking to stimulus intensity. For some stimuli (e.g., taste, odor stimuli) the task is easy. Panelists report no difficulty. For other stimuli there is some thought involved (for instance, the panelist who rates liking of the appearance of different patterns). For still other stimuli (e.g., sounds of different tones), panelists report that they have a great deal of difficulty making the judgments, but can do so once they have thought about the task and have had some experience.

These studies of "hedonic intensity functions" are quite instructive for both the product developer and marketer. First, panelists report differences in overall liking as the stimulus changes. Both the intensity and the hedonic value of the stimulus change as the physical magnitude increases. Second, the hedonic changes do not necessarily parallel the sensory changes. As the stimulus increases, perceived intensity continues to increase and may eventually flatten out, no matter what the stimulus continuum may be. However, liking ratings do not continue to increase *ad infinitum*. Rather, liking often increases at first (initially paralleling the increase in sensory intensity). Soon, however, liking reaches its maximum level. Any increase in the physical stimulus intensity beyond the optimum level (where liking reaches its maximum) produces a decrease in liking, even though perceived sensory intensity continues to increase. Figure 2 shows this prototypical liking curve.

There are three implications of these two patterns for the formulator and for the marketer:

1. If the hedonic pattern is relatively flat, and if the sensory intensity pattern is also relatively flat, then the formulator can change the physical concentration (intensity) at will, without worrying that the consumer will either perceive the difference between the "current" and "reformulated" product or care about the difference. The consumer is simply too insensitive to the physical changes for anyone to worry. Examples of this type include viscosity of a product or the fragrance intensity of a musk. Within the typical region of usage, the consumer is quite insensitive to modest changes around the current product. Changes of 5–10% can hardly be perceived. This "best of all possible worlds" appears in Figure 3A. Both curves are fairly flat. Consumers do perceive differences and do assign different liking ratings, but the range of ratings assigned is relatively small.

2. If the hedonic pattern is relatively flat but the sensory intensity pattern is steep, then the consumer will perceive the differences, but will not care (at least in terms of overall liking). Figure 3B shows this pattern which often occurs when the physical continuum is visual or textural. For instance, depth of color or size of product (within reasonable ranges) will describe this type of

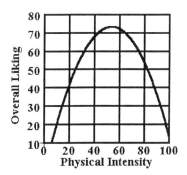

Figure 2 Relation between stimulus intensity (independent variable) and overall liking.

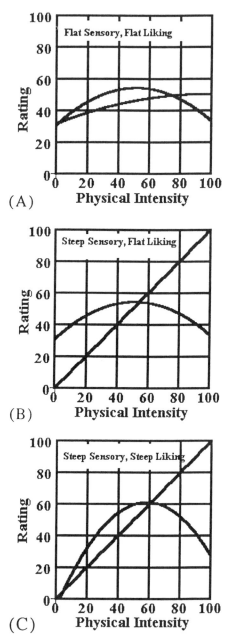

Figure 3 How sensory intensity and liking vary with stimulus level. The straight line corresponds to perceived intensity, the parabola corresponds to liking. (A) Flat sensory curve, flat liking curve, (B) Steep sensory curve, flat liking curve, and (C) Steep sensory curve, steep liking curve.

pattern. If the package designer increases the depth of color of a yellow package, then the consumer easily perceives the color to change dramatically. The curve relating perceived depth of color to physical saturation is steep. On the other hand, within a relatively large range of color saturation, many consumers simply do not care. If asked to rate how much they like the package, many will assign the different packages similar numbers, showing that they are indifferent to the changes made by the designer. Many researchers believe (albeit incorrectly) that if a consumer perceives a change in the current product, then this change is "bad," and so these researchers report back to the developer or package designer that the revised product must be changed again until no perceivable difference is reported between "current" and "revised" products. In reality, consumers are faced with product changes all the time. It is only when the product changes in a branded product lead to a loss in liking that the marketer should worry.

3. If both the sensory curve and the liking curve are steep, or if the liking curve reaches an intermediate optimum, then the marketer must worry about maintaining product quality in the face of changing levels. Figure 3C shows this pattern. Not only are consumers able to perceive differences between the stimulus levels, but these differences substantially affect how much they like the product. Examples of this include the grit of toothpaste (where consumers are sensitive to grit size and exhibit a narrow band around which their acceptance optimizes). Other examples come from the world of taste (e.g., the taste of salt; of sugar, etc.). This pattern does not typically occur for most visual stimuli, but does occur for textural and kinesthetic stimuli. One example is the effort involved in opening a package. Both acceptance and perceived effort increase as the physical effort involved increases. However, beyond a certain point, as the force required to open the bottle top increases, perceived effort increases and the consumer reports that it becomes less pleasant to open the container. Consumers can perceive small differences in effort needed to open containers, thus yielding a steep curve relating force vs. perceived effort. Furthermore, their hedonic functions are also steep, with an intermediate optimum.

There is a fourth pattern, namely, insensitivity to physical differences as they generate sensory responses, but strong sensitivity to these physical differences as they generate hedonic responses. This pattern is difficult to interpret. If consumers do not perceive sensory differences on an attribute as the physical intensity changes, then how can they validly assign differences in overall liking? It might simply be that the sensory attribute being rated does not covary with the physical stimulus being changed. Consumer panelists may perceive differences with changes in the physical stimulus, but are given the wrong attribute with which to report these perceived changes. Given the correct attribute, the consumers might report a noticeable change in perceived sensory intensity.

6. Laws of Sensory Mixture

Although traditional psychophysics (and especially scaling) deals with stimuli which vary along a single dimension, in nature and specifically in product development, most stimuli comprise combinations of components. Thus, for psychophysics to provide the product developer with even more insight, it has to deal with the sensory (and hedonic) response to mixtures. Fortunately, over the past 35 years, researchers have investigated how the sensory system transforms mixtures of stimuli into sensory and hedonic responses.

When a psychophysicist studies mixtures, the objective is to understand how the sensory system works. In contrast, when the developer studies mixtures, the objective is to understand how the components of the mixture interact with each other to produce a product. The psychophysicist is interested in laws of perception. The developer is interested in the product itself, using the consumer as a "bio-assay device." It is important to keep this distinction in mind, because many of the psychophysical studies will seem rudimentary and simplistic at best. Yet, these studies can be very instructive.

What Visual Research Teaches

We can divide visual research into two aspects—color and surface. Both of these are governed by a variety of rules (or at least patterns).
1. Studies of color vision show that there are three basic primary colors whose mixtures in various proportions give rise to the colors of the rainbow. All colors can be created by different proportions of the three basic primary colors. Furthermore, color science has made the prediction of color quality and color intensity virtually exact, given the quality and intensity of the components.
2. When the colors are embedded in liquids or in solids rather than being presented as simple "monochromatic" (single color) lights, matters become more difficult. It is possible to create a limited set of liquids, each with its own characteristic color. However, blending liquids is not the same as blending lights. Liquids have surface properties that affect color blending. The resultant color of the liquid can be predicted, but the rules of mixture are entirely different from those of simple lights.

What Taste Research Teaches

The original work on mixtures was done with taste materials, such as sugar, salt, acid and quinine (a bitter tasting alkaloid). Early speculations on the sense of taste suggested that it comprised four different basic tastes or "primaries"—sweet, salty, sour and bitter. Indeed, there was speculation (at least in the 1920's) that all taste sensations could be produced by the appropriate combinations of chemicals corresponding to these four basic primaries (Von Skramlik, 1921).

Once consumer panelists assign numbers to represent the perceived intensity of a stimulus, the natural next step is to have the same consumer assign numbers to "complex stimuli." The consumer can "profile" the taste sensation by breaking the sensation into its components. (For instance, the consumer might rate the amount of sweetness, the amount of saltiness, the amount of sourness (tartness) and the amount of bitterness in the same taste perception.)

Modern day research on taste mixtures has investigated whether one taste enhances or suppresses another, or has no effect at all. Attribute profiling enables the researchers to answer this question quite easily. The panelist is presented with liquids comprising either single taste stimuli (e.g., solutions of pure sucrose, pure sodium chloride) or mixtures (e.g., the same concentration of sucrose and sodium chloride as before, but mixed together into one liquid). The panelist does not know that some of the stimuli are "pure" and that other stimuli are physical blends. All the panelist needs to do is to rate the various liquids on taste intensity. The panelist might simply rate all of the solutions on sweetness, or even profile the intensity of the four taste sensations in each liquid (i.e., strength of the sweet taste, salty taste, sour taste, and bitter taste).

This exercise can be easily done by anyone who can mix two stimulus solids together and dissolve them in a liquid diluent. The results are surprisingly simple to understand:

1. When the components are dissimilar in taste (e.g., sweet sugar + salty tasting sodium chloride), then each taste suppresses the other taste. In the mixture, the taste intensity of both sweet and salt will be lower than the taste of the sweet stimulus without the salt being present. (The researcher knows the strength of the unblended sweet taste because the same panelist also evaluates a pure sugar solution of the same concentration.) If one taste quality (e.g. salty) is substantially more intense than another taste quality (e.g., sweet), then the stronger taste quality may entirely *mask* the weaker taste quality so that the weaker taste disappears. This is suppression. However, the stronger taste quality will be diminished as well (although it will not disappear).

2. When the components are dissimilar the suppression may not be equal on both sides. The salt taste suppresses the sweet taste more than the sweet taste suppresses the salt taste. The magnitude of the suppression and the degree of asymmetry in suppression is a function of two factors:
   a. What pair of stimuli are being tested.
   b. The concentration (and ultimately the taste intensity) of each component of the mixture (tested alone).

3. When the components have similar tastes (e.g., two sweeteners, two acids) the taste intensity of the mixture is stronger than either component alone. However, for the most part the taste intensity is not as strong as the arithmetic sum of the taste intensities. That is, if sample A (e.g., an artificial sweetener)

has a sweetness of 25 on the magnitude estimation scale, and if sample B (e.g., sucrose) has a sweetness of 35 on the same scale, then a blend of both stimuli in the same liquid at the same concentration will achieve a sweetness less than the simple sum of 60 (25 + 35). If the sensory system simply "added" the separate taste intensities of the components we would expect to have a rated sweetness around 60. We do not. Some researchers have ascribed this lack of additivity to a simple scaling artifact. For many scales the underlying scale is not linear. If the scale has end points at 0 and 100, and if two samples score 75 and 75 each, then the mixture cannot score 150 (75 + 75) because 150 is not an allowable scale point. However, the same lack of simple additivity occurs whether the stimuli both have low sensory intensities (e.g., 15 and 15 on a 100-point scale) or the stimuli are scaled on magnitude estimation, which has no end points and thus is not subject to the "end effect."

    4. It is rare for a taste mixture to be more intense (synergistic) than the arithmetic sum of its components. There are some combinations for which this synergism occurs, especially for mixtures of the so-called "taste enhancers" (monosodium glutamate) and other taste components (e.g., acids). For these combinations the taste intensity is sometimes equal to or slightly greater than the taste intensity of the arithmetic sum of the components.

    5. Taste mixtures may exhibit one of two behaviors in terms of their sensory "quality." We know from basic research that there are four primary tastes (sweet, salty, sour, bitter). When two taste compounds are mixed, the result is a taste sensation in which the components either blend together harmoniously or clash. Examples of a harmonious mixture include sweet and sour, and sweet and bitter. An example of a clashing taste is the blend of sweet and salty.

    6. Often the pleasantness rating assigned to a taste mixture is higher than the pleasantness rating assigned to the single components. People like the quality of taste blends far more than the quality of the single components. Furthermore, sometimes one can add a disliked taste material (e.g., the bitter compound quinine) to a pleasant-tasting material (e.g., the sweet compound sucrose) and generate a far more pleasant sensation than from either component alone. In fact, the negative-tasting quinine enhances the liking of the sweet taste, to produce a sensation that is significantly better liked. (We might have expected the panelist to assign a liking rating somewhere in the middle, between the positive taste of sweet and the negative taste of quinine. Surprisingly, the panelist assigns to the mixture a higher number than was assigned to the pure sweet taste alone.)

What Olfactory Research Teaches

Odor perception is significantly more complex than taste perception. There is no simple set of fundamental smells as there is of fundamental tastes or colors.

The qualitative range of odor sensations is significantly greater than the qualitative range of taste sensations.

The scientific study of olfactory mixtures reveals at once some surprisingly simple rules, and some extraordinarily complex phenomena.

1. When similar smelling odorants are mixed, they add together to produce a more intense smell. Again, however, the smell intensity of the mixture is less intense than the arithmetic sum of the smell intensities of the components.

2. When dissimilar smelling odorants are mixed together, they suppress each other. The intensity of each smell in the mixture is diminished. In many cases the stronger smell completely masks the weaker smell.

3. Sometimes the separate smell qualities can be recognized in the mixture, and sometimes they cannot. Smell qualities may or may not fuse when components are combined. There are no general rules at this stage in olfactory science which predict whether a pair of smell qualities will fuse to produce a new impression or will remain blended in the mixture, yet be quite perceptible as different entities. (Keep in mind that when one mixes these two odorants, both odor intensities will diminish but the odor qualities will be separable on a perceptual basis.)

4. By adding more odorants to the mixture, the researcher increases the likelihood that the odor qualities will blend into an unanalyzable whole or "gestalt." (This is the basis of perfumery.) Thus, it is harder to distinguish the separate components in a five-component mixture than in a two-component mixture.

5. By adding more odorants to the mixture the researcher creates mutual suppression between the various components. However, the total sensory intensity of the olfactory mixture does not drop to 0 (not perceptible) as more components are added to the mixture. Rather, the mixture intensity oscillates, soon settling down to a middle level of intensity. This oscillation and convergence towards a mid-level occurs because all of the components of the blend contribute their own respective intensities to the mixture as well as suppress the perceived intensities of the other components. A dynamic process occurs in which an odorant (e.g., odorant A) is both suppressed by other odorants directly (e.g., suppressed by B), and released from suppression (by B) because a third odorant (C) suppresses the suppressor (B). The interactions (suppression, addition) are instantaneous and quite complex.

6. It is very difficult to know what a mixture will smell like, given the odor qualities of the components. The mixture will radically differ from the smells of the components, but often a totally new, blended fragrance will emerge from the mixture. The quality of the blend will be similar to the qualities of the components, yet somehow will be quite qualitatively new. Furthermore, as more odorants are added to the blend, the quality will change. Consumers presented with the blend will not easily perceive the blend as a blend, but rather will perceive the quality of the mixture to be an emergent "gestalt" or unity.

7. A very low intensity component, when added to a mixture of odorants, may affect the quality, but not necessarily the intensity. These low intensity components will "round out" the fragrance, and reinforce the unitary sensory perception of the mixture. They will also move the fragrance in various directions (not necessarily strongly, but noticeably). Thus, the contribution of a component to the mixture may be greater than one would expect on the basis of its sensory intensity. Furthermore, the contribution will be more "qualitative" than "quantitative." The component will more likely change the nature of the olfactory blend, rather than the intensity.

8. The liking rating assigned to a mixture of odorants is often higher than the liking rating assigned to any of its components. Consumers like blends of odorants far more than they like pure odorants alone. Simple examples of this can be seen in the perfumery laboratory. Pure odorants with characteristic qualities can be significantly improved (at least in terms of liking) by the addition of very low intensities of other components. The quality shifts slightly so that the pure odorant does not have the noticeable chemical smell. It is more "rounded." The characteristic quality still remains, albeit in a tempered, modified form. The panelist often prefers that mixture, even though the panelist does not recognize the mixture as a blend of components. The blend simply smells better.

7. An Overview

Classical measures of perception as developed by psychophysics provide the researcher with four clear benefits:

1. Discrimination testing shows how panelists perceive product differences. However, for applied research and product development (especially with consumer goods), the concept of discrimination must be replaced by the concept of "difference." Research should not look for consumer responses of "same versus different," but rather should measure the "degree of difference."
2. Intensity scaling allows the panelist to act as a measuring instrument. Basic psychophysical experiments show that there is a reliable, but non-linear relation between the physical intensity measured by the product developer and the sensory rating assigned by the panelist. This information is useful to set ingredient limits for product development and for quality control (where measured physical deviations from the normative level can be translated into consumer perceived changes in sensory magnitude).
3. Panelists can assign numbers to how much they like the stimulus. When panelists assign these numbers as well as assign numbers to represent degree of perceived intensity, the two resulting curves differ from each

other. Sensory intensity increases with physical intensity. Liking, however, increases with physical intensity, peaks, and then drops beyond the optimum. (This inverted U shaped curve will vary by stimulus, panelist, and intensity level of the physical stimulus.)
4. Most consumer products are mixtures. Psychophysics investigates mixtures, and shows how the consumer transforms mixtures of physical stimuli into complex sensory responses. There is both additivity and suppression. Stimuli with similar sensory characteristics add together so that the mixture is stronger than either component alone. However, it is rare that the mixture is as strong as the arithmetic sum of its components. Stimuli with different sensory characteristics (e.g., different qualities of tastes; different qualities of smell) suppress each other. Each component in the mixture suppresses the other components, so that the mixture intensity is weaker than most of the separate components. However, the laws of mixture are sufficiently complex and idiosyncratic to the components. It is difficult to predict the likely sensory profile of the mixture, given the sensory profile of the components.

Appendix: On the Expertization of Consumers

Introduction

Marketing researchers work with panelists who they believe to represent the population of consumers. In a market research study, one of the standard qualifications is that the consumer should not have participated in a research study in the past 3 months (or in some studies in the past 6 months). This qualification insures that the participating consumer does not become overly "used," and potentially cease to be representative of an actual consumer. With overuse of the panelist, the researcher worries that the panelist may no longer represent "the consumer." Alternatively it is feared that the panelist will turn into another type of individual (e.g., become somehow "expertized").

A Similar Problem—"Consumerizing Experts"

In a like fashion, many researchers use expert panelists to rate the sensory properties of products. These researchers begin by recognizing that the expert panelists are trained to act as objective "instruments"—that they are to assess the sensory properties of the product in the same way that an instrument assesses the physical characteristics of the product. There is to be no value judgment of "good vs. bad," nor any attempt to imitate the consumer. Yet, over time, the expert panelist often evolves into a new type of individual—not quite the original expert, but not quite a consumer. With increasing familiarity with the product, some expert panelists begin to incorporate judgments of "off-spec" vs. "on-spec." The expert panelists cease to be simply objective instruments to assess the specific characteristics of the products. Instead they begin to integrate the entire sensory profile of product characteristics into a unified whole, and add value judgments that the product is either "within spec" (within the standard target range) or "out of spec" (outside the target range).

How and Why These Subtle Shifts Occur—Overly Used Consumers

What is it about repeated testing of products in the same category that leads to the drift and change of consumer panelists into "experts" (or at least biased panelists)? In most studies there is no deliberate effort on the part of the researcher to create such an expertized individual. Certainly the panelist does not receive the feedback necessary to create an expert. In programs which train expert panelists there is a standardized procedure which provides feedback to the panelist, and which creates the expert by means of molding the reactions. Feedback works quite well in conventional training of experts.

Despite the lack of a formalized feedback system initiated by the researcher, there is always informal and unstructured feedback. Panelists who evaluate a product (e.g., a fragrance) again and again observe their own behavior. They observe the stimulus and their reactions. Panelists (and indeed all human beings) seek patterns and regularity in nature in order to deal with complexities. A panelist who evaluates the same type of stimulus again and again starts out evaluating a "blooming, buzzing confusion" of attributes. But, with repeated exposure the panelist begins to organize the confusion of attributes. The panelist does not need the researcher's feedback to organize. He merely needs the exposure to the stimulus in order to perceive recurrent patterns. Once the panelist realizes that there are patterns in the sensory attributes (e.g., a change in the product with time, such as initially vs. 30-second dry down), or once the panelist begins to recognize commonalties between some pairs of "similar products" which are absent in other pairs, the process of expertization has begun. The panelist looks for patterns, and reorganizes the percept.

As an example of the effect, consider the evaluation of simple, pure chemicals. In a study of different chemicals, each of which was evaluated multiple times, Moskowitz & Gerbers (1974) analyzed the complexity of the sensory rating profile assigned by a group of consumers. The sensory profiles for the same chemicals could be compared over different evaluations (e.g., the first set of evaluations, the second set of evaluations.) With each successive set of evaluations, panelists became more discriminating in their ratings. They recognized more and more sensory characteristics in the chemicals, as if they were first able to home in on the basic characteristics and then with increasing experience proceed to perceive the more subtle nuances. Such an effect can also be demonstrated by the evaluation of common spices and flavorings from other sources, such as vanilla. At first all the smells seem similar. They are all merely small, hard-to-describe variations around the common theme of vanilla. But, with repeated evaluations of the same vanilla samples, the panelist soon perceives and tries to label the fine nuances that differentiate one product from another. Panelists recognize more "notes" or characteristics in these vanilla fragrances, even without feedback.

Can There Ever Be a "Pure" Consumer Reaction?

If consumers are self-organizing systems which seek to create patterns and impose patterns on stimuli automatically, then can there ever be a pure consumer rating? All consumers have some degree of familiarity with the various products that are tested. Therefore, all consumers must be in different states of "organizing" the sensory characteristics of the products.

There probably is no pure consumer reaction to a product. What is to be avoided, however, is the forced acceleration of this perceptual organization. At the very least one must take this acceleration into account (e.g., in evaluating well-known, widely used products). It is probably best for the researcher to follow the two conventional market research guidelines listed below:

1. Use a spectrum of different consumers, including those who are heavy users versus those who are light or infrequent users. By all means, check the sensory profile ratings assigned by these different groups to the same product in order to determine the degree of difference. For instance, do heavy or frequent users of the product perceive the product to have a substantially different sensory profile than do lighter users?
2. Avoid having the same panelist evaluate the product on a repeated, frequent basis. It is inevitable that the panelist will become expertized. Panelists cannot help organizing their perceptions of a product after repeated exposures. This fact strongly suggests that an in-house testing facility should make every effort to expand its consumer base. The in-house employees, although not truly experts in the conventional sense of the word, have become expertized.

References

American Society for Testing and Materials, 1991. Determination of odor and taste thresholds by a forced-choice ascending concentration series method of limits. ASTM Annual Book Of Standards, 15.07, End Use Products, pp. 34–39.

Aust, L.B. 1989. Applications of sensory science within the personal care business: Part 1. Journal of Sensory Studies, *3*, 181–186.

Aust, L.B. 1989. Applications of sensory science within the personal care business: Part 2. Journal of Sensory Studies, *3*, 187–192.

Basker, D. 1980. Polygonal and polyhedral taste testing. Journal of Food Quality, *3*, 1-10.
Berglund, B., Berglund, U. and Lindvall, T. 1973. On the principle of odor interaction. Acta Psychologica, *35*, 255-268.
Berglund, P.T., Lau, K. and Holm, E.T. 1993. Improvement of triangle test data by use of incentives. Journal of Sensory Studies, *8*, 301-316.
Blakeslee, A.F. and Salmon, T.N. 1935. Genetics of sensory thresholds: Individual taste reactions for different substances. Proceedings of the National Academy of Sciences, *21*, 84-90.
Butler, G., Poste, L.M., Wolynetz, M.R., Agar, V.E. and Larmond, E. 1987. Alternative analyses of magnitude estimation data. Journal of Sensory Studies, *2*, 243-257.
Cain, W.S. 1969. Odor intensity: differences in the exponent of the psychophysical function. Perception and Psychophysics, *6*, 349-354.
Cain, W.S. 1975. Odor intensity: mixtures and masking. Chemical Senses and Flavor, *3*, 339-352.
Draper, N.R. and Smith, H. 1981. Applied Regression Analysis. John Wiley & Sons, New York.
Engen, T. 1964. Psychophysical scaling of odor intensity and quality. Annals of the New York Academy of Sciences, *16*, 504-516.
Engen, T. 1971A. Psychophysics 1. Discrimination and detection. *In:* Woodworth and Schloserg's Experimental Psychology, Third Edition, J.W. Kling & L.A. Riggs (eds.), pp. 11-46. Holt, Rinehart and Winston, New York.
Engen, T. 1971B. Psychophysics 2. Scaling Methods. *In:* Woodworth and Schloserg's Experimental Psychology, Third Edition, J.W. Kling & L.A. Riggs (eds.), pp. 47-86. Holt, Rinehart and Winston, New York.
Ennis, D.M. 1993. The power of sensory discrimination methods. Journal of Sensory Studies, *8*, 353-370.
Fechner, G.T. 1860. Elemente der Psychophysik. Breitkopf und Hartel, Leipzig.
Fisher, R.A. 1956. Mathematics of a lady tasting tea. *In:* The World Of Mathematics, Vol. III, pp. 1512-1520. Simon and Schuster, New York.
Giovanni, M.E. and Pangborn, R.M. 1983. Measurement of taste intensity and degree of liking of beverages by graphic scales and magnitude estimation. Journal of Food Science, *48*, 1175-1182.
Givon, M. 1989. Taste tests: changing the rules to improve the game. Marketing Science, *8*, 281-290.
Gordon, J. 1965. Evaluation of sugar-acid-sweetness relationships in orange juice by a response surface approach. Journal of Food Science, *39*, 903-907.
Gridgeman, N.T. 1959. Sensory item sorting. Biometrics, *15*, 298-306.
Gridgeman, N.T. 1970. Re-examination of the two-stage triangle test for perception of sensory differences. Journal of Food Science, *35*, 87.
Johnson, J.L., Dzendolet, E. and Clydesdale, F. 1983. Psychophysical relationship between sweet and redness in strawberry flavored drinks. Journal of Food Protection, *46*, 21-25.
Kroeze, J.H.A. 1982. Reduced sweetness and saltiness judgments of NaCl-sucrose mixtures depend on a central inhibiting mechanism. *In:* Determination Of Behavior By

Chemical Stimuli, J.E. Steiner & J.R. Ganchrow (eds.), pp. 161-174. IRL Press Ltd., London.

Lawless, H.T. 1977. The pleasantness of mixtures in taste and olfaction. Sensory Processes, *1*, 227-237.

Lawless, H.T. 1984. Flavor description of white wine by expert and nonexpert wine consumers. Journal of Food Science, *49*, 120-123.

Lawless, H.T. and Klein, B.P. (eds.) 1991. Sensory Science Theory And Applications In Foods. Marcel Dekker, New York.

Lawless, H.T. and Malone, G.J. 1986. The discriminative efficiency of common scaling methods. Journal of Sensory Studies, *1*, 85-98.

Marks, L.E. 1968. Stimulus range, number of categories, and the form of the category scale. American Journal of Psychology, *81*, 467-479.

McBride, R.L. and Finlay, D.C. 1989. Perception of taste mixtures by experienced and novice assessors. Journal of Sensory Studies, *3*, 237-248.

McBride, R.L. and Johnson, R.L. 1987. Perception of sugar-acid mixtures in lemon juice drink. International Journal of Food Science and Technology, *22*, 399-408.

McBurney, D.H. 1984. Taste and olfaction: sensory discrimination. *In:* Handbook Of Physiology - The Nervous System III, J. Brookhart and V.B. Mountcastle (eds.), pp. 1067-1086. Williams and Wilkins, Baltimore.

Morrison, D.G. 1981. Triangle taste tests: are the subjects who respond correctly lucky or good? Journal of Marketing, *45*, 111-119.

Moskowitz, H.R. 1971. Intensity tastes for pure tastes and for taste mixtures. Perception & Psychophysics, *9*, 51-56.

Moskowitz, H.R. 1972. Perceptual changes in taste mixtures. Perception & Psychophysics, *11*, 257-262.

Moskowitz, H.R., 1981A. Sensory intensity versus hedonic functions: classical psychophysical approaches. Journal of Food Quality, *5*, 109-138.

Moskowitz, H.R. 1981B. Relative importance of perceptual factors to consumer acceptance: linear versus quadratic analysis. Journal of Food Science, *46*, 244-248.

Moskowitz, H.R. 1984. Cosmetic Product Testing: A Modern Psychophysical Approach. Marcel Dekker, New York.

Moskowitz, H.R. and Fishken, D. 1979. What effects do repeated evaluations play in fragrance perceptions? Perfumer and Flavorist (May), 45-52.

Moskowitz, H.R. & Gerbers, C.M. 1974. Dimensional salience of odors. Annals of the New York Academy of Sciences, *237*, 3-16.

Moskowitz, H.R., Jacobs, B. and Firtle, N. 1980. Discrimination testing and product decisions. Journal of Marketing Research, *17*, 84-90.

O'Mahony, M. 1992. Understanding discrimination tests: a user-friendly treatment of response bias, rating and ranking R-Index tests and their relationship to signal detection. Journal of Sensory Studies, *7*, 1-48.

Pangborn, R.M. 1960. Taste interrelationships. Food Research, *25*, 245-256.

Pearce, J.H., Korth, B. and Warren, C.B. 1986. Evaluation of three scaling methods for hedonics. Journal of Sensory Studies, *1*, 27-46.

Powers, J.J. 1988. Mathematics of a lady tasting tea revisited. Journal of Sensory Studies, *3*, 151-158.

Riskey, D.R., Parducci, A. and Beauchamp, G.K. 1979. Effects of context in the judgments of sweetness and pleasantness. Perception & Psychophysics, 26, 171–176.

Shepard, R.N. 1962. The analysis of proximities: Multidimensional scaling with an unknown distance function. Psychometrika, 27, 219–246.

Stevens, S.S. 1953. On the brightness of lights and the loudness of sounds, Science, 118, 576.

Stevens, S.S. 1975. Psychophysics: An Introduction To Its Perceptual, Neural And Social Prospects. John Wiley & Sons, New York.

Thurstonian, R. 1971. On the exponents in Stevens' law and the constant in Ekman's law. Psychological Review, 78, 71–80.

Von Skramlik, E. 1921. Mischungsgleichungen im Gebiete des Geschmacksinnes, Zeitschrift fur Sinnesphysiologie, 5(3B), 26–78, 142–143.

6

Individual Differences: Conventional Subgroups, Preference Segmentation and the Construction of a Product Line

1. Introduction

Individual differences pervade the world of consumer products. One need only look at the vast number of products of the same type in the market to recognize the effect of individual preferences on the array of products offered by a manufacturer. Dozens of fragrances are marketed annually to appeal to all sorts of consumers. These fragrances vary in their sensory character, packaging graphics, image, and pricing. On a more mundane level individual preferences for lipstick colors and mascaras force manufacturers to offer many dozens of shades in order to please the consumer. Advertising agencies recognize these individual differences as well. The advertising for some products is geared to convey an "upscale" image, which advertisers know to attract some consumers but scare away others. Other advertising stresses the functional benefits of the product and is often geared to individuals who are more price conscious.

Scientists and marketers view individual differences from two perspectives. To the scientist interested in the basics of sensory perception, individual differences are annoying factors which cloud otherwise easy-to-understand rules about consumer perception of products. To the marketer, however, these individual differences signal that the marketplace can accommodate several different products, not just one. Each consumer segment can be satisfied by a different product.

2. Current Ways to Divide the Consumer Population

There are a variety of ways to divide the population of consumers depending upon one's interest. Marketers and marketing researchers divide consumers on the basis of demographic breaks (e.g., market, age, income) or on the basis of purchase behavior (e.g., purchasers vs. non-purchasers of the category; brand used most often, ever use the brand, or have used the brand within the past 3 months). The rationale here is that consumers may differ in their preferences based upon their gender (e.g., males like some products whereas females may dislike the same products), or purchase behavior. When individual differences are based upon purchase behavior, the unwritten assumption is that individuals who show different purchase behavior will also exhibit different product preferences, and vice versa. According to the marketer, individual differences for the most part should be traceable to the demographic or purchase intent patterns.

Researchers, both basic and in R&D, approach individuals from a different point of view. Many researchers recognize that there is no simple accounting for person-to-person differences. Perhaps there is some correlation between a person's age and product preferences, but that correlation is tenuous and at best only suggestive of a deeper reason for the individual differences.

Over the past two decades, basic researchers in marketing have looked for personality or lifestyle bases underlying individual differences. The assumption is that consumers with different personalities or with different outlooks on the world will exhibit different product preferences. Thus, a person who is serious may have different product preferences than a person who is lighthearted, carefree, and young.

The jury is out on these methods for segmenting consumers. Although personality and lifestyle segmentation is popular among marketing researchers, it is not clear that this way of dividing the consumer population really uncovers product preferences. Consumers can be defined as belonging to a segment, but it is not clear that people in a segment like the same products.

Ten years ago the technique of "sensory preference segmentation" (Moskowitz et al., 1985) was introduced as an approach to divide consumers. The rationale for sensory preference segmentation was that individuals differ on the basis of the sensory aspects of the products that they like and dislike. Individuals falling into the same sensory preference segment show similar patterns of likes and dislikes. For instance, in fragrances there may be two preference segments—one segment liking fruity/flowery smells and another segment liking spicy smells. These preference segments are based upon responses to actual products, and how sensory attributes drive overall liking. (These preference groupings may or may not correlate with a person's age, income, or brand usage.) We will deal with this concept in detail later on in this chapter.

3. Individual Differences—Scale Usage and Sensory Perception

Individuals can differ in a variety of ways. Some researchers assume that individual differences manifest themselves in sensory perception. For instance, individual differences may emerge because some consumers rate a product high on sensory magnitude (e.g., perceived graininess of a powder) whereas other consumers rate the same product low. The individual data look quite different. The naïve interpretation is that the panelist who assigns high numbers on the scale perceives the product to be strong, whereas the panelist who assigns low numbers on the same scale to the same product perceives the product to be weak on that attribute.

The foregoing variability is one minor form of individual difference. The difference manifests itself in scaling behavior. The two panelists may perceive the product in an identical fashion. However, the panelists use the scale differently. One individual may prefer to use the full range of the scale, and reserve the low end of the scale for stimuli that are weak. Another individual may prefer to collapse his or her ratings to the middle of the scale, and not venture out toward the extremes unless the attribute being rated is clearly and unmistakably perceived to be very strong or very weak. Only in those extreme cases does the panelist assign a high number at the top end of the scale or a low number at the bottom end of the scale.

Scaling differences are interesting but not particularly productive. We are not interested in the way consumers use numbers. If we were, then we would find that in many test situations the variability between consumers in the use of numbers far exceeds the variability between the products being tested. This variability in scaling behavior is simply an artifact of scaling, pervasive, and differing by individual. The variability does not provide the researcher with a true measure of sensory or hedonic variability across consumers in the population.

Individual Differences—Sensory Perception

Beyond the level of scaling there are individual differences in perceived sensory intensity. Researchers recognize that some individuals possess greater sensory acuity than others. It has long been a problem to demonstrate these individual differences in perceived sensory intensity without confounding the data with artifacts due to scaling. (If two panelists assign different numbers to the same stimuli and one panelist assigns higher numbers, then does this automatically mean that the panelist assigning the higher numbers also perceives the stimulus to be more intense?)

In the simplest of cases the researchers who recognized this confounding between sensory intensity and scale have resorted to threshold measurement.

(The threshold is the lowest physical intensity that a panelist can perceive. Below this physical level the panelist reports no perceived sensory intensity at all.) According to those who use threshold measures, individual differences consist of demonstrating that individuals exhibit dramatically different thresholds. Some individuals will exhibit high thresholds so that they require a high physical intensity in order to detect the stimulus. Other panelists demonstrate low thresholds so that they require a low physical intensity to detect the stimulus. The threshold is a physical measure, and thus avoids bias due to scale usage (although the threshold measurement is subject to attitudinal and payoff biases). It is assumed (albeit perhaps incorrectly) that panelists exhibiting a low threshold are "more sensitive" to the stimuli and thus perceive the stimuli to be stronger, whereas panelists exhibiting a high threshold are "less sensitive" and thus perceive the stimuli to be weaker.

There is no scientific basis on which to assume that there exists a relation between perceived sensory intensity at above-threshold levels and the threshold level needed for perception. Some of the strongest refuting evidence comes from the chemical senses of taste and smell. Saccharin and aspartame have thresholds for the perception of sweetness many hundreds of times lower than the threshold of sugar. Yet, at above-threshold levels saccharin and aspartame are not perceived to be "hundreds of times sweeter." (If one liquid is perceived to be even 50 times sweeter than another, that difference in sweetness would be so dramatic as to be overwhelming.) A similar observation holds for different odorants. Some odorants have thresholds thousands of times lower than another (e.g., musk can be detected at thresholds tens of thousands of times lower in absolute concentration than can alcohols or the fruit fragrance of esters). Yet, consumers perceive these odorants to have the same intensity at supra-threshold concentration.

Despite the lack of a clear relation between threshold level and supra threshold intensity, researchers who study intensity perception have not given up. They recognize that some consumers are impaired in their sensory perception whereas others are acute. These researchers have been stymied by most scaling procedures. However, some investigators (e.g., Marks et al., 1986) have suggested that the panelists rate two sets of stimuli varying in their physical intensity. One set of stimuli comprises the reference stimuli. These are stimuli (e.g., sounds, lights) for which one believes there to be little, if any, interpanelist differences. The second set of stimuli comprises the test stimuli, which the researcher believes to be subject to interpanelist differences. The researcher can index each individual's ratings against the set of stimuli that are presumed to be equivalent across consumers. Thus, it is assumed that people differ in their sensory responsiveness to stimuli. The commonality to a reference set allows the researcher to compare the panelists to each other, after ratings from each panelist have been separately indexed against the reference set stimuli.

4. Inter-Individual Differences in Hedonics—Where the Gold Really Lies

Individual differences really reveal themselves when it comes to hedonics. Although individuals can be shown to have similar sensory acuities and similar scaling behavior, they may differ radically as to whether or not they like the stimulus. Individual differences are most noticeable in the chemical senses (taste and smell), and especially with regard to actual fragrances and foods. Fewer individual differences exist when it comes to appearance (vision), hearing (audition), or kinaesthesis (touch).

Individual Differences in Liking of Simple Stimuli

The earliest examples of individual differences in hedonics go back to the beginning of the twentieth century. Researchers at that time were interested in the constancy of taste and smell perceptions across the population. In some of the simplest studies the researcher would present the panelist with a stimulus and instruct the panelist to respond whether the stimulus was liked or disliked. What was interesting about the results was that virtually no stimulus received universal acceptance or rejection. The most acceptable fragrance was the smell of oil of roses, the most objectionable was the smell of carbon bisulfide. However, even these two odorous stimuli were not universally liked or universally disliked. Some people disliked the smell of roses, other people liked the smell of carbon bisulfide.

Individual Differences in the Importance of Sensory Inputs—Lotion

People differ from each other in the attention that they pay to stimuli. Specifically, individuals differ in their opinion of the relative importance of appearance, smell, and touch as those sensory inputs drive product acceptance. We can demonstrate this variation between consumers quite easily if each of the consumers evaluates a set of complex products. These products must excite many different sensory inputs.

Our example here is lotion. The panelists comprised lotion users. Each of the panelists rated overall liking as well as liking of appearance, liking of fragrance, and liking of feel of the product when it was applied. The full set of products comprised 26 formulations that ranged in appearance, fragrance, and skinfeel. Panelists used seven of the products for one week each. The products were rotated so that each panelist was assigned a random set of seven. Panelists rated their reactions to the product on each day. The data here will be presented for the first day's use only. (One could do even more in-depth analyses on the products, looking at the day-by-day ratings, to determine whether ongoing experience with a product changes the relative importance of sensory inputs.)

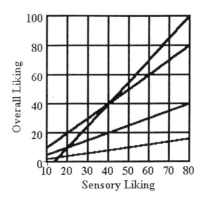

Figure 1 Example of leverage (or linear slope) analysis. The independent variable is the attribute liking rating. The dependent variable is the overall liking rating. Each line corresponds to a different sensory attribute. Steep slopes correspond to important attributes. Flat slopes correspond to unimportant attributes.

Importance of a sensory input can be measured by the slope relating overall liking to attribute liking. Figure 1 shows four slopes, ranging from 0.2 to 1.5. The independent variable is attribute liking, the dependent variable is overall liking. The slope of the straight line measures relative importance. The higher the slope the more important the sensory input. For instance, if the slope is 0.2, then a 1-unit increase in attribute liking corresponds to a 0.2-unit increase in overall liking. Clearly the attribute liking is not a strong driver of overall liking since large increases in the attribute liking do not correspond to equally large (or even larger) increases in overall liking. In contrast, when the slope is high (e.g., around 1.0 or higher) then the attribute liking is important because it covaries with a strong effect on overall liking.

We can look at group data to find out the relative importance of each sensory input. But, we can also go one step further and look at individual data. Each individual separately scaled the three attribute likings (appearance, fragrance, skinfeel), along with overall liking. Thus, we can compute the slopes for each individual for each attribute liking. The distribution of these slopes shows us the range of relative importance.

Figure 2 (A-C) shows the distributions of the three slopes. There is a great deal of interindividual variability in the slopes, showing different relative importances.

An even more instructive analysis considers percentage data. Each panelist generates three numbers—the relative slopes for the three sensory inputs. These three slopes can be percentaged so that they add up to 1.0. In this way all the panelists are put on the same scale. We can then plot the relative slopes

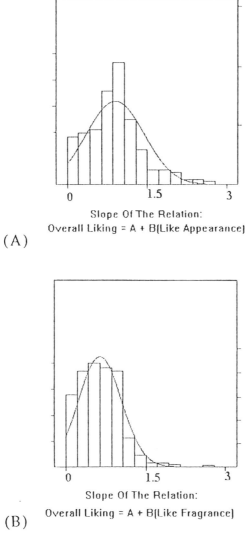

Figure 2 Distribution of slopes from individual panelists for lotion. (A) Liking of appearance (independent variable) vs. overall liking, (B) Liking of fragrance vs. overall liking, and (C) Liking of skinfeel vs. overall liking.

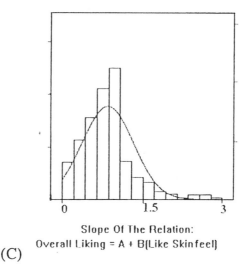

(C)

Slope Of The Relation:
Overall Liking = A + B(Like Skinfeel)

Figure 2 Continued

for lotion on triangular coordinate scales, as shown in Figure 3A. The location of the panelist as a point in the triangular coordinates tells us what sensory input is important to the panelist. Points very close to one vertex indicate that the panelist pays a great deal of attention to that sensory input, and pays little attention to any other input. Points lying on the line joining two vertices mean that the panelist pays attention to those two sensory inputs, but ignores the third. Finally, points lying in the middle mean that the panelist attends to all three sensory inputs.

Figure 3A suggests that the vast majority of the panelists pay attention to appearance and skinfeel, and slightly less attention to fragrance. However, individuals differ. They do not all pay equal attention to appearance and skinfeel.

Individual Differences in Sensory Inputs—Anti-Perspirants

Let us look at another product category—anti-perspirants. The same procedure used earlier in the lotion study was followed here. The panelists evaluated six products, each over a 6-week period. The products were in-market anti-perspirants. Panelists rated the products on a series of liking and sensory characteristics for each day. Every product was used for 5 days. On days 6 and 7 of each week, the panelist used a "blank" or washout product containing no active ingredient.

Figure 3B shows the results. Figure 3B shows the individuals plotted in triangular coordinates (after the slopes for appearance, fragrance, and skinfeel

Individual Differences

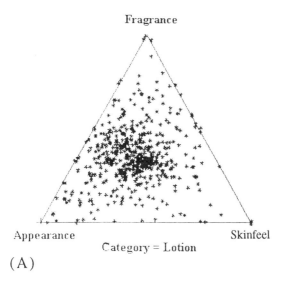

(A)

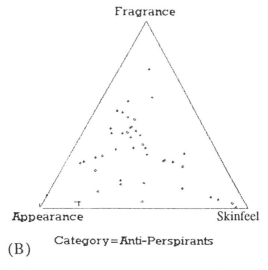

(B)

Figure 3 Distribution of relative slopes in triangular coordinates. Each vertex corresponds to attribute liking. Each point corresponds to a particular panelist. Points close to a vertex mean that the attribute is most important. Points close to the middle of the coordinate system mean that all attributes are equally important in driving overall liking. (A) Category = lotion, vertices are liking of appearance, fragrance and skinfeel, respectively. (B) Category = anti-perspirant, vertices are appearance, fragrance, and skinfeel.

were converted to percentages). We see the same pattern, but this time appearance and fragrance are significantly more important. We see again that the panelists do not use the same criteria. Some panelists are more visually oriented, others more smell oriented, and still others more touch oriented.

5. Different Cognitive Worlds: Do We Look at Products Differently?

Introduction

Individuals differ in the way that they respond to the attributes of a product. We can uncover the nuances of those individual differences by asking consumers to tell us what different attributes mean to them. We will soon find that individuals live in different "cognitive worlds." The worlds do intersect, however. For instance, most people will agree on the definition of a harsh smell, or at least agree that one fragrance smells harsh whereas another fragrance does not. The interindividual agreement is high, but as in every other aspect of product testing the agreement is not perfect.

A very instructive analysis looks at the correlations between attributes for a single panelist who rates many products on many attributes. Which attributes covary and which vary independently of each other? This analysis leads naturally to a panelist-by-panelist "factor analysis" of attributes. The factor analysis shows the underlying structure of the different attributes, on a panelist-by-panelist basis. It quickly reveals whether consumers use the attributes similarly.

Study 1—Anti-Perspirant Fragrances

Fragrances are the easiest stimuli to work with in terms of linguistic differences because they excite only one sense input—smell. We don't have to worry about the panelists paying different degrees of attention to appearance, fragrance, and touch.

In this study the fragrances were selected to be appropriate for an antiperspirant. The stimuli were therefore functional fragrances. (The original objective of the study was to select a line of different fragrances for a new antiperspirant product.) The panelists participated in a central evaluation test in which each panelist rated 15 different fragrances on a variety of sensory and liking attributes, shown in Table 1A. Some of the attributes are purely sensory in nature (e.g., sweet smell). Other attributes are image attributes (e.g., appropriate for use during sports).

The factor analysis of the sensory data for the total panel (averaging individuals, but looking at differences on a product-to-product basis) shows that there are three different factors along which the sensory attributes fall (Table 1B). However, on an individual-by-individual basis this simplistic pattern no longer

Table 1A Attributes Used for Evaluation of Anti-Perspirant Fragrance

Liking of fragrance
Strength of fragrance
Clean smelling
Fresh smelling
Harsh smelling
Refreshing smell
Unique smell
Medicinal smell
Masculine vs. feminine smell
Smells like it will stop odor
Smells like it will keep me dry
Smells like it will make me feel dry
Fruity smell
Citrusy smell
Sweet smell
Light smell
Spicy smell
Crisp smell
Musky smell
Smells like it will last long
Smells invigorating
Smells acrid
Smells stale
Smells like it will leave no residue on skin
Appropriate for younger vs. older person
Appropriate for regular use
Appropriate for after sports

holds. Each panelist generates a factor structure. Some individuals show more dimensions, and others show fewer dimensions when we factor analyze the individual data. (The data from each panelist was treated similarly—extraction of all "roots" with eigenvalues greater than 1.0, and rotation to a simple quartimax solution. The quartimax solution creates an easy-to-interpret factor structure.) Table 1C shows this factor structure. The same attribute may fall into different factors when we look at individual data. On a consumer-by-consumer basis some individuals generate one general factor, whereas other individuals use the attributes in distinct ways so they generate multiple factors. This latter group which generates more sensory factors may live in a more complex world of fragrance perception (or at least a more complex world of fragrance description by attributes).

Table 1B Factor Structure for the Full Panel of Consumers—Anti-Perspirant Fragrance

Attribute	Loading On		
	Factor 1	Factor 2	Factor 3
Harsh	0.73	−0.24	0.08
Medicinal	0.65	−0.29	0.07
Acrid	0.63	0.14	0.01
Spicy	0.59	0.33	−0.11
Stale	0.53	−0.38	0.16
Fresh	−0.14	0.86	0.14
Clean	−0.12	0.84	0.10
Crisp	0.24	0.75	0.01
Fruity	0.07	0.02	0.85
Citrusy	0.21	0.09	0.78
Sweet	−0.05	0.18	0.73
Feminine	−0.22	−0.15	0.55
Light	−0.27	0.23	0.25
Musky	0.06	0.13	−0.11
% Variance	18%	18%	16%

Note: Loading = correlation of attribute with the factor.

Study 2—Shampoos

The panelists used six shampoos at home, each for 1 week. Each panelist rated every shampoo on a wide variety of attributes. Table 2A shows the list of attributes. It is more instructive to limit the factor analysis to groups of similar attributes (e.g., all sensory attributes) than to mix classes of attributes (e.g., combine sensory and liking attributes).

The first analysis of the shampoo data is a factor analysis of the average ratings (across panelists) for the sensory ratings of the products. The factor structure appears in Table 2B. We see from this structure that the sensory attributes yield one factor.

The next analysis consists of an individual-by-individual factor analysis. Table 2C shows the results in an easy-to-read form. We can see that the simple three-factor structure we obtained by averaging across the panelists reappears, but in different forms across people. People use attributes differently. Some individuals show factor structures comprising only one dimension. Most individuals show factor structures comprising two or more dimensions. However,

Table 1C Factor Structures for Individuals—Anti-Perspirant Fragrance

	Percent of Panelists with a Given Factor Structure—Anti-perspirant Fragrances							
			1 Factor	4%				
			2 Factors	28%				
			3 Factors	60%				
			4 Factors	8%				
			Significant Loadings on Factors					
	F1+	F1−	F2+	F2−	F3+	F3−	F4+	F4−	None

	F1+	F1−	F2+	F2−	F3+	F3−	F4+	F4−	None
Factor 1									
Harsh	14	4	8	3	1	0	0	0	0
Medicinal	10	6	4	1	4	0	0	0	25
Acrid	7	5	7	4	8	2	3	0	14
Spicy	7	4	6	4	7	3	0	0	19
Stale	7	5	6	3	1	4	1	0	23
Factor 2									
Fresh	14	9	10	4	4	0	1	1	7
Clean	13	10	7	4	4	0	0	0	38
Crisp	10	4	7	2	3	0	3	1	20
Factor 3									
Fruity	12	2	8	1	7	1	1	0	18
Citrusy	17	3	7	3	5	1	1	0	13
Sweet	13	3	6	2	5	0	1	0	20
Feminine	9	8	3	0	4	2	0	0	24
No Loading									
Light	5	5	1	5	2	0	2	0	30
Musky	8	8	6	2	2	5	0	0	19

Numbers = number of panelists for whom the attribute loads either positively (+) or negatively (−) on the factor

even if two individuals show the same number of dimensions, each individual will show different attributes falling into each category.

We can do a separate analysis for the liking ratings. Although most marketing and sensory researchers believe that consumers can separately evaluate the liking of appearance, fragrance, and feel, it is not clear whether these dimensions truly differ from each other. They may correlate highly, meaning that the consumer has a hard time responding to each of them independently, either because it is truly difficult to separate the sensory inputs and rate the liking of each input separately, or because of response biases and the tendency to uprate a product on all attributes if the product is uprated on one attribute. This is called the "halo effect."

Table 2A Attributes Used for the Shampoo Test

Like the shampoo
Like the fragrance
Strength of fragrance
Perceived cleanliness of hair after shampoo
Like the appearance of the liquid
Like the color of the shampoo
Thickness of the liquid
Like thickness of the lather
Quickness with which the shampoo lathers
Amount of lather
Ability of the shampoo to lather
Amount of shampoo used
Ease of rinsing out the shampoo and lather
Clean feeling of hair
Amount of overcleaning of hair by the shampoo
Ease of combing the hair when wet
Gentleness of the hair by feel
Ease of combing the hair when dry
Shininess of the hair when dry
Like the shininess of the hair
Manageability of the hair
Amount of body possessed by hair

Table 3A shows the results for the factor analyses of the liking ratings. Again we see that, whereas the total panel shows only one major factor (Table 3A), in fact there is a distribution of response patterns across the different consumers. Many consumers do, in fact, fail to behaviorally distinguish between the various liking attributes so that their liking ratings really yield only one

Table 2B Factor Structure for the Full Panel of Consumer Sensory Attributes—Shampoo

Attribute	Loading on factor
Amount of lather	0.86
Pours quickly	0.82
Thickness of lather	0.72
Shininess when hair dry	0.44
Thickness of shampoo	0.41
Strength of fragrance	0.18
Amount used	−0.48
Amount of residue	−0.52

Individual Differences

Table 2C Factor Structure for Sensory Attributes—by Individual—Shampoo

Percent of Panelists with a Given Factor Structure—Shampoo Sensory Attributes

	1 Factor	5%
	2 Factors	65%
	3 Factors	10%
	4 Factors	15%
	5 Factors	5%

Significant Loadings on Factors

	F1+	F1−	F2+	F2−	None
Amount of lather	28	3	4	0	0
Pours quickly	18	4	7	0	6
Thickness of lather	22	2	5	2	4
Shininess when hair dry	6	6	9	1	13
Thickness of shampoo	11	1	9	1	13
Strength of fragrance	8	4	9	7	7
Amount used	4	7	10	8	6
Amount of residue	5	7	6	3	14

underlying dimension. Yet there are other panelists who clearly differentiate between the different attributes. These other panelists show more complex factor structures, so that their separate liking scales, in fact, represent truly different criteria for judging acceptance. They do not lump all of the sensory inputs together when making their liking ratings.

An Overview

Consumers live in different cognitive worlds. Researchers know that consumers exhibit significantly different preferences when it comes to fragrances and other personal products. However, the differences are not only in terms of the

Table 3A Factor Structure for the Total Panel of Consumers Liking/Performance Attributes—Shampoo

Attribute	Loading on factor
Like appearance	.81
Like the hair shine	.71
Like color	.68
Like fragrance	.61
How well it cleans	.56
Like thickness	.42
How well it lathers	.21

Table 3B Factor Structures for Individuals—Shampoo (Liking and Performance Attributes)

Percent of Panelists with a Given Factor Structure—Shampoo Liking and Performance Attributes

1 Factor 66%
2 Factors 34%

	Significant Loadings				
	F1+	F1-	F2+	F2-	None
Like appearance	18	0	7	1	9
Like the hair shine	12	4	6	1	12
Like color	14	1	7	6	7
Like fragrance	12	2	6	4	11
How well it cleans	15	2	6	3	9
Like thickness	11	1	8	2	13
How well it lathers	6	1	10	1	17

ratings of liking. There are substantial person-to-person differences in the way that the individual organizes and uses descriptive terms in questionnaires. This aspect of individual differences straddles perception and cognition, and needs more investigation. Those who train expert panelists know that individuals use terms differently. This data reaffirms that the consumers live in different worlds, in terms of the complexity of their perception of product characteristics as reflected in rating attributes.

6. Individual Differences in Reactions to Products vs. Brands or Packages

Introduction

Up to now we have considered individual differences in responses to simple stimuli or products. We know, however, that in the actual marketplace, hedonic response (liking/disliking or purchase intent) is affected both by the product characteristics and by the brand name and packaging. Quite often the effect of packaging and brand identification is greater than the effect of products, especially with regard to fragrances. We can ask two new questions about individual differences:

1. Are individual differences greater with blind products than with branded products?
2. Can we identify consumers who are more responsive to product characteristics, versus consumers who are more responsive to label (brand) or

package characteristics? That is, can we identify consumers who have different criteria for overall product acceptance? In this second example we look for differences among consumers in terms of what is important to them—product or brand label.

Individual Differences in Liking of Blind vs. Branded Products

To examine the range of individual differences in blind versus branded products, we have the same panel of consumers evaluate a set of products, first "blind" (without any identification) and then in packaged or identified form. The physical stimuli are identical (although coded with different identification numbers). However, the first set is presented in an "unromanced" format, as simple stimuli, whereas the second set is presented in marketed format.

The key question we want to answer is whether we obtain greater variability of the products on a blind basis or on a branded basis. The criterion attribute is overall liking. How do individuals differ from each other? For instance, on a blind basis we might find a narrower range of liking ratings for the total panel, whereas on a branded basis we might find a wider range of liking ratings. We can go one step further and analyze the data by analysis of variance. The analysis of variance breaks out the variability in the data into three components—variability due to the product, variability due to the panelist, and error variability due to unexplained causes. As we will see, the structure of variability in ratings changes when we identify the products.

In the example below the panelists rated six fragrances twice, once on a blind basis and once on a branded basis. The blind ratings comprised evaluations of the fragrance on a blotter. The branded ratings comprised evaluations of the same fragrance on a blotter, but with the bottle (and thus the brand) clearly in front of the panelist so that the panelist had no problem identifying the fragrance. The evaluation with 60 consumers took 3 hours, with approximately 10 minutes between each fragrance evaluation. The regimen maintained sensitivity during the course of the evaluation.

The results from the study (mean ratings of liking, analysis of variance tables) appear in Tables 4A–4C. It is clear that, overall, the products differed more when the fragrances were identified than when the fragrances were tested blind. [We see this from the high F ratio for "product"—in this case, fragrances]. However, the people differed less! Branding the fragrance reduced the interpersonal variation betweem consumers (lowered the F ratio of the panelist), but increased the product-to-product differences (raised the F ratio of the product).

Individual Differences in Reactivity to Product vs. Package/Label

The second area of interest is the degree to which consumers differ in their reaction to product versus brand. What drives consumer interest? We know from

Table 4 Results from Fragrance Evaluation: Blind vs. Branded

A—Effects of Fragrance and Branding Condition on Attribute Ratings

Attribute	Condition	F-ratio panelist	Interpretation	F-ratio product	Interpretation
Liking	Blind	1.21	No effect	1.84	No effect
Liking	Branded	0.89	No effect	5.45	Significant
Sweet	Blind	0.98	No effect	1.04	No effect
Sweet	Branded	1.17	No effect	9.53	Significant
Spicy	Blind	0.21	No effect	1.06	No effect
Spicy	Branded	0.87	No effect	3.04	Significant
For Older	Blind	1.14	No effect	2.02	Slight significance
For Older	Branded	0.99	No effect	7.77	Significant

B—Sum of Squares for Fragrance Data
(Total Variability of Ratings Partialed Out by Product, Panelist, Error)

Attribute	Condition	Product sum of squares	Panelist sum of squares	Error sum of squares	Total sum of squares
Liking	Blind	2278	9276	38480	50034
Liking	Branded	3465	2836	20648	26949
Sweet	Blind	992	5777	29497	36266
Sweet	Branded	5151	3920	16756	25827
Spicy	Blind	1500	10779	44041	56320
Spicy	Branded	1982	3509	20197	25688
For Older	Blind	2265	7924	34628	44817
For Older	Branded	4581	3599	18270	26450

C—Source of Variation in Fragrance Ratings on a Percentage Basis

Attribute	Condition	Product % SS*	Panelist % SS*	Error % SS*	Total % SS*
Liking	Blind	5%	18%	77%	100%
Liking	Branded	12%	11%	77%	100%
Sweet	Blind	3%	16%	81%	100%
Sweet	Branded	20%	15%	65%	100%
Spicy	Blind	3%	19%	78%	100%
Spicy	Branded	8%	14%	78%	100%
For Older	Blind	5%	18%	77%	100%
For Older	Branded	17%	14%	69%	100%

* % SS = Percent of total sum of squares.

Individual Differences

the research studies above that individuals differ in what they like. But, how strong is that individual-to-individual difference in acceptance of products? Is it stronger than individual differences due to the package or the brand? Do people differ more in what sensory characteristics they like or in what brands they like?

To answer this question we do a "crossover" experiment. Each panelist must test at least four stimuli, as follows:

1. Two products under one package/label condition
2. The same two products under a different package/label condition.

(The label or brand condition may be appropriate for the product, or may simply be a label or brand that is tested because it is convenient.)

Sunscreen Study. In this study a group of 115 panelists participated. The panelists evaluated four products as follows:

1. Two sunscreens under Brand A (in Brand A's plastic bottle)
2. The same two sunscreens under Brand B (in Brand B's plastic bottle).

The sunscreens differed substantially from each other in terms of their sensory characteristics. Panelists used the four products at home during the summer, applying one product per day on each of four days. The panelists rated the product on a variety of characteristics just after use.

The results from this study can be analyzed either in aggregate or on an individual-by-individual basis. The results for any individual can be represented by an equation, as follows:

Rating = Constant + k_1(Product B) + k_2(Brand B)

The equation states that the rating for a product can be represented as the effects of three contributions:

1. Constant (which is the estimated rating of liking assigned to product A, presented in package A)
2. The effect of switching from product A to product B (given by the value k_1), and
3. The effect of switching from Brand A to Brand B (given by the value k_2).

By itself the foregoing equation is simply a model which shows the net effect of switching from one product to another (keeping the brand the same) or switching from one brand to another (keeping the product the same). However, we can add a level of interpretation to the results. If the absolute value of the coefficient k_1 is higher for the product than it is for the brand ($|k_1| > |k_2|$), then we conclude that the panelist is more sensitive to product changes relative to brand changes (at least for this particular pair of products and brands/packages). On the other hand, if the absolute value of coefficient k_1 is lower for the product than it is for the brand ($|k_1| < |k_2|$), then we conclude that the panelist is more sensitive to changes in the brand than to changes in the product. (The absolute value of the coefficient is more instructive than the actual signed value of the coefficient, because we are interested in the reactivity to products versus packages, not in which product or brand/package is preferred.)

Figure 4 shows the scattergram for the absolute values of k_1 vs. k_2. Each point in the scattergram corresponds to a panelist. For instance, when k_1 and k_2 are both high, then the panelist is sensitive to changes both in the product and the package. When k_1 is high and k_2 is low, this indicates a panelist who is sensitive to the product, but insensitive to the package. From Figure 4 we see that there are substantial interindividual differences in sensitivity to product vs. package/label. In this category there is no single pattern of sensitivity. An individual could be easily insensitive to both the product and the package, sensitive to both, or sensitive to one and insensitive to the other.

Individual Differences in the Sensory-Intensity vs. Liking Curve

Individual differences are also easy to demonstrate when the same stimulus varies in physical intensity. Scientific research shows that as stimulus intensity increases, liking first increases, peaks and then drops. Figure 5A shows the typical inverted U shaped curve. The specific parameters of the curve will vary across products. (Some of the curves will go straight up, others will drop straight down, whereas some of the curves will actually appear as a flat line.)

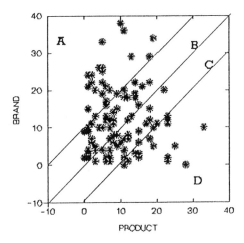

Figure 4 Partwise contribution of brand/label and product to overall liking. The contributions are shown in terms of absolute values. There are four regions: (A) brand far more important than product, (B) brand slightly more important than product, (C) product slightly more important than brand, (D) product far more important than brand.

Individual Differences

Figure 5A applies to the total panel, but the total panel data can be misleading. Where the average data suggests one simple curve, the individual data may suggest an entire family of curves. Some curves will go straight up, others straight down, and so forth. The average data results from combining these patterns, thus obscuring individual data.

A better feel for the range of individual differences can be obtained by looking at the curves relating sensory intensity to individual liking ratings. The sensory intensity data comes from the total panel; the liking rating comes from the individual liking ratings. The data is fitted so that the curve appears smooth. Figure 5B shows a variety of these individual curves. The panelists rated both

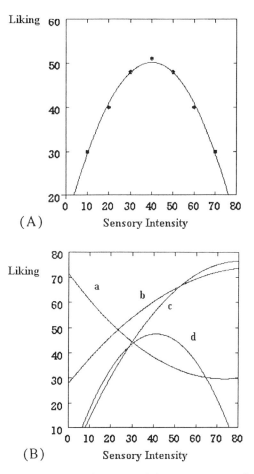

Figure 5 Inverted U shaped curve for overall liking versus sensory intensity. (A) Total panel and (B) Individual liking curves vs. total panel rating of sensory intensity.

sensory intensity and overall liking for a variety of products in a category. Each panelist's data can be related to the total panel rating of a sensory attribute to generate the curves shown in Figure 5B. (Keep in mind that the actual data in Figure 5B varies around the curve.)

7. Global Principles—Sensory Preference Segmentation

As we have seen above, researchers have established the existence of individual differences in a variety of ways. Most of the variability between individuals comes from their liking ratings, not their intensity ratings. However, is the researcher or marketer stuck with this individual variability as a "fact of life," and something with which he must contend? Or, is there a deeper, more profound organizing principle that can be used to gain both scientific insight and marketing advantage?

Instead of looking at individual variability on a person-by-person basis, we can hypothesize that in the population of consumers there exist different groups of individuals who exhibit similar preferences. That is, within the group of consumers, the preference patterns are similar. The preference patterns differ across groups. Armed with this organizing principle the researcher could then analyze data from a group of consumers in order to determine whether there exist such segments in the population, and how these segments differ from each other (both in terms of behavior, and in terms of attitudes, demographics, and purchase behavior).

Sensory segmentation follows a specific analytic protocol, described in detail in Table 5. There are seven steps and assumptions that underlie the segmentation approach. The assumptions are made in order to guard against incorrectly segmenting consumers on the basis of the numbers that they use (scaling behavior).

1. The basic organizing principle is that as we change a sensory attribute, liking changes as well.
2. The change in liking may describe an inverted U curve (the prototypical type of curve in these studies), or a linearly increasing or linearly decreasing function. The function may even be flat. We saw examples of these individual differences in Figure 5B.
3. The height of the liking curve shows how much a consumer likes or dislikes the product. However, in reality, that degree of liking is subject to many biases, including the individual's preferences for numbers on the scale, whether the individual is positive towards everything or negative towards everything. Little can be gained from an analysis of the degree of liking per se.

Table 5 Schematic for Sensory Preference Segmentation

Step 1—Define the set of products to be tested, and the sensory attributes. Test at least six products, and preferably eight or more products for the full set, and at least five products on an individual basis.

Step 2—Define the key liking attribute (e.g., overall liking). Optional: define specific attribute likings, such as liking of appearance, liking of fragrance, etc.

Step 3—For each product, develop its sensory profile, defined as the mean rating of that product on each sensory (non-evaluative) attribute (as determined in Step 1).

Step 4—Choose the first panelist. Identify which products that panelist rated, and identify the panelist's liking ratings for that product.

Step 5—For each sensory attribute (selected in Steps 1 and 3) and for the first panelist, relate that panelist's liking ratings (L) to the sensory intensity ratings (S) of a specific attribute. Use the quadratic equation:

$$L_i = A + B(S_j) + C(S_j)^2$$

The foregoing equation relates individual liking of panelist i (L_i) to the sensory level (S_j) of attribute j. The sensory levels do not change—they are the same from panelist to panelist, because the stimuli are the same. What differs (by panelist) are the liking ratings (L) assigned to the same stimuli.

Step 6—For panelist i, identify the sensory attribute level at which liking maximizes. Look at each sensory attribute separately. Each panelist will generate a vector or series of optimal sensory levels, one for each attribute. The optimum for the panelist will not depend upon the degree of liking, but only upon the sensory level at which that panelist's liking peaks.

Step 7—Repeat Steps 4-7 for each panelist, populating a rectangular matrix of rows (panelists) × columns (attributes). The numbers in this rectangular matrix are the sensory levels at which the panelist's liking reaches its maximum.

Step 8—Factor analyze the columns (sensory attributes) because they are redundant. (There are many attributes that mean the same thing.) Save the factor scores. For instance, for 100 panelists and 12 attributes, the starting matrix is 100 (rows) × 12 (columns). Factor analyzing the 12 columns and saving the factor scores for each panelist reduces the matrix to 100 (rows) by 3 or 4 (columns), depending on the number of factors.

Step 9—Cluster the rows (panelists). Individuals in the same cluster have similar optimal sensory profiles, where they like the products most. Furthermore, the analysis is independent of the degree to which a panelist likes or dislikes the products. The analysis simply depends upon the level at which liking peaks.

4. When a consumer rates a set of products in the category (e.g., lotions of various types), the consumer generates a set of curves. Each curve relates the consumer's rating of "overall liking" to a sensory attribute. Every sensory attribute, in turn, generates its own individual curve for that person. And, furthermore, every sensory attribute for that person generates a curve with a defined optimum level. Thus, if the panelist rates the lotion on five sensory attributes, then each attribute generates its own curve. Each curve has its own optimum. Therefore, that particular panelist is defined by five numbers (one per sensory attribute). Table 6A shows results for 20 panelists, each of whom generates five optimal levels (one for each sensory attribute of a product).

What is nice about the data in Table 6A is that the optimal levels on a specific sensory attribute are comparable across panelists. For instance, the first attribute is "fragrance intensity." Within the range of the products tested, some panelists liked stronger fragrances, other panelists preferred weaker fragrances, etc. In fact, across the range of panelists participating, there could be an entire spectrum of different levels of optimum fragrance intensities.

Table 6A Optimal Sensory Levels for Shampoo Data

Panelist	Fragrance intensity	Thickness of liquid	Quickness of lather	Amount of lather	Residue
1	52	67	86	73	33
2	59	60	86	87	11
3	62	78	78	77	18
4	52	78	85	83	33
5	45	60	83	86	11
6	55	68	83	73	33
7	55	78	86	79	11
8	49	61	83	86	11
9	48	61	79	86	33
10	62	78	86	80	11
11	52	70	79	72	22
12	59	79	83	72	30
13	62	79	84	83	16
14	54	60	82	72	33
15	62	78	84	86	33
16	62	79	78	78	23
17	62	60	83	81	20
18	62	60	87	86	11
19	52	68	86	86	33
20	53	68	87	86	33

Table 6B Intercorrelations Among Sensory Attributes (Based upon Optimal Sensory Levels)

	Fragrance intensity	Thickness of liquid	Quickness of lather	Amount of lather	Amount of residue
Fragrance intensity	1.00	—	—	—	—
Thickness of liquid	0.45	1.00	—	—	—
Quickness of lather	0.02	−0.10	1.00	—	—
Amount of lather	−0.10	−0.27	0.36	1.00	—
Amount of residue	−0.21	0.17	−0.19	−0.28	1.00

5. Whenever panelists rate a product on a multitude of sensory attributes there is always redundancy. Attributes correlate with each other. They share aspects in common. In fact, if the questionnaire were set up to concentrate primarily on lather, then on feel, and very little on fragrance, we would find that many of the attributes dealing with lather really involve the same basic factor. This redundancy or intercorrelation of attributes is a fact of life, well-known to all researchers. Despite intense efforts to find the "best" and most "parsimonious" or limited set of attributes, the attributes often correlate with each other (see Table 6B).

6. We remove redundancy by factor analysis. The factor analysis reduces the set of attributes to a limited number of basic, statistically independent dimensions. Each panelist, in turn, is now assigned a set of factor scores. Rather than having five numbers (one per each of the five attributes), we now locate the same panelists in the two-dimensional factor space. The benefit of the factor space is that the two dimensions are limited and statistically independent of each other. (Table 6C and Figure 6 (A–B) show the location of the 20 panelists in the factor space.)

7. The final step in the analysis consists of clustering the points in the factor space (panelists), based upon their geometric location. Points close together lie in the same cluster (see Figure 6B). Points far away from each other in this space lie in different clusters. Clusters now comprise consumers with similar patterns of optimal sensory levels. The definition of the cluster (naming the cluster) is left up to the researcher as an analytic step. Naming the clusters is not part of the cluster program.

The number of clusters is left up to the discretion of the researcher. The greater the number of clusters, the more homogeneous the composition of the cluster, but also the less general the finding. The fewer the number of clusters, the easier it becomes to make generalizations about the cluster and to understand the population of consumers in terms of a limited number of "rules." Table 6C shows the cluster to which each panelist belongs.

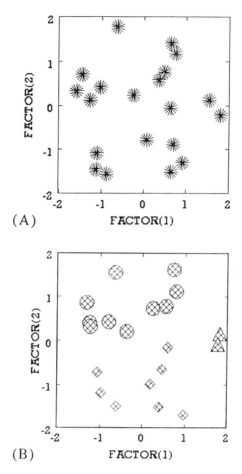

Figure 6 Plot of 20 panelists in factor space. (A) Location of each panelist, before clustering (B) Location of each panelist, after clustering into three segments. The three shapes (circle, diamond, triangle) refer to the three segments.

Understanding the Segments

By itself the sensory segmentation method is merely a statistical analysis tool by which to divide the consumer population into subgroups. It has no intrinsic "meaning" per se. However, the researcher can understand the nature of the segments in a very easy way. Once the segments have been identified, the researcher simply computes the liking by segments (Table 6D) and plots overall liking assigned by each of the segments against sensory intensity. The data used

Individual Differences

Table 6C Location of 20 Panelists in a Two-Dimensional Factor Space and the Sensory Segment to Which Each Panelist Belongs*

Panelist	F1	F2	Cluster
P1	−0.809	0.637	2
P2	1.693	0.138	3
P3	−0.807	−1.364	1
P4	−0.274	0.168	2
P5	0.795	1.618	2
P6	−1.166	0.253	2
P7	0.624	−0.723	1
P8	0.891	1.126	2
P9	−0.716	1.717	2
P10	0.902	−1.453	1
P11	−1.426	0.186	2
P12	−1.173	−0.947	1
P13	0.628	−1.340	1
P14	−1.296	0.836	2
P15	0.116	−0.840	1
P16	−0.967	−1.295	1
P17	0.428	−0.115	1
P18	1.833	−0.205	3
P19	0.277	0.855	2
P20	0.447	0.748	2

* Note - the clustering algorithm revealed three distinct groupings or sensory segments.

in the plotting comes from the liking ratings assigned to the product by the total panel and by each segment. For three segments, there will be four curves (total panel, Segment 1, Segment 2, Segment 3, respectively). Each of the four curves will use the sensory ratings assigned by the total panel to the products, but will use a different set of liking ratings (means from Segment 1, means from Segment 2, and means from Segment 3, as well as the means from the total panel). The sensory intensity scale is the same for all segments, but the liking ratings will change by segment. The plots appear in Figures 7A–7B for perceived lather vs. overall liking.

Sensory preference segmentation often leads to substantially greater differences between consumers in terms of responsiveness to products than do other methods for dividing consumers (see Table 6D). Furthermore, sensory preference segmentation is directly related to the characteristics of products, rather than to some external variable (e.g., "lifestyle") that the researcher feels "might possibly be" associated with product preference (albeit in an unknown fashion).

Table 6D Means and Standard Deviations for the Five Products: Data from 20 Panelists in the Study

	Product							Standard deviation
	106	102	104	103	101	105	Mean	
Non-evaluative sensory attributes								
Strength/fragrance	60	47	61	54	54	60	56	6
Thickness	75	60	79	73	69	76	72	7
Quickness/lathering	83	84	86	81	87	78	83	3
Amount/lather	85	87	80	72	83	83	82	5
Amount/residue	27	12	16	17	33	18	20	8
Shiny/on drying	61	64	60	54	51	54	57	5
Body/on drying	53	66	59	55	50	57	57	6
Overall liking ratings								
Liking/total panel	66	62	61	59	59	55	60	4
Liking/segment 1	68	44	64	45	44	44	52	11
Liking/segment 2	65	71	57	71	68	64	66	5
Liking/segment 3	61	89	74	55	75	58	69	13
Liking/use brand A	63	66	62	62	58	52	58	5
Liking/use brand B	67	59	61	57	61	57	63	4

Some Observations on Sensory Preference Segmentation

Through a decade of work segmenting consumers on the basis of their sensory preferences, the author has discovered six general trends:

1. The preference segmentation usually makes intuitive sense. That is, the patterns are interpretable based upon the curve relating overall liking to sensory intensity (produced on a segment-by-segment basis). There are usually no great surprises.
2. Much of the segmentation falls along two distinct axes. One axis is the degree of sensory intensity. There is usually one segment that likes a weak stimulus, whereas there are often one or more segments that like stronger stimuli. The second axis divides the groups by product character. Sometimes the differences are based on texture preferences (e.g., likers of rough or viscous materials versus likers of smooth or thin materials). Other times the differences are based on fragrance or appearance preferences. There are no set rules as to the nature of this second axis.
3. Although the segmentation is not based upon degree of liking (but rather only on sensory-liking patterns), the segments often differ in their degree of liking for a range of products. The total panel typically shows a nar-

Individual Differences

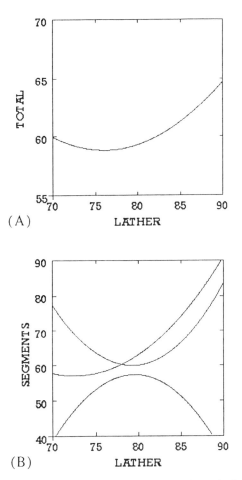

Figure 7 Overall liking versus perceived sensory intensity. (A) Overall liking versus lather—total panel and (B) Overall liking versus lather—by segments.

row range of means. In contrast, sensory preference segments often show a wider range than the total (or at least one or more of the sensory segments show this wider range). This expansion in liking range occurs because the individuals in different segments may exhibit opposing preferences for the products. The mean from the total panel encompasses the ratings of those who like the product together with those who dislike the product, and therefore tends toward the middle of the scale. In contrast, the mean from the individual segments comprises ratings from individuals who are more similar in what they like. Thus, within a sensory segment there are fewer "opposing" consumers (who would force the mean to converge toward the center of the scale).

4. There is often a low-impact segment. This segment is the most indifferent to the products. This indifference emerges when we look at the level and the range of the liking ratings. The low-impact segment often rates the products lower in absolute level than do the other segments. Furthermore, the low-impact segment often exhibits a narrower range of liking across the products than do the other segments.
5. It is rare to have a product that is truly beloved by the total panel, except in the most unusual of circumstances. Beloved products (with very high liking scores, means of 70 +) emerge much more clearly within sensory segments. Often some of the higher impact segments will like products far more than would have been guessed on the basis of data from the total panel. Such a high degree of liking, if reliable, means that the panelist will tend to repurchase the product because the product meets the panelist's expectations (assuming, of course, that all other factors such as price, promotion, packaging, distribution, etc., do not hinder the purchase).
6. The sensory segments transcend demographics. In any category, there may be a limited number of these segments. Even on a country-to-country basis, the segmentation may hold. If the researcher tests the same set of products worldwide, and segments the consumers independently of their country of origin, then the results will yield a worldwide segmentation. Furthermore, only a limited number of segments will emerge. Each country will then differ from every other country primarily in the percentage distribution of consumers in each of the segments. This way of thinking provides an organizing principle for companies who wish to market products worldwide, but who do not want to create an unduly large number of products, each tailored to a specific country. Sensory segmentation provides a way to reduce the number of products one needs to create.

8. A Case History—Fragrances

Cappello Inc. markets a line of fragrances for mass consumption. During the past four years Ginny Cappello, chief executive officer, developed a strong development and marketing staff for her 12-year-old company. The corporate vision was to manufacture and market fragrances of high quality to consumers through department stores. However, in order to increase revenues, Cappello Inc. also packed fragrances for other manufacturers, especially those in mass merchandise outlets which Ginny Cappello determined not to be in competition with their main business.

Testing Fragrances—The Experiment

Testing fragrances for acceptance can be accomplished in two ways. Some practitioners prefer to test the fragrances on blotters. A panelist can evaluate many fragrances in a session (up to 20 +), provided there is a reasonable rest period between samples (e.g., 5 minutes minimum), and provided the panelist is sufficiently motivated to participate and to provide honest answers. (Motivation is increased by a cash reward for participation, payable at the end of the fragrance evaluation.) Other researchers, considering themselves "purists," insist that the panelist must actually wear the fragrance on the skin for an extended period of time before evaluating it. According to the purists, each fragrance interacts in a unique fashion with every person's skin, so that it is the skin-fragrance combination that is critical to measure. The purists aver quite strongly that the researcher incorrectly measures fragrance acceptance when testing the fragrance on a blotter.

In the interests of collecting extensive data from panelists on many fragrances, the marketing group decided to test the fragrance on the blotter, and abandon the purist ideal of testing the fragrance on the skin. That ideal would have been unduly costly, and would not have provided the necessary ratings on many products by many consumers.

The marketing group selected 22 fragrances representing a diversity of different sensory impressions. These fragrances comprised both in-market fragrances (both upscale and mass-marketed entries), as well as a number of prototype fragrances for consideration. The marketing group at Cappello planned to introduce two fragrances that year, and wanted to obtain measures of consumer acceptance.

Field Setup and Execution

Panel Considerations. The panel comprised 89 female category users, selected to be Cappello's likely target group. These were middle class consumers who purchased fragrances. There were no screening requirements to participate in the project, other than the participant must have purchased a fragrance at either a department store or a mass market outlet. In other studies where the outlet is important, one might restrict participation to individuals who purchase fragrances at a specific class of outlet, such as department stores. Here, however, the research objective was more general and less tightly focused. Therefore, the screening was less rigid. The marketing group did not want to include women who received fragrances, but would not purchase fragrances for themselves. Fragrance purchasers were assumed to purchase on the basis of both image and sensory preference.

Field Execution. The panelists reported to a central field location comprising a large, well-ventilated hall, with numerous exhaust fans placed strategically in the room and next to the windows. (The exhaust fans insured that the fragrances, when applied, would not flow into the evaluation area, but rather would be evacuated out the window.)

The panelists reported in groups of 20–25. None of the panelists knew each other. A group setting, in which each panelist goes through a set of evaluations (but without talking to other panelists) insures a rapid, high quality, well-controlled set of interviews.

The panelists followed the sequence of activities listed in Table 7. The sequence ensured that the panelists understood what to do in terms of smelling

Table 7 Steps in Evaluating a Fine Fragrance on Blotters

Room Setup

Application room

- Exhaust fans to the outside
- Fragrance application in front of exhaust fans to eliminate fragrance
- Door to test room

Test room

- 10 tables, each with three consumers and one table interviewer
- The test room is well-ventilated, with positive pressure so that the air is always pushed outwards

Evaluation Procedure

Step 1—Re-screen consumers to insure that they are who they said they are. (Consumers originally screened by telephone, qualified and invited to participate.)

Step 2—Orientation by the study "monitor" (head interviewer) describing the study, and method for evaluation.

Step 3—Orientation exercise (with a fragrance, data for which is eliminated).

Step 4—Each consumer evaluates the first fragrance according to an individualized rotation sheet. The individual rotations are set up so that each fragrance appears equally as often in each of 15 test positions.

Step 5—The consumer rates the fragrance on the attributes.

Step 6—The interviewer at the table checks the ratings by asking the consumer two questions. The consumer must substantiate the ratings. This procedure improves the quality of the data, by making the consumer pay attention to the task. Neither the interviewer nor the consumer knows the "right answer."

Step 7—The consumer disposes of the used blotter in a garbage can, sealed except for a small hole to admit used blotters. The garbage can is located outside and away from the test room.

Step 8—The consumer waits at least 5 minutes and gets the next fragrance.

Steps 4–8 are repeated until the panelist completes 15 of the 22 fragrances (randomized block).

Step 9—The consumer fills out an attitude and usage questionnaire.

Step 10—The study is complete, and the panelists are dismissed after their data is checked one more time. The panelists are paid for their participation.

Total Time = Approximately 3.5 to 4 hours

the fragrance, and what the attributes meant. Often, a short introduction to the process and a practice exercise with a product whose data will be discarded in the analysis increases the quality of the results. Many panelists are unfamiliar with the specifics of interviews. Often their ratings assigned to the first product will be uncertain, and will differ from the ratings that would have been assigned to the same product if tested later, after the panelist had some experience in evaluating.

The procedure can last anywhere from 15 minutes to 4 hours. To many traditional researchers accustomed to working with fragrances, 4 hours seems to be an excessive amount of time for product evaluation. However, keep in mind that the majority of the time in the evaluation is spent waiting for the next product, or performing specific activities (e.g., filling out rating cards, getting the data checked, as well as waiting for the product). Most panelists spend no more than 5–10 seconds actually smelling each fragrance. With a minimum of a 10–15 minute wait between samples, any loss of sensitivity due to the previous evaluation is soon recovered. (The only time that this wait period may be ineffective occurs when the fragrances are strong musks, which adsorb into the lining of the nose, and desorb during the waiting period. Other precautions must be taken with musk, and the waiting times must be lengthened to insure that the desorption has stopped.)

The Questionnaire. Questionnaires for fragrance ratings are simple. There is only one sense involved—smell, so that the number of attributes is limited. Although there are many lists of descriptor terms for fragrance, in most consumer work, panelists usually evaluate a list of relatively simple ones, including overall strength, sweetness, floweriness, etc. Consumers do not have an extensive vocabulary for fragrances, although they can be taught to use many more descriptor terms with some instruction prior to the fragrance evaluations. The questionnaire appears in Table 8A. All of the attributes in the questionnaire were scaled on an anchored, 0–100 point scale. Panelists find the anchoring helpful because it reduces ambiguity.

Analyzing the Data

What is the Best Data to Use—The Product Tried First, or All of the Product Ratings (Independent of Order of Trial)? Many researchers prefer to use ratings assigned to the first product, and to stop the interview at that point, after the panelist has rated only one sample. Other researchers gather data from more products, but place the greatest emphasis on the ratings assigned to the first product. The rationale of both points of view is that the first product represents the "truest" measure of a consumer's response to the product. However, in reality the first product is often the most variable and prone to error. When rating a product for the first time, panelists have no frame of reference. They may not be clear about what

Table 8A Questionnaire for Woman's Fine Fragrance

Please smell the fragrance on the blotter, and then rate the fragrance. You may smell the fragrance on the blotter as often as you wish, but please make sure that you wait at least 3 seconds between smells. Hold the blotter at least 2 inches from your nose when smelling.

A. How much do you like the fragrance overall? (0 = hate → 100 = love)
B. How strong does the fragrance smell? (0 = very weak → 100 = very strong)
C. How clean does the fragrance smell? (0 = not clean → 100 = very clean)
D. How fresh does the fragrance smell? (0 = not fresh → 100 = very fresh)
E. How harsh does the fragrance smell? (0 = not harsh → 100 = very harsh)
F. How unique does the fragrance smell? (0 = common, everyday → 100 = very unique)
G. Describe the quality of the fragrance. (0 = cosmetic-like → 100 = functional/utilitarian–like)
H. Describe this fragrance. (0 = masculine smelling → 100 = feminine smelling)
I. Describe this fragrance. (0 = not at all sensuous → 100 = extremely sensuous)
J. When would you wear this fragrance? (0 = during the day → 100 = during the evening)
K. Describe this fragrance. (0 = not at all fruity → 100 = extremely fruity)
L. Describe this fragrance. (0 = not at all citrusy → 100 = extremely citrusy)
M. Describe this fragrance. (0 = not at all sweet → 100 = extremely sweet)
N. Describe this fragrance. (0 = heavy smelling → 100 = light smelling)
O. Describe this fragrance. (0 = not at all spicy → 100 = extremely spicy)
P. Describe this fragrance. (0 = not at all crisp → 100 = extremely crisp)
Q. Describe this fragrance. (0 = not at all musky → 100 = extremely musky)
R. Describe this fragrance. (0 = dissipates quickly → 100 = lasts a long time)
S. Who would use this fragrance? (0 = a younger person → 100 = an older person)
T. When would you wear this fragrance? (0 = sports/informal occasion → 100 = formal occasion)
U. How interested would you be in buying this fragrance if it were priced at the same price as your normal fragrance?
   1 = definitely not buy
   2 = probably not buy
   3 = might/might not buy
   4 = probably buy
   5 = definitely would buy

to do with the product even though an interviewer has explained the process, and may be uncertain about answering the questions. The first product brings with it many biases and uncertainties. These uncertainties rapidly disappear after the first product has been evaluated.

Products tested in the first position yield results that are significantly more variable than the same products tested in the second, third, . . . n^{th} position (see Table 8B). This can be shown by evaluations of samples which vary systematically in one characteristic, such as fragrance concentration (so that the characteristic is clearly evident). When directly compared to each other, the fragrances clearly demonstrate an increasing range of intensities. Now let panelists rate the degree of intensity, assessing the products in a randomized order. Look at the ratings assigned to fragrance intensity for the products tested in the first position. They do not "track" the known physical intensities. Look at the products tested in the second, third, fourth, etc. positions. They do track the physical intensities. Ratings in the first position, and primarily that position, cannot easily reproduce the known changes in the physical levels of the product. Similar results can be shown for products that vary in an even more obvious attribute, such as darkness, or sounds that vary in loudness. Ratings of intensity in the first position often fail to mirror the known physical levels of the product. The researcher must test hundreds of panelists in order to obtain a stable estimate of sensory intensity if the only sample to be considered in the analysis is the sample tested in the first position.

The Fragrance Database. The database comprises two parts. One portion of the database comprises the mean ratings on all of the sensory attributes (see Table 8C). The other section of the database comprises the liking ratings assigned by the total panel, as well as key subgroups. These subgroups for fragrances include age, income, type of fragrance used most often, where the consumer purchases the fragrance, and sensory preference segment. For the sensory preference segmentation, data suggested two subgroups. Table 8C shows the results for two pairs of subgroups—product user subgroup and sensory preference segment.

Key Observations from the Database. Table 8D, showing the liking ratings, supports the fact that the fragrances differ from each other on a blind basis. This differentiation clearly occurs both for sensory attributes and for liking.

The database can be mined for more insights. For instance, one relevant analysis is the variability of liking ratings across the means. If the segmentation (of any type) divides consumers into groups having different preferences, then we may expect different ranges of liking *across the products*, as a function of the subgroup. For instance, if the panelists in a segment are similar to each other in their fragrance likes and dislikes, then we might expect the pref-

Table 8B How Ratings of Perceived Fragrance Intensity Level Change as a Function of Physical Intensity and Order of Testing

Mean Ratings of Perceived Intensity by Position in the Sequence (12 Positions: P1 → P12)

Physical intensity	P1	P2	P3	P4	P5	P6	P7	P8	P9	P10	P11	P12
10	17	3	7	12	12	11	7	14	14	11	15	15
40	14	27	7	18	7	10	14	10	15	16	8	12
160	23	23	19	18	18	18	17	18	14	16	19	14
640	20	33	41	32	37	45	32	41	32	36	39	40
2560	43	64	48	68	58	43	51	39	43	57	52	69

Running Average of Ratings:
How the Cumulative Average Changes as the Ratings for the Same Product in Different Positions Are Combined to Yield an Average

Physical intensity	P1	P2	P3	P4	P5	P6	P7	P8	P9	P10	P11	P12
10	17	10	9	10	10	11	10	11	11	11	11	12
40	14	20	16	17	15	14	14	14	14	14	13	13
160	23	23	22	21	20	20	20	19	19	18	19	18
640	20	26	31	31	33	35	34	35	35	35	35	36
2560	43	54	52	56	56	54	54	52	51	51	51	53

Fragrance = one fragrance, systematically varied over a 256:1 range in liquid diluent.
P1 ... P12 = Position Tested.
Numbers in body of the table = average ratings of perceived fragrance intensity from each of 11 consumers (11/position, 132 consumers in total).
Note: Each panelist tested 12 fragrances, of which five were the systematically varied concentrations shown above.

Individual Differences

Table 8C Example of Basic Data from Woman's Fine Fragrance—Best Two Fragrances vs. Worst Two Fragrances

	Product			
	Best #1 # 21	Best #2 # 17	Worst #2 # 15	Worst #1 # 9
Liking				
Total panel	60	59	45	31
Buy (% Top two box)	58	52	44	23
Image				
Clean	63	67	63	41
Fresh	59	64	60	35
Refresh	54	61	52	31
Everyday—unique	43	54	51	51
Cosmetic—functional	31	36	33	39
Feminine—masculine	31	42	48	45
Dissipates—lasts	67	60	58	38
Younger—older	59	49	51	50
Informal—formal	52	50	41	27
Sensory				
Strength	71	66	65	44
Harsh	38	37	43	39
Fruity	14	24	35	23
Citrusy	22	28	40	25
Sweet	26	37	46	31
Light	39	41	44	52
Spicy	54	42	34	28
Crisp	56	57	51	28
Musky	51	37	26	20
Liking by subgroups				
Total panel	60	59	45	31
Sensory segment 1	55	61	53	35
Sensory segment 2	65	57	36	25
Use/product A	63	59	50	35
Use/product B	60	60	40	27

erence segments to show higher than average ranges and variabilities for the 22 mean ratings since the consumers in a segment like or dislike the same fragrances. There is no damping effect that would occur by combining consumers with different preferences in the same subgroup—such as likers and dislikers. That combination would diminish the range of liking ratings, and bring the mean toward the center of the scale. For any fragrance there would be both likers and dislikers. Consequently, the fragrance means would tend towards the center of the scale. Table 8D shows that the conventional way of segmenting the data (by user subgroup) yields a smaller range of liking across fragrances than does sensory segmentation. This suggests that the conventional methods to segment consumers do not really make the panelists more homogeneous in their preferences than would be the case for the total panel. The sensory preference segments behave differently. The segments show larger variation between means than does the total panel, suggesting that these individuals in a single segment have more similar preferences.

Do the Subgroups Show the Same Pattern of Fragrance Likes/Dislikes? A second analysis looks at the fragrance means to determine whether these segments are saying the same thing. We have ratings from each of the fragrance subgroups (demographic, sensory preference) on the set of 22 fragrances. Do these subgroups really show different patterns of liking? Or, is the way that we divide the panelists pretty much irrelevant (winners are winners, losers are losers)? Table 8D shows that, for some, fragrances stand out as highly acceptable for a particular segment, even though the same fragrance is only moderately well accepted for the total panel. However, this can be merely accidental. Figure 8A shows a scatterplot for the liking ratings of the two user subgroups. The two subgroups plot similarity. In contrast Figure 8B shows a scatterplot for the sensory segments. There is no clear pattern, suggesting that these two segments behave independently.

Individual Differences in Fragrance Acceptance—An Overview

Fragrance likes and dislikes are perhaps the major area where individual differences are a pervasive fact of life. Differences in fragrance likes have been noted in the scientific literature going back more than a half century. The approaches presented above show that these differences are not clearly linked to age, market, or type of fragrance used most often, but rather result from innate differences between consumers in the types of fragrances that they like.

9. Individual Differences in a Complex Product— Cinnamon-Flavored Toothpaste

Fragrances are relatively simple. They excite only one sense modality—smell. Individual differences in fragrance occur only because of differences in one's liking or disliking of the smell of the fragrance. Most products excite several senses, including appearance, smell, and touch. To what extent can we discover similar types of individual differences in these more complex categories?

Individual Differences

Table 8D Summary Statistics for Total Panel vs. Two Sets of Subgroups: Product Users, Sensory Preference Segments

Fragrance	Total	Use prod A	Use prod B	Abs diff* user subgroup	Sen seg 1	Sen seg 2	Abs diff sensory segments
21	60	63	56	7	55	66	11
17	59	59	59	0	61	57	4
20	58	60	55	5	65	49	16
18	56	60	50	10	69	40	29
12	55	58	51	7	65	43	22
13	54	60	45	15	59	48	11
14	54	59	47	13	62	44	18
22	53	50	58	8	36	74	38
19	53	56	49	8	61	43	18
8	52	57	45	13	61	41	20
16	51	53	48	5	61	39	22
7	51	53	48	5	58	42	16
2	50	52	47	5	60	38	22
10	50	56	41	15	54	45	9
5	49	51	46	5	56	40	16
1	49	50	48	3	51	47	4
6	48	52	42	10	61	32	29
4	48	52	42	10	52	43	9
11	47	46	49	3	52	41	11
3	47	49	44	5	55	37	18
15	45	50	38	13	53	35	18
9	31	35	25	10	35	26	9
Mean	51	54	47	8	56	44	17
Std. dev.	6	6	7		8	10	
Maximum	60	63	59	15	69	74	38
Minimum	31	35	25	0	35	26	4
Range	29	28	34	15	34	48	33

*Absolute difference between the ratings assigned by the two subgroups.

The next category is cinnamon-flavored toothpaste. These products vary considerably in appearance, flavor intensity, and mouthfeel. What is of interest here is the range of differences between consumers, and the existence of potential sensory preference segments.

Evaluating toothpaste is more complex than evaluating fragrances because there are more aspects to the product. Toothpastes call into play responses to

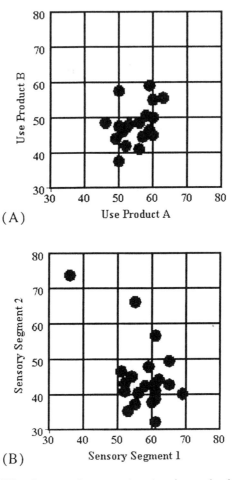

Figure 8 Overall liking for two subgroups, plotted against each other. Each point corresponds to a specific product. (A) By user subgroup and (B) By sensory preference segment.

multisensory inputs. They also create end benefits. In this study, eight different toothpastes were purchased in the market, at department stores, specialty shops, and by mail from specialty companies that sold their products by catalog. These eight products were selected to be reasonably different from each other. The purpose of the original study was to identify the key drivers in the category.

Test Procedure

Each of the eight commercial products was "repackaged" to disguise the brand. The products were then randomized, so that each panelist would try all eight products, one product per week, over an 8-week period. The total study comprised 90 consumers.

A toothpaste has to be tested over time and at home, rather than at a central location. There are many attributes to evaluate, including appearance, flavor, and feel as the product is extruded and then applied to the teeth. Table 9A shows the questionnaire. As in the fragrance evaluation, the panelists used an anchored 0–100 point scale to rate every attribute.

Field Specifics

Participating consumers comprised individuals who were defined as category users by means of a screening questionnaire, and who agreed to participate in the test for the 8-week period. The panelists were told that they would have to test eight products, and rate each product at the end of each 1-week period. Panelists were told that they would receive one product each week (on a Monday evening), would use it for the week as they normally use a toothpaste, and would then return with their completed questionnaire the next Monday to re-

Table 9A Questionnaire for Cinnamon-Flavored Toothpaste—Home Use Test

Use the product for 1 week, brushing the way you ordinarily do. At the end of 1 week, on Sunday evening, please rate the product on the questions below:

Please apply the the toothpaste to the toothbrush.
Now look at the toothpaste, but don't brush with it yet.

A. How much do you like the way the toothpaste looks?
 (0 = hate → 100 = love)
B. Describe the way the toothpaste looks.
 (0 = dry → 100 = watery)
C. How dark is the color of the toothpaste?
 (0 = very light → 100 = very dark)
D. How uniform looking is the toothpaste?
 (0 = not uniform at all → 100 = very uniform)
E. How strong is the aroma of the toothpaste?
 (0 = very weak → 100 = very strong)
F. How much do you like the aroma of the toothpaste?
 (0 = hate → 100 = love)

(continued)

Table 9A Continued

Now please brush with the toothpaste in the way that you ordinarily do. When you are finished, please answer the questions below.

G. How much do you like the toothpaste overall?
   (0 = hate → 100 = love)
H. How smooth is the feel of the toothpaste?
   (0 = very rough → 100 = very smooth)
I. How strong is the flavor of the toothpaste?
   (0 = very weak → 100 = very strong)
J. How much do you like the flavor of the toothpaste?
   (0 = hate → 100 = love)
K. How hot is the flavor of the toothpaste?
   (0 = mild → 100 = hot)
L. How cinnamony is the flavor of the toothpaste?
   (0 = not at all → 100 = extremely cinnamony)
M. How easy is it to apply the toothpaste to your teeth?
   (0 = very difficult → 100 = very easy)
N. How quickly can you cover your teeth with the toothpaste?
   (0 = takes a very long time to get the teeth covered, ready to brush → 100 = takes a short time from when I apply the toothpaste until I am ready to brush my teeth)
O. Describe the feel of the toothpaste.
   (0 = very smooth → 100 = very gritty)
P. How much did you like the feel of the toothpaste as you applied it?
   (0 = hate → 100 = love)
Q. How strong an aftertaste did the toothpaste have?
   (0 = very weak → 100 = very strong)
R. How much did you like the aftertaste of the toothpaste?
   (0 = hate → 100 = love)
S. What type of a job did the toothpaste do in *cleaning* your teeth?
   (0 = very poor job → 100 = very good job)
T. What type of job did the toothpaste do in *shining* your teeth?
   (0 = very poor job → 100 = very good job)
U. What type of job did the toothpaste do in *protecting* your teeth?
   (0 = very poor job → 100 = very good job)

ceive a new product. This sequence of field events insured a high degree of control over the test process, because panelists realized that they would have to show their completed questionnaires in person, before they could proceed with the evaluation. This setup insured cooperation and maintained panelist motivation, without compromising the quality of the data. Because panelists knew that their answers would be checked at the time that they handed in the question-

Individual Differences

naire, it made sense for the panelist to answer the questionnaire honestly. Furthermore, because the panelists completed the questionnaire at home, during the period of use, the data was not based on memory, but on their reactions to recent use. Finally, since the panelists filled out the questionnaire themselves rather than telling an interviewer their answers, the possibility of covert demands by the interviewer was eliminated. (Covert demands are implicit demands made by an interviewer during the course of an interview, such as an unspoken demand for the panelist to be "positive" about the product. These demands bias the data, because they are induced by the interview process. In a self-administered questionnaire such demands are significantly reduced.)

Basic Results—Product × Attribute Matrix

As was done for the previous example on fragrance, the researcher first creates a basic summary matrix, showing the average ratings on all attributes, on a product-by-product basis. This matrix overlooks the individual differences, but gives the researcher a sense of the differences between the toothpaste products. Table 9B shows some of the data.

Differences in Sensory Ratings—Relevant or Irrelevant?

In an assessment of individual differences it is always tempting to look at individual differences in all of the attributes, not just overall liking (as had been done for the fragrance). Although it can be interesting to study individual differences

Table 9B Partial Data Matrix for Cinnamon-Flavored Toothpaste (Ranked in Order of Overall Liking)

	Product							
	8	7	6	2	4	3	1	5
Liking								
Overall liking	64	64	57	57	57	55	48	48
Like appearance	54	62	42	67	43	55	51	48
Like flavor	56	57	50	57	56	52	51	39
Like feel	48	56	53	52	61	53	53	49
Sensory								
Darkness	45	42	51	44	53	56	44	42
Aroma	41	46	45	39	42	44	36	43
Flavor	73	42	68	51	53	48	53	39
Spiciness	65	30	62	53	50	38	43	37
Hotness	56	31	67	53	42	45	39	37

in the scaling of sensory characteristics, most scientific research suggests that on a sensory basis individuals are fairly similar to each other. That is, the differences that one encounters in the scaling of individual differences in the perception of sensory attributes probably result from scaling effects, rather than differences in sensory attribute perception. Thus, it is probably not particularly productive to study individual differences in the scaling of sensory attributes. This would turn out to be a study of response biases (interesting in itself, but actually a second order problem that does not help the manufacturer to develop a salable product).

For those researchers interested in individual differences in the scaling of sensory attributes (vs. liking), the most instructive analysis is an analysis of variance. The data for the analysis of variance are the ratings by the same set of panelists, rating the same set of products, on a variety of attributes (some of which are sensory attributes, others of which are liking or image, or even performance attributes).

Differences Between Products in Terms of Liking

As we did for fragrances, we can look at the ratings of key subgroups. The key rating is overall liking. Table 9C shows the results of this analysis, and includes data from the total panel, user subgroups, demographics, and sensory preference segments (created by the segmentation procedure described in Table 5). It is clear from the data that the biggest differences come from the sensory preference segments, and substantially lower inter-individual differences come about when we classify the individuals by other, more conventional procedures.

Do Interindividual Differences Carry Over to Attitudes?

We have seen that there are interindividual differences in responses to these lotion products, at least on a blind basis. These differences are primarily the result of different preferences for the sensory characteristics of products. Do these differences carry over to attitudes about products?

One of the easiest ways to measure attitudes towards products is by means of a questionnaire. However, it is difficult to measure relative importance of characteristics directly. That is, how important is the attribute "grainy" to a consumer? That question is ambiguous. Do we mean "How important is it that a product be grainy?" Or, do we mean "How important is it that a product have the right amount of graininess?" When the researcher wants to establish the relative importance of a sensory attribute (where too much or too little is not optimum), the question of "importance" is ambiguous. It is not clear what "importance" means to the consumer. In contrast, for other attributes, where the optimum level is always at the highest or lowest amount (e.g., value for the

Individual Differences

Table 9C Liking Ratings and Summary Statistics for Eight Cinnamon-Flavored Toothpastes: Total Panel and Different Subgroups

	Product												
	8	7	6	2	4	3	1	5	Avg	Std dev	Max	Min	Range
Total	64	64	57	57	57	55	48	48	56	6	64	48	16
Gender													
Males	62	67	60	58	60	57	50	50	58	6	67	50	17
Females	64	61	55	57	54	53	46	45	54	6	64	45	19
Absolute difference	2	6	5	1	6	4	4	5	4				
Market													
East	66	62	60	57	57	52	56	46	57	6	66	46	20
Midwest	64	64	57	59	61	57	49	47	57	6	64	47	17
West	61	64	55	55	47	53	39	49	53	8	64	39	25
Max absolute difference	5	2	5	4	14	5	16	3	4				
Brand Used													
User/A	57	60	53	60	59	59	50	46	56	5	60	46	14
User/B	69	67	61	55	54	51	46	49	56	8	69	46	22
Absolute difference	12	7	8	5	5	8	4	3	0				
SensorySegment													
Seg/A high impact	79	60	76	50	56	48	40	38	56	15	79	38	41
Seg/B low impact	53	66	44	62	57	59	54	54	56	7	66	44	22
Absolute difference	16	6	32	12	1	11	14	16	0				

Note: Std dev = standard deviation of the eight means.
Avg = Average.
Max = Maximum.
Min = Minimum.
Range = Maximum − Minimum.
Absolute difference = Between ratings assigned by two subgroups.

money, efficacious in preventing tooth decay), relative importance questions, directly asked, are no problem. There is no ambiguity.

We can ask panelists in the toothpaste study to assign ratings of "annoyance" to product or packaging (or even usage, positioning, and pricing) defects. A defect is something that is wrong with the product. The higher the annoyance rating assigned the more important the attribute is to the panelist. Table 9D shows examples of the questions asked for the toothpaste. Table 9E shows results from the different subgroups of panelists. We see from Table 9E that on an attitudinal basis the segment differences we discovered for product preference disappear. This is especially remarkable when we look at the data assigned by the sensory preference segments—groups of consumers with established differences in sensory preference profiles. These groups are quite similar to each other attitudinally in terms of what they feel is important. These groups radically differ in their actual likes and dislikes, however, based on product use.

An Overview to the Toothpaste Study

These results again show that there are interindividual differences in responsiveness to products. But, very importantly, they show that these differences are not simply the result of the conventional subgroups that marketers use. Consumers are remarkably similar to each other in what they like, even though they fall into different demographic breaks. What is important, however, is that these individual differences can be captured by other procedures, such as sensory preference segmentation. Individual differences do exist in responsiveness to stimuli. These differences give rise to sensory preference segments, or groups

Table 9D Annoyance Questionnaire for Toothpastes to Identify What Is Important on the Basis of Attitude

Instructions

Sometimes a toothpaste can have defects associated with it. These may be in the product, the packaging, or the performance. Please rate the degree to which you personally would feel annoyed were you to have purchased and used a toothpaste with the defect listed below. Please use the following scale:

0 = Not annoyed at all → 100 = Extremely annoyed

You may use any number on the scale to show the degree of annoyance that you personally would feel:
A. The toothpaste color was too dark.
B. The toothpaste color was too light.
C. The aroma was too strong, before you brushed with the toothpaste.
Etc.

Table 9E Results from Annoyance Questionnaire for Toothpastes

Defect	Total	User		Sensory segment		Gender	
		Use/A	Use/B	Seg/A	Seg/B	Males	Females
Color too dark	32	35	31	34	31	35	30
Color too light	41	40	42	41	42	39	42
Aroma too weak	23	24	22	25	21	23	23
Aroma too strong	67	68	66	70	65	66	67
Flavor too weak	32	36	29	34	35	30	34
Flavor too strong	57	57	57	60	54	55	58
Flavor not hot enough	34	37	33	36	31	33	34
Flavor too hot	67	65	68	64	68	65	68
Feel too smooth	23	22	24	24	23	25	20
Feel too gritty	57	56	57	58	55	56	57
Aftertaste too weak	23	22	23	26	21	24	23
Aftertaste too strong	68	66	69	73	65	66	69
Did not clean teeth	86	88	85	87	86	87	86
Did not protect teeth	88	88	88	89	87	86	88
Hard to open	68	69	68	70	67	69	68
Hard to close	66	65	68	67	66	66	66
Messy to use	78	79	77	77	78	80	76

of individuals with similar patterns of sensory likes and dislikes. The results hold for responses to blind products (without benefit of brand—e.g., both fragrances and toothpastes), although the sensory preference segmentation demonstrable for products may not generalize to attitudes about the sensory attributes of the products (by annoyance ratings).

10. Finding New Individuals in a Sensory Segment

Introduction

Dividing a group of consumers by conventional demographic and product usage criteria is straightforward. One need only instruct the panelists to fill out an attitude and usage questionnaire, and divide the panelists by gender, age, market, and product used most often. As we have seen, these conventional methods of dividing consumer panelists do not divide the consumers into homogeneous preference subgroups. In contrast, the sensory preference segmentation divides the consumers into groups with similar sensory preferences. However, the preference segmentation does not divide the panelists into easily accessible subgroups. We have seen from the current data on sensory segmentation and from a variety of previous studies (unreported here) that sensory preference segments transcend demographics. There is no simple method to identify which consumers will fall into a specific sensory segment. Gender, age, market, income and even brand used most often do not allow an easy prediction.

The problem facing the marketer is straightforward: If these sensory preference segments exist, then how can the researcher identify a member of the segment without having to repeat the entire study? Simply defining a segment in terms of the probability that a consumer will fall into the segment as a function of other usage criteria is not the answer. What is needed is a short, easy-to-administer questionnaire or printed stimulus, from which responses can be used to identify consumers from different segments. The questionnaire or printed stimulus should be easy to produce, and easy to analyze (e.g., using a key provided to the researcher).

A Worked Example—New Style Feminine Napkins

J&L Inc. (named for its founders, Jan and Laura), manufactures sanitary napkins under technology licensed from France. Recent technological developments in the category have created a major opportunity for a new product introduction. The patented technology provides a variety of end benefits, including comfort, security and privacy.

Some early research studies with the new technology suggested that it might segment the consumers in a way that was different from the traditional segmentation. Up to that point the market research community, both at J&L and at their competition, had segmented consumers on the basis of heaviness of flow, and general type of napkin used. However, that method of segmenting the consumers did not correlate with acceptance of napkins fabricated with the new technology. Many of the consumers who reacted positively to the concept rejected the product after wearing it. Other consumers who reacted negatively to the concept reported that they actually liked the product after wearing it.

Using Sensory Segmentation

Because it was unclear why consumers accepted or rejected the product, except on the basis of wear, the researchers at J&L hypothesized that it might be a sensory- or product-based segmentation. Clearly, reactions to the product concept did not predict acceptance. The napkin was sensorially different from the competition. This may have engendered reactions that could not be easily predicted.

The R&D researchers developed a variety of alternative napkins, including several with the new technology. The results were analyzed according to the sensory segmentation algorithm described in Table 5. The result showed three distinct segments, as demonstrated in Table 10A:

Individual Differences

Table 10A Acceptance Ratings for Eight Feminine Napkins: By Total Panel and Three Segments

Short description	Product code	Total panel (100%)	Seg 1 indif-ferent (20%)	Seg 2 high tech (35%)	Seg 3 secure (45%)
New cover	101	51	44	68	41
New absorber	102	56	42	67	53
New adhesive	103	50	38	61	47
Small, thick	104	53	43	44	65
Small, thin	105	53	41	42	68
Small, thin	106	51	43	45	60
Small, bulky	107	47	46	40	52
Large, bulky	108	38	47	40	33

1. Traditionalists, who did not like napkins created with new technology.
2. Those who liked the products created from the new technology, perhaps because of interest in "technology-based" characteristics. They liked products with visible features representing the technology.
3. Those who liked the products because of interest in privacy and security afforded by the products with the "new technology." They liked the products that were small, discreet, but worked well.

The segmentation clearly demonstrated the existence of three segments. The researchers were convinced that these three segments reflected true groups of consumers in the population. Furthermore, the researchers believed that they could create a significantly improved product using the new napkin technology, but they had to be able to identify individual consumers as belonging to the three different segments. The researchers wanted to be able to eliminate consumers from Segment 1. There appeared to be two different opportunities—one each for Segments 2 and 3, respectively. The key problem was how to identify consumers in each segment. Demographics and responsiveness to the basic concept were not helpful. Table 10B shows that consumers in the different segments do not clearly belong to any single or key demographic subgroup. There is no "magic bullet" which can identify a consumer as belonging to a subgroup.

Strategy 1—Product Usage as a Segment Identifier

The first strategy to identify consumers as belonging to different segments consisted of taking a look at the most polarizing products. It is clear from Table 10A that by choosing products 103 and 106 one can identify consumers as belonging to one of the three groups:

Table 10B Distribution of the Total Panel and Sensory Segments for Napkins on Demographic and Usage Variables

	Total panel	Seg 1 indifferent	Seg 2 high tech	Seg 3 secure
Median age—years	31	33	29	32
Median flow*	5.4	5.5	5.3	5.4
Magazine reading—frequency**	2.4	2.1	2.4	2.5
Physician visits—per year	3.9	3.1	4.6	3.7
Use Brand A (%)	46	48	43	54
Use Brand B (%)	54	52	57	46
Market - East (%)	36	32	37	37
Market - Midwest (%)	34	37	34	33
Market - West (%)	30	31	29	30
Income - low (%)	45	47	42	46
Income - medium/high (%)	55	53	58	54
Education - high school (%)	30	34	33	26
Education - college (%)	48	42	47	51
Education - graduate school (%)	22	24	20	23

* Relative number based upon a pre-defined flow scale (1 = low flow → 9 = heavy flow, over the cycle of the menstrual period).
** Defined as number of magazines read per week at home.

1. If both products are disliked to neutral on a rating scale (less than 50), the odds are that the consumer belongs to Segment 1 (traditionalist).
2. If Product 103 is liked and Product 106 disliked (or liked substantially less), then the odds are that the consumer comes from Segment 2.
3. If Product 106 is liked and Product 103 disliked (or liked substantially less), then the odds are that the consumer comes from Segment 3.
4. If both Products 103 and 106 score highly and equally, then it is not clear from which segment either comes. There is no differentiating pattern here. However, the odds are that the consumer does not fall into Segment 1 (traditionalist).

This strategy of product usage is attractive for small-scale research projects. It classifies consumers as belonging to a segment, but requires product usage. The method is expensive to implement, however, because it does require a small product use study. Consequently, the method requires a high initial screening cost (because of the requirement of usage). The researcher must repeat part of the product evaluation simply to identify the panelist as belonging to a segment.

Strategy 2—A Small Questionnaire

The foregoing approach of actually product testing a limited number of samples in order to determine to which segment a participant may belong is not practical if the screening has to be done by telephone or in a mall. It is infeasible because the interviewer cannot go around giving out samples of the products to try, and then phoning the panelists after a few days' usage. It is necessary to do something different. This "something" is to create a small screening questionnaire that can accurately identify to which segment a panelist belongs.

Requirements. The screening questionnaire must be an efficient tool to identify the category to which a consumer belongs. The screening questionnaire has two distinct uses:

1. *A tool to identify the distribution of segments in the larger population*: The initial study used to discover the segments is typically "small scale." In most developmental product research where there are many products to test, the researcher usually tests products with relatively few panelists. As a consequence, the estimates of the size of the segments may not be accurate. The product use test does not involve a sufficiently large number of panelists, nor is it typically conducted in many cities with a wide geographical distribution. Therefore, if the researcher can develop a good screening questionnaire then it may be feasible to better estimate the distribution of these segments in the population at large.
2. *A tool to rapidly screen potential panelists in order to identify the appropriate participants in additional, follow-up research*: If the researcher wants to create improved prototypes for each segment, then it is important that the participants in the study be of the appropriate segment. The researcher does not want to go through the entire segmentation process each time.

The Approach—Create a Large Questionnaire and Find Predictors. The approach is straightforward. It creates a variety of different phrases that can be easily read to consumers, either in person or over the phone. This questionnaire should be tailored to the category being evaluated. There may be up to 50 different questions or statements.

The questionnaire is presented to the consumer in the initial product evaluation phase (where the segments were first identified). The panelist who evaluates the products also fills out the questionnaire, using a specific and easy-to-use rating scale. [One scale is a 5-point agree/disagree scale; another scale is a 5-point interest scale]. The key point is that the scale be easy to administer.

Each panelist in the study first completes the questionnaire before participating in the product evaluations. This is necessary because, in subsequent recruitment, the panelists will be screened and assigned to segments without ever having experienced the particular product. Therefore, it is vital that the screening questionnaire be completed prior to the actual product use, in order to parallel what will happen later on.

Analysis of the Questionnaire Data. After the panelists have been assigned to the segments we can look at the profile of their answers in the screening questionnaire. The objective is to create a set of predictor equations which allow the researcher to classify a panelist into one of the three segments. The objective is also to limit the number of questions to 10 or less. (This limited number of questions is critical if the recruitment is to be done by telephone. The fewer the number of questions to ask the more likely it will be that the interviewer can complete the interview without irritating the potential panelist or interviewee.)

Table 10C shows a subset of the mean ratings for the screening questionnaire, for total panel, and for the three segments. By using "multiple discriminant analysis," the researcher can create a short questionnaire of no more than eight items which maximizes the number of correctly classified panelists. This analysis is shown in Table 10D. By using the eight questions and the weighting factors shown in Table 10D, the researcher can correctly assign the panelists to segments at least 73% of the time.

Table 10C Screening Questions for Sanitary Napkins and Mean Ratings of Total Panel and Three Segments—Agree (5) → Disagree (1) Scale

Q	Phrasing of the question	Total	Seg 1 indiff	Seg 2 hi-tech	Seg 3 secure
1	I feel that my personal physician knows what is best for me and my health.	3.2	3.3	2.5	3.6
2	I am fascinated by the way physical things work and like to learn about them.	3.1	3.1	3.9	2.4
3	I have a burning desire to explore new things in the world around me.	2.9	2.1	4.1	2.3
4	I am a very private person in most of the things that I do.	3.7	3.5	3.1	4.2
5	I shop primarily for what is on sale, as long as it is of good value.	3.7	4.4	3.6	3.5
6	I like things the way that they were, in the "good old days."	3.4	4.1	2.4	3.9
7	I would describe myself as ultra-feminine in what I do and what I like.	4.0	3.9	3.8	4.3
8	I would like to benefit from advances in technology.	4.3	4.0	4.5	4.3
9	I am brand loyal, and always buy my favorite brand because I trust it.	3.8	3.2	3.4	4.4

Table 10D Approach for Classifying an Individual as Belonging to a Segment Based upon the Pattern of Responses to Eight Classifying Questions (A-H)

Approach Explicated

1. For the people who have already been classified in the study as belonging to the three specific segments, look at the pattern of their responses on the 50 screening questions.
2. Use the method of multiple discriminant analysis to create two equations that best predict this three-segment classification. Typically, the discriminant analysis will only consider a limited number of the 50 predictors. (In this case it used eight of the predictors, which we call A-H. Predictors A-H are questions, which the panelist answers from 1 = strongly disagree → 5 = strongly agree. See Table 10C.)
3. The two discriminant functions which emerge are shown in the table below, and labelled F1 and F2. These are two independent or orthogonal axes. In fact they can be plotted against each other (F1 as the abscissa, F2 as the ordinate).
4. For the panelists who have participated, estimate the discriminant function. The estimation is easy. Simply multiply the rating for the question (A-H) by the appropriate coefficient in the function. Sum up the products, and add in the additive constant. The sum is the expected value on the discriminant function. Each person in the original napkin study has two numbers → expected value of F1 and expected value of F2. Each person in the original study can be located on the discriminant plot.
5. Each panelist has already been classified as belonging to sensory segment 1, sensory segment 2, or sensory segment 3, respectively.
6. For each segment, estimate the average value of F1 (across all panelists in that segment), and the average value of F2 (again, across all panelists in that segment). The result is the centroid of each segment, on each discriminant function. This centroid can be located on the discriminant plot.
7. Now bring in a new panelist (e.g., panelist 101). That panelist will have answered all eight screening questions, as shown in the table above. Estimate panelist 101's value for discriminant function 1 (F1) and discriminant function 2 (F2). It is 1.21 for F1, and −1.13 for F2.
8. Compute the distance to the centroid of segment 1, segment 2, and segment 3. The distance is a geometrical distance. (This can be done easily with a spread sheet.) The centroids for the three segments are as follows:

	Seg 1	Seg 2	Seg 3
F1	0.23	−2.01	1.57
F2	−2.25	0.36	0.49

9. In the simplest of cases, assign the panelist to the segment whose centroid is closest to the location of the panelist on the discriminant plot.

(continued)

Table 10D Continued

10. In some more sophisticated cases one can compute the probability that a person falls into a segment, rather than simply assigning the person to the nearest segment.

	F1	F2	Pan 101 rating	Pan 102 rating	Pan 103 rating
Constant	-1.6	-5.4			
A	0.3	0.1	3	3	4
B	-0.8	-0.2	1	4	2
C	-0.4	0.5	4	1	6
D	0.3	0.4	2	1	3
E	0.2	-0.8	4	1	0
F	0.5	0.7	3	6	5
G	-0.2	0.3	2	3	4
H	0.4	0.2	6	2	5
		Panelist	101	102	103
		Expected F1	1.21	-0.66	-0.18
		Expected F2	-1.13	0.14	4.83
		Segment #	1	2	3

Strategy 3—Developing a Print Ad for Advertising or Sampling

From time to time the marketer may want to attract individuals in each segment through print media. One objective may be to entice consumers in a specific segment to purchase a product. Another objective may be to distribute small amounts of each product (e.g., one sample per segment type) to a large number of consumers, with instructions about how to identify the segment to which they belong, and then which product to try.

    This third strategy cannot use either actual product usage or a short number of screening questions to identify a panelist as belonging to a segment. The strategy must use print advertisements. The print ad medium is necessary because the objective is no longer research. Rather, the goal is to create consumer interest in trying the product, or at least responding to the advertisement as an actual ad, competing with other advertisements in the same category. Keep in mind that unlike the previous two strategies, which used dozens to hundreds of consumers, this strategy must work inexpensively with hundreds or thousands of consumers.

The Approach. The approach uses full concepts rather than simple phrases or products. The same panelists who participated in the product study (from which the segmentation was derived) also evaluate a variety of different print advertisements. We are looking for a concept/print advertisement which appeals selectively

Individual Differences

Table 10E Ratings of 10 Print Advertisements by Total Panel and by Segment

Ad	Total panel	Seg 1 indifferent	Seg 2 high tech	Seg 3 secure
1	62	54	50	75
2	60	51	68	58
3	56	54	67	51
4	54	53	63	47
5	52	50	41	61
6	43	40	38	48
7	41	40	46	38
8	40	38	42	39
9	38	41	35	39
10	37	45	38	33

Strategy:
Seg1—No ad "pulls" or "attracts."
Seg2—Use "Ad 3."
Seg3—Use "Ad 1."

to the segment under consideration, and not to the other segments. We are not looking at this point to classify the panelist, but simply to attract consumers from one of the three segments, without attracting the other consumers.

The strategy for this approach follows. The panelists are presented with a set of 10 different print advertisements, stressing different features and benefits of the product. The panelists pick the top two from the set. Each of these top two is assigned a rating of 100. The remaining eight advertisements are assigned a value of 0. The objective is to find the advertisement that is chosen by one segment, and not by another.

Table 10E shows the analysis of this strategy, and presents the final selections of the print advertisements which maximize the likelihood of attracting the correct consumer group.

11. An Overview to Sensory Segmentation

Sensory segmentation is a powerful procedure to identify consumers as belonging to different groups, with different product preferences. The problem, however, is that these sensory segments do not fall into neat little bundles of consumers, easily identifiable from classification data. The segments are robust and reliable. They all spread through the population. There must be other methods to get at the segments, beyond the simplest methods (demographics, usage) that may not work.

This section presented three different approaches:

1. Product usage
2. Screening questionnaire
3. Broad scale appeal of print ads.

No one approach is perfect. Each approach misclassifies some of the consumers. The key, however, is to perform significantly better than chance. If the researcher believes that the answer to improved product performance consists of finding consumers in the appropriate preference segments, then the next logical step in the process is to develop a way to identify the segment to which a given consumer belongs. Only then can the researcher proceed with confidence using the sensory segmentation procedure to gain an advantage in the marketplace.

12. Creating a Line of Products

Individual differences in product preference can be a source of annoyance to the individual who is interested in a simple world where everything is alike, and where everyone plays by the same rules. Individual differences can also be a source of tremendous opportunity. What one consumer disdains another consumer may adore.

How does the marketer use individual differences to develop a line of products? Are there standard procedures?

The Conventional Approach—Demographic Differences

The conventional approach is to assume that each consumer segment (e.g., males/females, age, even brand usage) corresponds to a specific niche in the marketplace. Each niche can be filled by its own set of products. A significant amount of effort is expended by researchers who divide up the market into finer and finer differences, based upon conventional segmentation. The segmentation may be based upon income, market, or the most frequently used product. There is enough data available on consumers to enable the adventurous researcher to divide the population in as many ways as he likes.

The problem with conventional approaches is that they do not lead to truly new products. As we saw above for both fragrances and toothpastes, there is no clear link between product preferences and demographics. That is, consumers in different demographic segments clearly exhibit different characteristics. Their incomes, and even their usage of products, differ. But, it is hard to demonstrate that the segments or cohesive clusters of consumers developed by these conventional approaches signify opportunities for products. The products preferred by one segment are often just as acceptable to consumers in another segment.

Looking at Individuals "in the Raw"—TURF Analysis

A productive way to create a line of products considers each individual separately, without reference to a subgroup or segment of consumers. The rationale of this approach is that a line of items should be maximally satisfactory (satisfies the greatest number of consumers).

Consider a line of fragrances for a category. Assume for the sake of argument that the manufacturer wishes to market a line of four fragrances. Which fragrances should the manufacturer choose? Should the line comprise only the most acceptable fragrances? If so, then there is the ever-present possibility that consumers who like one fragrance will like all of the fragrances. Actually, by selecting the four best fragrances for the line, the manufacturer may have inadvertently selected items which "cannibalize" each other. That is, an individual in the population who chooses fragrance A may also choose fragrance B, C or D. An individual who does not choose fragrance A may not choose fragrances B, C or D. This strategy satisfies some of the consumers, but not the greatest number who could be satisfied by another selection of four fragrances.

An alternative method to develop the line of four fragrances considers different sets of fragrances, four at a time, from the available set. (These fragrances may be blind or branded, or even described by a concept/package without the physical fragrance being present.) If all of the panelists originally evaluated each fragrance, then we have data for each panelist on every one of the fragrances. Therefore, for each set of four fragrances that the line could comprise, we can estimate the number of panelists who accept either one, two, three, or all four fragrances. (By acceptance we mean the panelist rates the fragrance higher than a certain criterion value on the liking scale.) Each set of four fragrances has a "line-value" associated with it. That line-value could be the proportion of panelists who find at least one fragrance of the four acceptable to them. The optimal line would then consist of a set of fragrances that maximizes the number of "satisfied" consumers, with a satisfied consumer defined as an individual who likes and would choose at least one of the fragrances offered.

This brute force approach is known as "TURF" analysis. (TURF is the acronym for "Total Unduplicated Reach and Frequency.") The objective of TURF analysis is to identify promising lines of products without necessarily understanding the underlying product preferences in a more profound way. TURF is first and foremost a computational scheme which estimates the range of consumer satisfaction with a given line of items.

A Case History: Shaving Cream

We can see the TURF approach in action in the following example. The manufacturer of a line of shaving creams knows that there is substantial person-to-person variation in the acceptance of the shaving creams. Furthermore, the manufacturer knows that to some extent this variation is due to fragrance, as well as to the physical feel of the cream. The issue facing the manufacturer is to design a line of new shaving creams to replace an old line. The old line comprised many different fragrances and a limited number of different "textures" of the shaving cream. How many entries should the new line comprise? What should they be?

Step 1—Creation, Submission, Selection of Prototypes. The manufacturer opted to use the TURF analysis procedure to create the line of products. The first stage in the experiment consisted of a briefing to R&D to create a variety of shaving creams with different textures and fragrances. Marketing was not specific in this early stage regarding the particular fragrances and textures desired. That is, there was no compelling reason to choose a specific selection of fragrances over and above another selection. Furthermore, marketing as well as R&D wanted to learn from this experiment, rather than conduct an experiment to validate the decisions that had already been made.

R&D received well over 4 dozen submissions from its regular fragrance suppliers, each of whom had been given the same brief—come up with different fragrances for a variety of shaving creams. Three shaving cream bases (varying in creaminess and in density) had been previously developed by the R&D group, and were provided to the fragrance suppliers who were then instructed to incorporate their own submissions into the base.

Step 2—Screening the Submissions Down to a Reasonable Set. The 53 submissions received in Step 1 above were simply too many to test with the reduced budgets provided to R&D. As a group, R&D screened the 53 submissions down to 16 promising submissions. Among the group rules of screening was that the final candidates for the TURF analysis would have to represent both different types of fragrance and different textures. Therefore, R&D made every effort to balance the 16 final submissions in order to have a reasonable number of each texture type and general fragrance type.

Step 3 Executing the Evaluations. The proper execution of the shaving cream test is difficult. It is tempting to have the panelist evaluate each product at home, shave with the product and experience the product in its natural setting. This would provide a better "read" of the consumer reactions. On the other hand, the procedure is cumbersome and error-prone. Ideally, each consumer should evaluate all 16 products. The test would take over a month, longer than marketing wanted to wait.

An alternative method, which seemed at the time more attractive to marketing, was to pre-recruit the consumers to participate in an evening session lasting 3 hours. During that session the panelist would smell the shaving cream

Individual Differences

in a cup, and apply some of the cream to the face. The panelist would then rate the cream, and wash it off in a sink provided for that purpose. The panelist would not shave. Then the panelist would wait 12 minutes and rate the next product.

The purists in the company objected strenuously to this approach because they believed strongly that the central location test was inappropriate for a product that was designed to be used at home, for actual shaving. However, when they recognized that they could significantly decrease the cost of the project and obtain the data far more rapidly, the dissenters to the research plan agreed to go ahead. They insisted, however, on being present at the test to insure that it went well.

The test execution comprised 80 respondents, all users of the category (but not necessarily users of the brand). The panelists evaluated each of the 16 samples in two sequential 3-hour sessions, rating each sample on a series of attributes. The key attribute was "overall liking," although the attributes dealt with appearance, fragrance, feel, and a series of terms dealing with performance. Overall liking was rated on a 0–100 point scale (0 = hate → 100 = love).

Results—Matrix of Liking × Product × Panelist. The starting analysis consists of a matrix of product × panelists. The numbers in the matrix correspond to the liking rating. If the data is not complete (if each panelist only rated a subset of the products, rather than a complete set), then we complete the matrix by replacing a panelist's missing data by the mean rating assigned by that same panelist to actual products. (This completion or filling in of missing data would not change the mean rating assigned by any panelists. Furthermore, the mean of a panelist's rating is the best estimate of missing data for that panelist.)

Identifying Consumer Responses to Different Lines of Fragrances. The next analysis consists of selecting different combinations of fragrances, either 1, 2, 3, or 4 at a time. For each selection we determine the proportion of panelists who would be satisfied with at least one selection. By "satisfied" we mean that the panelist would assign to a product a rating of at least 65 on the 100-point scale. A rating of 65 on the scale corresponds to a product that is quite well liked. (We could also make the criterion stricter—e.g., a liking rating of 75, etc. In any case we must decide ahead of the analysis what criterion defines a "satisfied" panelist.)

Table 11A shows the results of this analysis. Only some of the 2-, 3-, and 4-line combinations are shown. [For three items there are (16*15*14)/(1*2*3) or 560 combinations. The number of combinations grows geometrically with the number of items in the fragrance line]. Depending upon the specific combination of items, the manufacturer can satisfy most of the consumers or just some of the consumers. Clearly the more items in the line the closer the manufacturer will be to satisfying any consumer with at least one item in the line.

Table 11A Examples of TURF Analysis to Develop a Line of Shaving Creams: Performance of Various Lines of Items

Items in line				%Satisfaction				
901				37%	Products	=	1	
911				37%	Highest reach	=	37%	
903				36%	Mean reach	=	30%	
902				33%	Lowest reach	=	23%	
907				33%				
912				33%				
917				33%				
909				31%				
906				30%				
914				30%				
921				30%				
904				28%				
905				28%				
908				28%				
910				28%				
919				28%				
915				27%				
920				25%				
916				24%				
918				24%				
913				23%				
901	906			61%	Products	=	2	
906	911			61%	Highest reach	=	61%	
907	917			56%	Mean reach	=	51%	
917	918			50%	Lowest reach	=	39%	
901	902	906		73%	Products	=	3	
901	902	903		66%	Highest reach	=	78%	
904	906	915		62%	Mean reach	=	65%	
918	920	921		59%	Lowest reach	=	51%	
915	916	918		55%				
905	910	912	918	82%	Products	=	4	
917	918	919	921	82%	Highest reach	=	87%	
903	906	912	918	77%	Mean reach	=	76%	
901	902	909	916	76%	Lowest reach	=	62%	
901	918	920	921	72%				
903	906	913	918	69%				

Reach = Proportion of satisfied consumers (who would assign at least one shaving cream in the line a rating of 65 or higher).

Going One Step Further—Adding Fragrances to an Existing Line

TURF analysis can go beyond simply selecting the fragrances (or other entries) in the line. If there are existing items in the line that must be used, then this requirement will change the decision. The existing items may not be the optimal selection for the line because they may not please the greatest number of consumers (in concert with other items). However, the existing items may be forced into the line because of a desire for "continuity," and a desire not to disappoint current users of the line.

Let's now construct a line comprising either 1, 2, or 3 additional items. Assume again that the line comprises a total of four products. However, also assume that either 1, 2, or 3 items will be forced in. What additional items should be selected in order to maximize interest? Table 11B shows the result of this decision. The table shows which items of the test set are assumed to represent current products in the line and the optimal solution to the problem, given the requirement that the current items remain in the line. Note that the result is a less-than-optimal line in terms of the proportion of the consumers who are satisfied with the line.

Creating Lines with Defined Product Differences—Fragrance and Texture

Often the problem of creating a line of items becomes more complicated because the marketer wants to promote a specific variety of product characteristics in the line. TURF analysis by itself simply selects the best set of items which, in

Table 11B TURF Analysis for Shaving Creams—Results of Forcing in Specific Fragrance (#901)

Items in line	% Satisfaction		
901 906 911	70%	Products in line	= 3
901 903 905	67%	Highest reach	= 70%
901 906 914	66%	Mean reach	= 67%
		Lowest reach	= 53%
901 906	61%	Products in line	= 2
901 903	60%	Highest reach	= 61%
901 911	59%	Mean reach	= 54%
		Lowest reach	= 43%
901	37%	Products in line	= 1

Table 11C TURF Analysis For Shaving Creams—Results of Forcing in One Fragrance (#901) and One Texture (#910)

Items in line				%Satisfaction		
901	910	903	904	83%	Items in line	4
901	910	903	905	82%	Highest reach	83%
901	910	913	921	80%	Mean reach	77%
901	910	916	921	78%	Lowest reach	69%
901	910	908		65%	Items in line	3
901	910	912		64%	Highest reach	67%
901	910	921		62%	Mean reach	58%
901	910	913		60%	Lowest reach	53%
901	910			54%	Items in line	2

concert, maximize the proportion of satisfied consumers. TURF analysis does not specify the composition of the line (in terms of the range of characteristics).

Suppose, however, that the manufacturer wishes to insure that the line comprises a specific texture and a specific fragrance, but does not care what remaining products are used. The TURF analysis could be changed slightly to reflect this requirement. Table 11C shows the results of this type of analysis. Since the marketer specifies a given texture, the TURF analysis is not free to maximize overall acceptance across the entire set of products because that strategy would not generate the desired inclusion of necessary products. The resulting solutions in Table 11C are less than optimal, but satisfy the marketing criteria.

12. Sensory Preference Segments in Line Development

The previous analysis using TURF creates lines of items, but does not do so in a framework which involves understanding the true preferences of consumers. That is, TURF analysis is entirely atheoretical. There are no assumptions at all about the preferences of consumers. There is simply a brute force computation of consumer acceptance given different sets of products offered.

Sensory preference segmentation provides a more theory-based way to divide the population. According to sensory segmentation, the world of consumers comprises a limited number of different clusters of individuals, each cluster showing distinct preferences for product sensory attributes. Thus, it seems natural to divide the product offerings into sets that appeal to each sensory preference segment. That is, rather than blindly going out to market a line of items, just maximizing consumer acceptance, it may be better to select one, several, or even all sensory preference segments and design products for each

Individual Differences 383

segment. In this way the manufacturer can be sure to maximize the number of satisfied consumers.

To illustrate this approach, consider a line of lipsticks designed for young women (ages 12–17). Marketing research revealed that many of these consumers like "hot colors." However, the manufacturer did not know which specific colors to offer in the line. Store audit data wasn't particularly helpful because the number of items in various lipstick lines varied from store to store. Secondary purchase data also did not help, because it did not reveal what consumers wanted—it only showed what was available and what was purchased.

The manufacturer first ran four focus groups, two in each of two geographically dispersed markets in the U.S. The focus groups were of some help narrowing down the types of colors, but did not provide the necessary information about how many should be in the line, which specific colors, and the expected quantities of each color to manufacture. The manufacturer, who was new to the category, was also understandably nervous about the research because the results of the focus group were not immediately actionable.

The marketing research director hypothesized that consumer preferences for lipstick colors might be distributed according to sensory preference segments. The manager observed that certain pairs of colors were never selected by the same consumer. That is, from usage data (attitude surveys) and from the focus groups, it appeared that there were at least two groups of consumers that did not overlap—those who used rich, deep colors and those who used pale, pastel-type colors. There were probably other groups of consumers as well that fell into segments different from the two that the marketing research director hypothesized.

Because the company management was committed to the lipstick category and wanted to do things right, the marketing research director suggested that they ought to follow up the focus groups with a quantitative study to identify the nature of the segments. The research design was fairly straightforward, as the six-step sequence below shows:

1. Purchase all of the available lipsticks on the market.
2. Identify a reasonable set of different lipsticks. These lipsticks were to have distinct colors. (Management did not want to spend money testing two colors that were the same, even if these colors were offered by different manufacturers.) The screening exercise, held at the marketer's office and attended by 10 consumers, selected 40 different colors. The colors were evaluated by application to a white blotter (of the same type used by fragrance evaluators).
3. Recruit 150 panelists for a 2-hour test session. The panelists were recruited in four markets around the U.S. (to provide geographical distribution), and ranged in age from 12 to 49 (to provide age and usage distribution).

4. Each panelist rated the degree of liking for the product, using a scale from 0 (hate) to 9 (love), as well as rating every product on two additional attributes—lightness (from light to dark) and redness (from pink to red). The panelist also used a 0–9 scale for each attribute. Each panelist rated all 40 products in a randomized order.
5. The data was then subjected to the sensory segmentation analysis as described in Table 5. The independent variables were degree of lightness, and amount of redness. (These independent variables were taken from the average ratings assigned by the total panel.) The dependent variable was the individual's liking rating.
6. The segmentation analysis revealed three different segments, best defined as pink/white preferrers, deep red preferrers, and those who accepted a wide range of middle colors. A post hoc analysis of the data showed that the pink/white preferrers tended to be younger, whereas the deep red preferrers tended to be older. However, the classification by age was not identical with the sensory segmentation. (There were some pink/white preferrers who were fairly old.)

Table 12A shows the partial list of liking ratings for five of the 40 lipsticks. The ratings are shown for the total panel, the three sensory segments, and the two user groups.

Creating the Line

Since the primary purpose of the project was to develop the line of lipsticks, the marketing manager wanted to move forward with a recommended line. The marketing research director suggested that a reasonable approach would be to select one or more colors from each of the three preference segments. That is, it made sense to use the sensory segmentation to define the line. The marketing manager recognized the wisdom of this approach, and suggested that they offer a total of nine different colors, three from each of the three segments.

Table 12A Results from Lipstick Study—Partial Data for Liking—Total Panel, Sensory Segments, User Group

Product	Type	Total panel	Hot seg 1	Medium seg 2	Subdued seg 3	Use prod/A	Use prod/B
1	Medium	55	65	57	47	58	52
2	Hot	37	56	24	37	34	36
3	Medium	51	56	45	55	47	55
4	Subdued	43	47	26	56	37	49
5	Subdued	42	46	28	54	33	51

Individual Differences

Table 12B Comparison of Two Alternative Strategies for the Lipstick Line

Strategy 1—Select the Nine Most Acceptable Lipsticks for the Entire Panel

Product	Type	Total panel	Segment hot seg 1	Segment medium seg 2	Segment subdued seg 3	Use prod/A	Use prod/B
1	Medium	55	65	57	47	58	52
18	Medium	53	58	45	58	49	57
34	Medium	53	57	53	51	54	52
13	Medium	52	53	56	48	50	54
40	Medium	52	56	52	50	52	52
3	Medium	51	56	45	55	47	55
14	Medium	50	42	51	55	40	60
24	Subdued	50	35	48	62	36	64
25	Subdued	50	36	47	63	36	64

Strategy 2—Select Three Products for Each of the Three Sensory Segments

Product	Type	Total panel	Segment hot seg 1	Segment medium seg 2	Segment subdued seg 3	Use prod/A	Use prod/B
6	Hot	47	*70	45	35	59	35
27	Hot	38	*69	22	33	47	29
15	Hot	43	*68	37	35	51	35
1	Medium	55	65	*57	47	58	52
13	Medium	52	53	*56	48	50	54
34	Medium	53	57	*53	51	54	52
10	Subdued	47	39	35	*63	34	60
25	Subdued	50	36	47	*63	36	64
30	Subdued	45	31	35	*63	28	62

Furthermore, the marketing director suggested that each selection for each segment should be strongly targeted to that segment.

Table 12B shows the proposed line, and reveals that the selection of different colors did in fact appeal to the various segments. However, it is important to recognize that judgment entered into the equation as well.

What Is Better for a Line—TURF Analysis or Sensory Segmentation?

Given the two foregoing approaches, TURF analysis and Sensory Preference Segmentation, which is better? Both procedures have a great deal of merit, but also have problems as listed below:

TURF Analysis
  Positives:
  1. Looks at each individual
  2. Identifies the proportion of satisfied consumers
  3. No need for theory
  Negatives:
  1. No theoretical basis (organizing principle)
  2. Therefore, no way to predict reaction to new products that were not tested

Sensory Preference Segmentation
  Positives:
  1. Based upon theory
  2. Enables the research to use the organizing principle to predict responses to new products
  Negatives:
  1. Unable to estimate proportion of satisfied consumers because it works on aggregate data.

13. Going One Step Further in Line Development: Opting for a Fair Share of the Marketplace

Introduction

How can a marketer decide the optimum strategy for developing products, given the pervasiveness of individual differences? In a crowded market where there are many contenders for the consumer's pocketbook, which product should be introduced, and why?

The naïve answer to this question is to introduce the most highly-liked prototype into the market. The reasoning goes something like this:

1. The most-liked product, if it is liked sufficiently, should appeal to a large number of consumers.
2. Once the consumers discover that they like this product more than other products on the market, they will switch to this product and become loyal users.
3. It makes the most sense to introduce the most highly-liked product, because that single product should attract the greater number of new consumers.

This development and marketing strategy is faulty for the following reasons:

1. The consumers in a category are often segmented in terms of their preferences.

Individual Differences

2. If we consider the existing segments in a category to represent groups of consumers with different preferences, then we might discover that the product that has the "best shot" at the category will be most highly accepted by only one or two of the segments, and not by all consumers.
3. If the segments who like the prototype are sufficiently large in terms of total number of consumers, then it is likely that other manufacturers will have already discovered these segments. The other manufacturers may not realize that consumers comprise a limited number of sensory preference segments, but by trial and error the other manufacturers probably will have happened upon the appropriate products for the segments.
4. Consequently, the segment to which the new product appeals so highly may be already inundated with competitor offerings. Although the product is highly acceptable to consumers in the segment, it is likely that the competitors have introduced products that are also highly acceptable.

An Alternate Approach—Fair Share Analysis and Marketing Strategy

Let us approach the problem differently. Assume for a moment that the manufacturer has done his homework and has identified the competitor products in the marketplace, as well as the existence and composition of sensory preference segments.

For the sake of this example, let us assume that the product category is toothpaste. The manufacturer has tested 50 different toothpastes in the category, and has found that consumers divide into three segments as shown in Table 13A. Furthermore, it appears that there is an imbalance in the category. By imbalance we mean that there are segments in the category comprising many potential consumers, but relatively few products given the number of consumers. There are other segments, perhaps just as large, which have many products competing for the consumers in these segments. Table 13A shows how (on the average) consumers in each segment react to the toothpastes available to them.

In Table 13B we also see an example of a theoretical "fair share" analysis, without one's new product being considered. If all of the brands were equally advertised and equally used, then we would expect consumers in a sensory preference segment to consider purchasing primarily those products that they like very much on an absolute basis. (We define liking as a mean rating of 63 or higher on overall acceptance, when the product is tested "blind.")

Following this train of thought, we would expect that consumers would divide their purchases equally among the products that appeal to them. In Sensory Segment 1, we have four toothpastes which are highly appealing. We expect the consumers in this category to divide their purchases among these four. In fact, if there were no advertising or branding effects, we would expect each of the four highly accepted products for Segment 1 to be chosen by 1/4 of the consumer segment. That is, for Sensory Segment 1, only four toothpastes of

Table 13A Toothpaste—Liking Data by Total Panel and Three Sensory Preference Segments

Size of segment in the population	Total (100%)	Seg 1 (21%)	Seg 2 (28%)	Seg 3 (51%)
Number of toothpastes out of 50 tested with high scores >				
Total > 63	8	4	9	15
Total > 66	5	2	9	10
Total > 69	3	2	5	7
Total > 72	2	1	4	3
Scores of first 15 toothpaste products	Total	Seg 1	Seg 2	Seg 3
101	52	48	56	53
102	18	21	13	20
103	41	40	40	44
104	17	24	14	14
105	24	30	19	22
106	72	59	76	77
107	32	26	37	35
108	58	62	49	57
109	31	24	46	29
110	58	59	52	51
111	56	54	47	66
112	64	59	67	63
113	16	20	14	12
114	69	68	70	67
115	50	41	60	53

those tested should enter their "consideration set" on the basis of pure sensory acceptance. Furthermore, Sensory Segment 1 comprises only 21% of the total set of consumers. Thus, each product appealing to Sensory Segment 1 should have approximately 5.3% of the total consumers (21%/4), if advertising, promotion, brand history, etc. are equal.

Now consider the strategy of introducing a new product to appeal to Sensory Segment 1. If the product does not appeal to the consumers on a sensory basis (e.g., the liking rating is less than 63), we would count this product out of the consideration set. If the product does appeal to the consumers, however, then it must carve out its share from the 21% of the consumers in Segment 1. An acceptable entry makes five products which must share the 21% of the consumers in Sensory Segment 1. Each product now should have 4.2% of the consumers (viz., 21%/5).

Individual Differences

Table 13B Fair Share Analysis: Effect on Potential Share If Consumers Only Select Highly Acceptable Toothpastes (Liking ≥ 63 for Current and Proposed New Entry)

	Seg 1	Seg 2	Seg 3
Size of segment in population	21%	28%	51%
Total number of toothpastes scoring 63 +	4	9	15
Current fair share of each toothpaste:			
In segment	25.0%	11.1%	6.7%
In total population	5.3%	3.1%	3.4%
Effect on share of introducing a toothpaste scoring 63+:			
In segment	20.0%	10.0%	6.3%
In total population	4.2%	2.8%	3.2%

Assumption — all segments use toothpaste at the same rate of consumption.

The marketer can follow this train of thought through an analysis of all of the segments. It may turn out that a product designed for a small segment (e.g., one that has only 21% of the marketplace) may be more promising than one designed for a large segment (e.g., one that has 50% of the marketplace). The small segment may be more attractive and viable than the large segment when it has far fewer products appealing to consumers in that segment.

14. An Overview to Individual Differences

With the fracturing of conventional markets, with the emergence of just-in-time manufacturing, and with the breakup of mass marketing into niche marketing and even individual product tailoring, an understanding of individual differences is more important than ever before. Properly using individual differences for one's market advantage may, in the years to come, mean the difference between modest achievement and great success.

As we have seen in this chapter, there are many ways to understand and to use individual differences in commercial product development. Individual differences are a fact of life, both in model systems (e.g., pure tastes and smells) and in actual products. There is no accounting for individual tastes, nor need there be. Individual differences are pervasive.

Furthermore, for applied purposes there is no need to standardize people. The differences in consumers' product preferences are hints about the potential for different products that need to be created. Once the marketer realizes that these differences present opportunities rather than problems, the marketer is empowered to develop new products in an efficient way, and gain market share.

There are two ways to approach individual differences. One way is scientific, the other uses brute force. Scientific analysis (e.g., such as that typified by Sensory Preference Segmentation) looks for organizing principles underlying individual differences. Is nature sending the researcher a message? What is the pattern underlying these differences? And, can that pattern be used to further understand the processes of perception and hedonics? A brute force analysis (e.g., such as that typified by TURF analysis) does not look for general principles and underlying rules, but simply tries to maximize some objective criterion (e.g., overall proportion of satisfied consumers, given a proposed line of items). Both approaches have merit—although the scientific analysis builds an understanding of the structure of individual differences for future utilization by the marketer.

References

Allison, R.I. and Uhl, K.P. 1964. Influence of beer brand identification on taste perception. Journal of Marketing Research, *1*, 36-39.

Blom, G. 1955. How many taste testers? Wallerstein Laboratories Communications, *18*, 173-178.

Booth, D.A. 1987. Individualised objective measurement of sensory and image factors in product acceptance. Chemistry and Industry (London), *13*, 441-446.

Booth, D.A. 1988. Practical measurement of the strengths of actual influences on what consumers do: scientific brand design. Journal of The Market Research Society (U.K.), *30*, 127-146.

Booth, D.A. 1990. Designing products for individual customers. *In:* Psychological Basis of Sensory Evaluation, R.L. McBride and H.J.H. MacFie (eds.), pp. 163-192. Elsevier Applied Science, Barking, Essex, U.K.

Booth, D.A., Thompson, A.L. and Shaledian, B. 1983. A robust, brief measure of an individual's most preferred level of salt in an ordinary foodstuff. Appetite, *4*, 301-312.

Conner, M.T. and Booth, D.A. 1992. Combined measurement of food taste and consumer preference in the individual: Reliability, precision and stability data. Journal of Food Quality, *15*, 1-17.

Cowart, B.J. 1989. Relationships between taste and smell across the adult life span. *In:* Nutrition and the Chemical Senses in Aging, C. Murphy, W.S. Cain and D.M. Hegsted (eds.), New York Academy of Sciences, New York.

Desor, J.A., Green, L.S. and Maller, O. 1975. Preferences for sweet and salty tastes in 9 to 15-year olds and adult humans. Science, *190*, 686-687.

Dukes, W.F. and Bevan, F. 1952. Accentuation and response variability in the perception of relevant objects. Journal of Personality, *20*, 457-465.

Griffiths, R.P., Clifton, V.J. and Booth, D.A. 1984. Measurement of an individual's optimally preferred level of a food flavor. *In:* Progress in Flavour Research, J. Adda (ed.), pp. 81-90. Elsevier Science Publishers, Amsterdam.

Jacoby, J., Olson, J.C. and Haddock, R.A. 1971. Price, brand name and product composition characteristics as determinants of perceived quality. Journal of Applied Psychology, 55, 570-590.

MacFie, H.J., Bratchell, N., Greenhoof, K. and Vallis, L.V. 1989. Designs to balance the effect of order of presentation and first-order carry-over effects in halls tests. Journal of Sensory Studies, 4, 129-148.

Marks, L.E., Szczesiul, R. and Ohlott, P. On the cross-modality perception of intensity. Journal of Experimental Psychology, Human Perception and Performance, 12, 517-534.

Martens, M., Rodbotten, M., Martens, H., Risvik, E. and Russwurm, Jr., H. 1988. Dissimilarities in cognition of flavour terms related to various sensory laboratories in a multivariate study. Journal of Sensory Studies, 3, 123-135.

McBride, R.L. and Booth, D.A. 1986. Using classical psychophysics to determine ideal flavour intensity. Journal of Food Technology, 21, 775-780.

McConnell, J.D. 1968. The development of brand loyalty: an experimental study. Journal of Marketing Research, 5, 13-19.

Miller, A.J. 1987. Adjusting taste scores for variations in use of scales. Journal of Sensory Studies, 2, 231-242.

Moskowitz, H.R. 1981. Relative importance of perceptual factors to consumer acceptance: linear versus quadratic analysis. Journal of Food Science, 46, 244-248.

Moskowitz, H.R. 1984. Cosmetic Product Testing—A Modern Psychophysical Approach. Marcel Dekker, New York.

Moskowitz, H.R. 1984. Relative importance of sensory factors to acceptance: theoretical and empirical analyses. Journal of Food Quality, 7, 75-90.

Moskowitz, H.R. 1985. New Directions for Product Testing and Sensory Analysis of Foods. Food and Nutrition Press, Trumbull, CT.

Moskowitz, H.R. 1986. Sensory segmentation of fragrance preferences. Journal of the Society of Cosmetic Chemistry, 37, 233-247.

Moskowitz, H.R. 1989. Sensory segmentation and the simultaneous optimization products and concepts for development and marketing of new foods. In: Food Acceptability, D.M.H. Thomson (ed.), pp. 311-326. Elsevier Applied Science, Barking, Essex, U.K.

Moskowitz, H.R. 1992. Importance of sensory factors for acceptance of seafood. Journal of Sensory Studies, 7, 147-156.

Moskowitz, H.R., Jacobs, B.E. and Lazar, N. 1985. Product response segmentation and the analysis of individual differences in liking. Journal of Food Quality, 8, 168-191.

Moskowitz, H.R. and Jacobs, B.E. 1986. The relative importance of sensory attributes for food acceptance. Acta Alimentaria, 15, 29-36.

Moskowitz, H.R. and Krieger, B. 1993. What sensory characteristics drive product quality? An assessment of individual differences. Journal of Sensory Studies, 8, 271-283.

Naes, T. 1990. Handling individual differences between assessors in sensory profiling. Food Quality and Preference, 2, 187-199.

Naes, T. and Solheim, R. 1991. Detection and intepretation with and between assessors in sensory profiling. Journal of Sensory Studies, 6, 159-177.

Neslin, A. 1981. Linking product features to perceptions: self stated versus statistically revealed importance weights. Journal of Marketing Research, 18, 80-86.

Pangborn, R.M. 1970. Individual variations in affective responses to taste stimuli. Psychonomic Science, *21*, 125–128.

Schlich, P. 1993. Uses of change-over designs and repeated measurement in sensory and consumer studies. Food Quality and Preference, *4,* 223–236.

Schutz, H.G. and Wahl, O. 1981. Consumer perception of the relative importance of appearance, flavor and texture to food acceptance. *In:* Criteria of Food Acceptance. How Man Chooses What He Eats, J. Solms and R.L. Hall (eds.), pp. 97–116. Forster Verlag, Zurich.

Sheen, M.R. and Drayton, J.L. 1989. Influence of brand label on sensory perception. *In:* Food Acceptability, D.M.H. Thomson (ed.), pp. 89–99. Elsevier Applied Science, Barking, Essex, U.K.

Shepherd, R., Farleigh, C.A. and Land, D.G. 1984. Effects of stimulus context on preference judgments for salt. Perception, *13*, 739–742.

Shepherd, R., Smith, K. and Farleigh, C.A. 1989. The relation between intensity, hedonic and relative-to-ideal ratings. Food Quality and Preference, *1*, 75–80.

Stevens, D.A., Dooley, D.A. and Laird, J.D. 1989. Explaining individual differences in flavour perception and food acceptance. *In:* Food Acceptability, D.M.H. Thomson (ed.), pp. 173–180. Elsevier Applied Science, Barking, Essex, U.K.

Vickers, Z.M. 1993. Incorporating tasting into a conjoint analysis of taste, health claim, price and brand for purchasing strawberry yogurt. Journal of Sensory Studies, *8*, 341–352.

Weitz, B. and Wright, P. 1979. Retrospective self insight about the factors considered in product evaluations. Journal of Consumer Research, *6*, 280–294.

Wheatley, J.J. and Chiu, S.Y. 1977. The effects of price, store image, and product and respondent characteristics on perceptions of quality. Journal of Marketing Research, *14*, 181–186.

7
Interrelating Data Sets:
Physical Measures, Expert Panel and Consumer Ratings

1. Introduction

During the past 60 years, researchers have correlated the instrumental measurements of products with consumer ratings. There are two objectives of these research efforts:

1. On a purely *scientific* basis, researchers are interested in the correlation between subjective and instrumental (often termed "objective") measures. Correlation analysis gives the researchers an ability to quantify consumer reactions (e.g., perceived intensity), even if in a roundabout way. That is, the researcher feels comfortable when the consumer rating of a characteristic can be correlated with a known physical property of the stimulus.
2. On a purely *applied* basis, researchers want to substitute inexpensive instrumental measures for the more expensive consumer measures. For instance, in quality control it is important to discover when a product is out of specification. Consumer panels that perform this function can become extraordinarily expensive when a routine consumer test is performed on a batch-to-batch basis. Some companies economize by training a panel of experts. These on-site expert panelists, generally employees, save money because they are already on the payroll. However, replacing the panelists with machines calibrated to consumer acceptance will reduce the cost far more.

2. Traditional Methods For Correlating Data Sets

The scientific literature is replete with attempts to correlate data from different sources. The conventional method uses correlation analysis. Typically, the researcher assesses several products in the same category on a variety of different measures. Some of these are instrumental measures, others are consumer ratings. Then the researcher correlates the data from the two sets. Table 1A

Table 1 Example of Data from Instruments, Consumers, Experts—Data from a Cleansing Hand "Mousse"

	A. Partial Data Base (Five Products from the Set)				
	Product				
	101	102	103	104	105
Physical measures					
Penetration	21	21	24	20	21
Color	25	24	22	24	23
Oil	51	51	50	52	50
Moisture	1.9	1.9	1.3	2.6	1.5
Consumer sensory					
Dark	51	46	53	48	60
Shiny	53	57	57	71	60
Ease of application	75	82	76	87	79
Smooth	75	79	75	86	75
Oily	61	65	62	77	66
Sticky	69	64	57	56	54
Firm	46	38	46	33	41
Creamy	63	67	69	75	70
How fast it rubs in	51	52	63	65	64
Expert sensory					
Dry	6.0	6.0	6.0	5.8	5.8
Easy to spread	9.0	7.8	7.8	7.0	8.8
Airy texture	0.5	0.3	0.3	0.5	0.2
Firm texture	14.6	14.0	14.0	14.5	14.4
Amount of particulates	0.0	0.0	0.0	0.0	0.9
Size of particulates	0.0	0.0	0.0	0.0	2.5
Air pockets	0.2	0.3	0.3	0.1	0.5
Density	11.0	10.0	8.0	7.5	8.3
Spreadability	8.5	8.5	7.5	7.5	10.3
Slipperiness	0.5	0.8	1.5	1.5	1.5
Adhesiveness	10.0	10.0	7.5	6.2	6.0
Ease of manipulating	10.5	8.5	8.5	7.8	6.5
Cohesiveness	12.9	13.8	12.5	13.0	10.0

Table 1 Continued

B. Correlations Between Physical Measures (Columns)
and Sensory Attribute Ratings (Rows)
(Data from All 19 Products)

	Penetration	Color	Oil	Moisture
Consumer sensory				
Dark	0.24	−0.81	0.05	−0.20
Shiny	0.05	−0.19	0.26	0.26
Ease of application	0.13	−0.14	0.82	0.04
Smooth	−0.14	−0.14	0.45	0.20
Oily	0.17	−0.23	0.68	0.07
Sticky	−0.32	−0.14	−0.44	0.32
Firm	−0.08	−0.01	−0.73	−0.04
Creamy	0.13	0.00	0.62	−0.07
How fast it rubs in	0.21	0.01	0.46	−0.31
Expert sensory				
Dry	−0.26	0.44	−0.13	0.06
Ease of spreading	−0.27	−0.22	−0.70	−0.11
Airy texture	0.59	0.22	−0.03	−0.32
Firm texture	−0.61	−0.22	−0.22	0.20
Amount of particulates	−0.06	0.07	0.08	0.14
Size of particulates	−0.05	0.22	−0.07	0.17
Air pockets	0.61	0.19	0.06	−0.34
Density	−0.47	0.28	−0.37	0.19
Slipperiness	0.34	−0.17	0.37	−0.11
Adhesiveness	−0.24	0.21	−0.50	0.24
Ease of manipulating	−0.45	−0.13	−0.43	0.33
Cohesiveness	−0.10	−0.01	0.03	0.10

shows some typical data. The first set of rows shows the instrumental measure, the remaining rows show the consumer subjective measures (including sensory and liking attributes). Table 1B shows the intercorrelations between the different rows.

Correlations tell the researcher that two variables are linearly related, but do not reveal the quantitative nature of that relation. If the two variables are related to each other linearly, then is the relation steep or flat? (See Figure 1.) Furthermore, what happens in those cases where there is a relation between an instrumental measure and a consumer measure, but the relation is not linear? For instance, it may be a parabola, or a curve with an asymptote (see Figure 2).

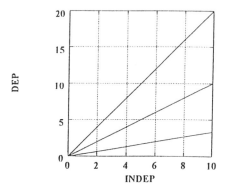

Figure 1 Linear relation between instrumental measure (independent) and sensory ratings (dependent).

Perhaps most research deals with correlations rather than with higher order analysis because correlations are easy to perform and to understand. Correlation "packages" on a personal computer make no assumptions about selecting the appropriate predictors for an equation. They simply compute a single statistic which links two variables. Furthermore, correlation packages often identify those correlations which are statistically significant (which by chance would occur only five times or fewer in 100 analyses). Finally, because researchers often measure many physical characteristics of a single product, the correlation analysis usually turns up one or two physical measures that appear to be related to the consumer attribute rating. (It is rewarding to discover that there are at least 1 or 2 significant results after the researcher has invested the time and the money. With correlation analyses, more often than not, some correlation statistics achieve statistical significance by chance alone.)

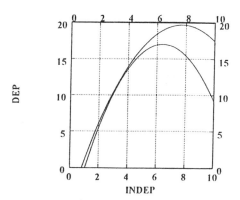

Figure 2 Non-linear (quadratic) relations between instrumental measures (independent) and sensory ratings (dependent).

Building Equations for Sensory-Instrumental Relations

Another way to relate the two domains of sensory and instrumental measures uses equations. The equation is *descriptive* because it describes how changes in one domain (instrumental) covary with changes in the other domain (subjective, consumer). The equation makes no pretense of describing the processes which actually transform the physical stimulus into the subjective response.

Equations are more satisfying than correlations because they enable the researcher to describe and to predict. If the equation is sufficiently accurate (fits the data quite well), and if the researcher estimates by interpolation (estimating a consumer response from instrumental data that lies within the range of instrumental data used to develop the equation), then the equation will often provide an accurate estimate. Correlations cannot perform this function because the correlation statistic simply measures the strength of a linear relation between two variables.

Equations have three other key benefits that are important to the researcher:

Equations can be linear or non-linear. A linear equation assumes that the dependent variable changes in a fixed fashion with unit changes in the independent variable. A non-linear equation does not make this assumption. With a non-linear equation, unit changes in the independent variable can correspond to small, medium or large changes in the independent variable.

Once the researcher decides to graduate from a simple linear equation to a non-linear equation there are many different ways to express the non-linear relation. Each of the five equations below looks different in graphical coordinates, and describes a different pattern of how the dependent variable changes with changes in the independent variable.

There is one primary way to express a linear equation, as shown below:

$$\text{Dependent Variable} = k_0 + k_1 (\text{Independent Variable}) \qquad (1)$$

There are many ways to write a non-linear equation. Four ways are shown below:

$$\text{Dependent Variable} = k_0 + k_1 (1/\text{Independent Variable}) \qquad (2)$$

$$\text{Dependent Variable } k_0 + k_1 (\text{Independent Variable}) + k_2 (\text{Independent Variable})^2$$

$$\text{Dependent Variable} = k_0 + k_1[\text{Log}(\text{Independent Variable})] \qquad (4)$$

$$\text{Dependent Variable} = k_0 + k_1[(k_2 - \text{Independent Variable})^2] \qquad (5)$$

Equations can contain one independent variable or several independent variables. Up to now we have limited our discussion to the simplest of all cases—where there is one independent and one dependent variable. However, it is perfectly plausible to have situations where two or more independent variables jointly determine the value of the dependent variable. In this case we need the more complicated equations, as written below:

$$\text{Dependent Variable} = k_0 + k_1(X_1) + k_2(X_2)\ldots \quad (6)$$

(X_1 and X_2 are independent variables)

Furthermore, the independent variables need not be present in their linear forms only, but can be present in non-linear forms as well (e.g., squares, cubes, etc.).

$$\text{Dependent Variable} = k_0 + k_1(X_1) + k_2(X_2)k_3(X_1)^2 + k_4(X_2)^2 \ldots \quad (7)$$

Equations can contain two or more independent variables interacting with each other. Up to now we have been dealing with relatively simple equations in which the independent variables were kept separate from each other. It is perfectly possible to have interactions between the independent variables so that they multiply each other, or are divided by each other. For example:

$$\text{Dependent Variable} = k_0 + k_1(X_1) + k_2(X_2) + k_3(X_1 * X_2) \quad (8)$$

There is a multiplicative term (or in statistical terms an "interaction"), $X_1 * X_2$. This means that the effect of X_1 is not simply a function of the value of X_1 only. If X_2 is high, then the effect of X_1 will be large. If X_2 is low, then the effect of X_1 will be small. The pairwise interaction term, $X_1 * X_2$ affects both independent variables, linking them together so that they are no longer mathematically independent of each other, and no longer separately contribute to the value of the dependent variable.

An Example with Soap Data

To demonstrate the approach of equation fitting, consider the data for soaps shown in Table 2. The data comprises various measures of product character-

Table 2 Partial Data for Six of 43 Soaps—Data from Instrumental Measures and Consumer Ratings

	Product					
	101	102	103	104	105	106
Physical measures						
Color	5	6	6	8	5	3
Density	59	35	41	63	48	61
Fat	64	60	59	61	62	61
Hard	20	6	21	7	8	24
Lather	56	60	59	50	60	49
Length	13	12	15	16	6	4
Melt	53	54	59	56	63	51
Roughness	32	34	14	14	15	12
Width	21	32	13	14	31	22

Table 2 Continued

	Product					
	101	102	103	104	105	106
Overall liking						
Segment 1	62	47	60	55	69	54
Segment 2	36	64	42	40	43	38
Males	48	53	49	54	47	36
Females	47	59	53	42	58	51
Use Brand A	39	56	44	48	43	41
Use Brand B	52	56	54	46	56	46
Young	42	53	54	48	56	44
Old	54	58	49	47	47	42
Low income	45	66	54	49	56	35
High income	50	52	50	46	50	48
Sensory liking						
Like appearance	47	58	45	37	47	46
Like color	51	54	47	38	45	44
Like shape	55	54	55	51	56	53
Like fragrance	56	58	48	49	53	52
Like lather	52	48	48	48	51	52
Like texture	46	51	43	38	42	44
Like skinfeel	43	53	42	47	54	40
Sensory attributes						
Uniform	63	67	56	49	46	62
Glistening	49	43	64	71	61	65
Dark	75	65	68	70	74	71
Large	56	50	41	46	54	54
Thick	67	51	48	63	47	60
Fragrance strength	52	50	48	48	53	53
Quick lathering	63	49	48	64	52	57
Amount of lather	58	47	46	50	49	59
Ease of lathering	59	48	38	71	33	65
Residue	46	42	48	56	52	50
Hardness	74	67	59	68	72	72
Smooth	37	49	46	51	46	40
Image						
Quality	45	57	43	41	47	36
Face vs. body	47	58	52	50	55	43
Female vs. male	48	51	43	38	50	39
Family vs. guests	44	55	37	41	52	40

istics, including color, lathering (by physical measures), density, size, physical measures of length, width, etc. The data was obtained from a collection of qualitatively different soaps, each of which was measured on a variety of physical measures. The same products were evaluated by a consumer panel, who used the soaps and rated them on a series of sensory, image and liking attributes.

The initial analysis comprises a correlation matrix (see Table 3). The results show the degree to which the relation is linear. As was discussed above, the correlation matrix does not show the nature of the relation, but simply the degree to which the relation is linear.

Table 3 Simple Correlations Between Physical (Instrumental) Measures and Consumer Ratings

	Physical measures								
	Color	Density	Fat	Hard	Lather	Length	Melt	Rough	Width
Sensory liking									
Appearance	−0.13	−0.22	0.18	−0.13	0.75	−0.22	−0.09	0.29	0.08
Color	−0.01	−0.19	0.13	−0.18	0.68	−0.09	−0.14	0.29	0.06
Shape	−0.30	0.13	0.03	−0.06	0.40	−0.07	−0.05	0.18	−0.06
Fragrance	−0.02	−0.08	0.26	0.07	0.62	−0.18	−0.27	0.44	0.16
Lather	0.12	−0.23	0.44	−0.13	0.78	−0.18	−0.36	0.39	0.07
Texture	0.11	−0.40	0.28	−0.27	0.80	−0.04	−0.16	0.35	0.05
Skinfeel	0.13	−0.32	0.39	−0.38	0.79	−0.18	−0.33	0.48	0.09
Sensory attribute									
Uniform appearance	−0.25	0.07	0.10	0.08	0.55	−0.08	−0.19	0.26	−0.03
Glistening	0.17	0.27	−0.11	−0.23	−0.50	−0.03	−0.09	−0.28	0.04
Dark	0.20	0.04	0.27	0.05	0.12	−0.11	−0.32	0.16	0.14
Large	−0.19	0.27	0.15	−0.13	0.41	−0.29	−0.28	0.31	0.06
Thick	−0.17	0.64	−0.09	0.17	0.03	0.14	−0.12	0.04	−0.09
Fragrance strength	−0.05	−0.04	0.15	−0.16	0.06	−0.37	−0.49	0.32	0.15
Quick lathering	0.30	0.02	0.18	−0.23	0.28	0.13	−0.42	0.18	0.01
Amount of lather	0.09	0.01	0.39	0.04	0.53	−0.08	−0.47	0.37	0.07
Ease of lathering	0.15	0.15	−0.02	−0.26	0.05	0.03	−0.40	−0.05	−0.13
Residue	0.24	0.13	−0.09	−0.17	−0.31	0.10	−0.02	−0.40	−0.04
Hardness	0.09	−0.07	0.34	−0.25	0.60	−0.30	−0.39	0.28	0.04
Smooth	0.32	−0.10	−0.29	−0.62	−0.13	0.20	−0.02	−0.03	0.07
Image									
Quality	−0.01	−0.38	0.34	−0.29	0.86	−0.11	−0.23	0.40	0.01
Face vs. body	0.12	−0.47	0.32	−0.31	0.84	−0.12	−0.23	0.37	−0.04
Male vs. female	0.00	−0.37	0.40	−0.12	0.86	−0.14	−0.18	0.46	0.05
Family vs. guest	0.04	−0.38	0.37	−0.22	0.89	−0.15	−0.26	0.40	0.10

Note: numbers in body of table are the simple correlations between the objective physical (column) measure and the consumer attribute rating (row).

Table 4 Slope and "Goodness-of-Fit" for Linear Equations: Dependent Variable = Amount of Lather, or Perceived Residue, Independent Variable = Physical Measure

	Subjective attributes					
	Amount of lather			Residue		
	M	r^2	S.E.	M	r^2	S.E.
Physical measure						
Melt	−0.67	0.22	5.33	−0.02	0.01	4.13
Rough	0.27	0.14	5.61	−0.21	0.16	3.78
Density	0.01	0.01	6.03	0.07	0.02	4.11
Fat	0.48	0.15	5.56	−0.09	0.01	4.12
Hard	0.04	0.01	6.03	−0.11	0.03	4.07
Length	0.39	0.28	5.12	−0.12	0.16	3.78
Lather	0.41	0.29	5.30	−0.16	0.09	3.94
Width	0.08	0.01	6.02	−0.03	0.01	4.13

S.E. = Standard error of regression.
Equation: Subjective sensory attribute = M(physical measure) + B
M = Slope
r^2, S.E. = Goodness-of-fit statistics.

Linear Functions. The next level of analysis creates linear relations between the instrumental measures (as independent variables) and two consumer attributes (as dependent variables). Table 4 shows the slope of the straight line, the intercept, and most importantly, the percent of the variability accounted for by the linear equation. As Table 4 shows, the linear equation accounts for some variability for certain sensory attributes (e.g., melt point vs. perceived amount of lather) but accounts for virtually no variability for other attributes (e.g., physical width vs. perceived amount of residue).

Quadratic Functions. Since linear equations do not necessarily fit the data, perhaps quadratic equations will do better. Rather than introducing other predictor terms, we opt for a more parsimonious quadratic equation:

Consumer Attribute = $k_0 + k_1$(Independent) + k_2(Independent)^2

Table 5 shows the goodness-of-fit statistics for the quadratic function versus the same goodness-of-fit statistics for the linear function to compare the two functions. In some cases the quadratic function adds additional explanatory power, suggesting that the relation between the additional response and the instrumental variable is not linear, but rather curvilinear. However, even with linear and quadratic terms (using the same predictor) we may not develop a good equation.

Table 5 Comparison of Linear and Quadratic Equations: Fitting Instrumental Measures to Perceived Amount of Lather, Perceived Amount of Residue

	Subjective attributes							
	Amount of lather				Residue			
	Linear		Quadratic		Linear		Quadratic	
	r^2	S.E.	r^2	S.E.	r^2	S.E.	r^2	S.E.
Physical measures								
Melt	0.22	5.33	0.27	5.19	0.01	4.13	0.01	4.19
Rough	0.14	5.61	0.39	5.62	0.16	3.78	0.18	3.81
Density	0.01	6.03	0.01	6.12	0.02	4.11	0.02	4.16
Fat	0.15	5.56	0.23	5.37	0.01	4.12	0.06	4.06
Hard	0.01	6.03	0.15	5.61	0.03	4.07	0.03	4.13
Length	0.28	5.12	0.37	4.87	0.16	3.78	0.16	3.84
Lather	0.29	5.30	0.31	5.09	0.09	3.94	0.09	3.94
Width	0.01	6.02	0.08	5.91	0.01	4.13	0.04	4.11

Using Multiple Linear Predictors. The next level of analysis creates more complicated linear equations, using multiple predictors. Part of the problem with these complex linear equations is that they use several predictors. But which predictors should be used, and when should the researcher stop adding predictors to the equation? Each additional predictor increases the percent of variability explained by the equation.

Fortunately, there are statistical techniques which enable the researcher to create linear equations which are statistically "parsimonious." The number of predictors is determined by statistical considerations. For example, only those predictor terms or independent variables are added which contribute significantly to the ability of the equation to describe the data. The statistical criterion consists of adding or deleting predictor variables in order to maximize the percent of variability accounted for by the equation, while at the same time guarding against adding too many predictors so that the equation is no longer parsimonious.

Table 6 shows several of these equations, developed by full regression (all predictors - Table 6A) and then by stepwise regression (Table 6B). Note that each objective measure has different predictor terms emerging from stepwise regression. The researcher does not know these terms ahead of time. Stepwise regression is a purely statistical procedure which attempts to maximize the "goodness-of-fit," without considering whether the predictor term makes sense (other than statistically accounting for variability in the data).

The final analysis shows what happens when the researcher performs regression analysis, this time with linear and square terms, as well as predictor

Table 6 Models Relating "Perceived Amount of Lather" to Physical Variables

A. Models Using All Eight Instrumental/Objective Variables

Multiple r^2:	0.51
Standard error of estimate:	4.80

Variable	Coefficient
Constant	23.753
Density	0.195
Fat	0.144
Hard	0.134
Lather	0.362
Length	0.031
Melt	−0.368
Rough	0.127
Width	0.028

B. Results from Stepwise Regression: Using Only the Most Significant Independent Variables

Multiple r^2:	0.40
Standard error of estimate:	4.71

Variable	Coefficient
Constant	60.462
Lather	0.329
Melt	−0.520

C. Predicting Perceived Amount of Lather from Physical Lather Alone

Multiple r^2:	0.31
Standard error of estimate:	5.09

Variable	Coefficient
Constant	−7.769
Lather	1.709
Lather*Lather	−0.012

D. Predicting Perceived Amount of Lather from Physical Melt Alone

Multiple r^2:	0.28
Standard error of estimate:	5.20

Variable	Coefficient
Constant	−183.931
Melt	9.028
Melt*Melt	−0.086

(*continued*)

Table 6 Continued

E. Predicting Perceived Amount of Lather from Physical Amount of Lather, Physical Melt, and Their Squares

Multiple r^2: 0.45
Standard error of estimate: 4.70

Variable	Coefficient
Constant	−158.118
Lather	0.342
Lather*Lather	−0.001
Melt	7.296
Melt*Melt	−0.069

F. Predicting Perceived Amount of Lather from the Interaction of Physical Amount of Lather, Physical Melt

Multiple r^2: 0.47
Standard error of estimate: 4.50

Variable	Coefficient
Constant	193.430
Lather	−1.996
Melt	−2.819
Lather*Melt	0.040

G. Predicting Perceived Amount of Lather from Linear and Square Terms, as Well as Interaction of Physical Amount of Lather, Physical Melt

Multiple r^2: 0.52
Standard error of estimate 4.50

Variable	Coefficient
Constant	106.900
Lather	−4.116
Melt	2.470
Lather*Lather	0.014
Melt*Melt	−0.053
Lather*Melt	0.050

Interrelating Data Sets

terms. If the researcher uses all linear, square, and cross terms, the equation could become extraordinarily unwieldy. (Here, with nine independent linear terms there are also nine square terms, and 36 cross terms from which to choose predictors for the equation.) We can simplify the process somewhat by using only the linear terms that appeared in Table 6, and then trying to add those square terms corresponding to the linear terms, and also trying to add the pairwise cross terms.

Table 6 (6C-6G) shows a variety of equations resulting from this strategy. The quadratic and cross terms add some additional predictability to the equation, but not as much as was added by the linear terms. This is always a problem facing the researcher. On the one hand, one wants to account for as much variability as possible in the data. On the other hand there is a true plethora of predictors. Which predictors should one use? Should one use quadratic terms as well? How about pairwise cross terms? When should the researcher stop?

A Point of View on Fitting Equations

Superficially, it appears quite easy to relate two domains—instrumental and consumer measures. There are correlations and regression analysis. Many predictors can be called into play. However, interpreting the equations is difficult. Correlation analysis shows the existence of a linear relation but gives no further insight into the nature of the relation. Model building by means of equations goes one step further by creating an equation to describe the relation between the instrumental measures and the subjective responses. But, even model building is problematic. What is the correct model to use? Should the researcher opt for a simple linear equation that is easy to interpret? Or, should the researcher create a better-fitting model by incorporating linear terms, square terms, and interactions?

A further problem arises when it comes to interpreting the data. What does the relation mean? When the researcher uses a simple linear equation, the meaning is simple. The equation describes how changes in an independent variable (the instrumental measure) covary with changes in the dependent variable (e.g., liking or sensory response). When the equation is more complex (e.g., with multiple predictors) it is difficult to interpret the equation. Certainly the equation fits the data, but does it explain anything beyond the simple statistical description? Can the data be used for additional insights? The answer to this is, unfortunately, probably not. The very strength of the statistical approach (equation fitting to describe relations in the data) does not rest on any theoretical model of what happens in the product.

3. Going One Step Further: Relating Profiles of Consumer Ratings to Profiles of Instrumental Measures

Introduction

There is an alternative approach to interrelating instrumental and subjective ratings—the method of "reverse engineering," described in some detail in the chapter on Category Appraisal. Reverse engineering relates a profile of instrumental measures to a profile of consumer responses. For any combination of instrumental measures, reverse engineering estimates the profile of consumer response.

A Worked Example—Abrasive Gel Product for Oral Care

The approach is illustrated below for a new gel used for toothpaste. The approach follows nine steps:

1. *Test many products.* For each product, collect both instrumental measures and consumer liking and sensory ratings. It is vital to test many products, not just one. Ideally, the products being tested should be related to each other in a meaningful fashion (e.g., the products might be systematic variations of a limited set of ingredients or processing conditions). However, the products need not be related to each other, except by belonging to the same category and by being reasonably "similar" to each other. (The more similar to each other the products in the set are, the better the data will be.)

2. *Divide the variables into two sets—the instrumental measures and the consumer responses.* Table 7 shows the schematic for an abrasive gel. The instrumental measures are obtained from physical "sensors." Each product generates a profile of instrumental measures, one measure per sensor. The more sensors used to "profile" the product, the better the analysis will work. The researcher should not reduce the number of instrumental measures. Although it is tempting to reduce these measures to a limited number that the researcher believes to correlate with consumer sensory attributes, it is best to collect as many different instrumental measures as possible.

3. *Develop a set of "independent predictors."* The ultimate analysis of reverse engineering is to create and use equations. The equations must use a limited number of independent variables. The instrumental measures themselves are not statistically independent of each other. However, we can create a set of "independent variables" by factor analysis. Factor analysis reduces the set of instrumental variables to a limited number of statistically independent variables. Each of the gels generates its own profile of factor scores. The factor scores now become the independent variables in the equation. Table 8 shows the factor scores.

4. *Create equations relating the factor scores to both instrumental and consumer ratings.* The model uses linear and quadratic terms, as well as cross

Table 7 Partial Data Set Showing Instrumental Measures (Sensors) and Consumer Ratings Used to Assess the Abrasive Gel Products

	Product			
	101	102	103	104
Instrumental				
Sensor 1	176	189	171	207
Sensor 2	187	292	254	288
Sensor 3	164	205	166	218
Sensor 4	51	72	67	94
Sensor 5	308	543	242	473
Sensor 6	172	191	174	200
Sensor 7	203	217	197	230
Sensor 8	118	142	122	139
Consumer ratings				
Darkness	51	46	53	48
Gritty	53	57	57	71
Easy to apply	75	82	76	87
etc.				

terms (if they add significant predictability). The models fit some of the attributes better than they fit others. The models fit the instrumental data best because it is from the instrumental measures that we derive the factor scores. The models do not fit consumer acceptance as well, perhaps because consumer acceptability is more complicated than even consumer-rated sensory attributes. Table 9A shows the parameters of two equations. Table 9B shows the goodness-of-fit statistics for all 14 of the equations (eight equations for the eight sensors, six equations for consumer ratings, respectively).

5. *Identify a "goal profile" to be matched.* The goal profile may be a set of instrumental measures for which one wishes to estimate the corresponding profile of consumer attribute ratings. Or, the goal profile may be a set of consumer attribute ratings for which one wishes to estimate the corresponding profile of instrumental measures. One of the nice features about reverse engineering and multiobjective goal programming is that the researcher can choose either the instrumental measures or the consumer ratings as the goal to be matched. The optimization technique does not care which profile is used, nor, in fact, do all of the consumer attributes need to be used. In a similar fashion, all of the instrumental measures do not need to be used. All that needs to be used is some set of dependent variables, whether these are consumer ratings, instrumental ratings, or even a mixture of the two.

Table 8 Factor Scores for the 19 Abrasive Gel Products (Factor Analysis of the Instrumental Measures Only)*

Product	Factor 1	Factor 2	Factor 3
1	-1.02	0.15	0.61
2	0.73	0.08	0.42
3	-0.58	-1.38	0.83
4	1.25	1.06	1.21
5	-0.71	-0.13	0.67
6	1.75	-1.58	0.58
7	-0.96	-0.10	0.01
8	0.63	-0.34	-3.37
9	-0.70	-1.98	0.01
10	0.41	-0.09	-0.65
11	-0.98	0.31	-0.13
12	0.87	2.31	0.48
13	-0.72	0.10	0.51
14	0.63	0.25	0.23
15	-0.88	0.35	-1.08
16	1.35	-1.11	0.55
17	-1.09	0.78	0.14
18	1.14	0.77	-0.82
19	-1.12	0.54	-0.20

*Another viable strategy is to do a factor analysis of the combined data sets—sensory rating assigned by consumers, and instrumental data obtained from "sensors."

Table 9 Representative Equations and Goodness-of-Fit Statistics for Abrasive Oral Gel (Using Three Factors)

A. Equations Relating Factor Scores ($F_1 - F_3$) to Instrumental Measures (Sensor #4), and Consumer Perception (Gritty)

Term	Instrumental sensor #4	Consumer gritty
Constant	59	55
F_1	8.19	0.76
F_2	4.93	-1.58
F_3	16.08	9.17
F_1^2	0.69	3.10
F_2^2	1.24	0.30
F_3^2	5.03	1.64
F_1*F_2	-0.53	1.15
F_1*F_3	1.98	3.02
F_2*F_3	-6.31	-0.08

Table 9 Continued

	B. Goodness-of-Fit Statistics of Equations	
	Multiple r^2	Standard error of regression
Instruments		
Sensor 1	0.91	3.4
Sensor 3	0.91	28.3
Sensor 3	0.90	10.2
Sensor 4	0.89	9.3
Sensor 5	0.94	59.2
Sensor 6	0.92	5.0
Sensor 7	0.92	4.0
Sensor 8	0.94	3.7
Consumer attributes		
Dark	0.59	6.8
Gritty	0.72	5.8
Easy to apply	0.87	7.1
Thick	0.70	9.2
Sticky	0.71	5.8
Firm	0.76	6.6

6. *Identify the allowable range of factor scores.* Reverse engineering identifies the combination of factor scores (independent variables) which in concert reproduce the goal profile as closely as possible. The values of the factor scores are systematically explored within a given range. The user must give that range of factor scores, upper and lower limits on each factor score.

7. *Impose constraints (e.g., on the consumer ratings or on the instrumental measures).* Reverse engineering is a mechanistic system. As such, it does not know when a mathematical solution makes sense. We can force the answer to yield results that make sense by setting up "implicit constraints." An implicit constraint is a constraint placed upon the dependent variable (e.g., consumer attribute rating). We can select one or several (or even all) dependent variables and require that the resulting profile remain within that implicit constraint.

Consider the following example. We may wish to find what profile of consumer attribute ratings corresponds to a given profile of instrumental measures. We know that in the study the rating of consumers for "darkness" lies between 50 and 60, across the full set of products that were tested. We want to insure that the range of "darkness" on the consumer's scale is not more than 60 nor less than 50. We want to constrain the answer from reverse engineering so that we never have a result with an expected darkness greater than 60 or less than 50.

Table 10A shows three scenarios, using instrumental measures as goals, along with a specified implicit constraint.

8. *Adjust the independent variables (factor scores) to maximize the closeness-of-fit between the goal to be matched and the expected profile.* Reverse engineering now systematically explores the different combinations of the independent variables (factor scores) within the range allowed (see Table 10B). At each iteration or "trial," the reverse engineering algorithm looks at the closeness of the goal profile to the estimated profile (estimated using the equation relating the factor scores to the attributes or instrumental measures). The algorithm adjusts the independent variables (factor scores) until it discovers a combination of factor scores yielding a profile as close as possible to the desired goal profile. Table 10C shows these results.

9. *Estimate the full profile of consumer ratings and instrumental measures.* Once the program has discovered the appropriate combination of factor scores, it is a straightforward task to estimate the full profile of values. Each combination of factor scores corresponds to a full set of expected consumer and instrumental measures (see Table 10C).

An Overview to Reverse Engineering and Relating the Instrumental and the Subjective Domain

The approach of reverse engineering presents a novel, alternative solution to the problem of relating the subjective and the instrumental measures. It differs from the conventional approaches because it assumes no theoretical relation between instrumental and consumer measures. It simply assumes that there exists a relation between a set of independent variables and a set of dependent variables. Furthermore, it recognizes that the independent variables may be correlated with each other, so it creates a usable set of independent, parsimonious predictors. From these predictors it creates models relating the predictors (and their squares and pairwise interactions) to the dependent variables. Each dependent variable generates its own equation. Finally, it uses the array of equations and a goal-fitting procedure to estimate the likely set of predictor variables corresponding to a given profile of dependent variables. From this procedure it estimates the likely profile of all consumer and instrumental variables.

This approach is atheoretical and mechanical. Many scientists spend years attempting to uncover the instrumental correlates of sensory perception. The efforts are rewarded with a deeper, more profound understanding of how specific physical measures covary with consumer ratings. But in applied work, time and money are limited. The efforts and time needed cannot be justified. A researcher in a corporate laboratory might need to learn how physical measures of a product correspond to consumer acceptance, because that information could help quality control efforts. But the researcher will not be able to justify the outlays required to solve the problem. Despite the interesting aspects of such a

Table 10 Example of Reverse Engineering by Goal Fitting
Goals → from Instrumental Measures—Category = Abrasive Oral Gel

Instrument	A. Goals to Be Matched		
	Goal 1	Goal 2	Goal 3
Sensor 1	168	180	180
Sensor 2	201	220	220
Sensor 3	200	180	180
Sensor 4	75	80	80
Sensor 5	600	560	560
Sensor 6	180	160	160
Sensor 7	217	200	200
Sensor 8	140	270	270
Implicit constraint			
Consumer rated darkness	None	None	50–55

B. Range of Factor Scores in the Study

Factor	Minimum	Maximum
F1	−1.00	1.25
F2	−1.00	1.00
F3	−1.00	1.25

C. Solution
(Factor Scores Generating the Match, and the Expected Profile of All Attributes Corresponding to That Factor Profile)

Factors	Solution Goal 1		Solution Goal 2		Solution Goal 3	
Factor 1	0.78		0.58		0.09	
Factor 2	0.93		−1.00		0.75	
Factor 3	0.30		0.24		−0.84	
Instrumental measures	Est	Goal	Est	Goal	Est	Goal
Sensor 1	189	168	183	180	162	180
Sensor 2	300	201	312	220	308	220
Sensor 3	193	200	180	180	176	180
Sensor 4	75	75	66	80	58	80
Sensor 5	515	600	515	560	476	560
Sensor 6	196	180	191	160	185	160
Sensor 7	226	217	215	200	220	200
Sensor 8	140	140	138	270	138	270
Estimated consumer rating						
Darkness	51		57		50	
Shiny	57		54		49	
Easy to apply	74		74		56	
Thick	72		64		65	
Sticky	65		63		67	
Firm	46		49		57	

Goal = Goal to be matched.
Est = Estimated rating or instrumental measure that will be achieved.

problem, and despite the future potential of the solution in terms of maintaining quality, the researcher will not receive the necessary funds nor management's blessing to proceed. The project will be shelved, relegated to the category of "blue sky" projects to be done "someday, but not just now." The more pragmatic approach described here for reverse engineering brings the problem and solution back into "real time" at an affordable price.

4. Interrelating Expert Panels and Consumer Data

Introduction

The growth of product testing has brought with it the development of in-house expert panels. These panels comprise individuals who have been trained to describe the sensory characteristics of products, such as appearance, fragrance, or texture characteristics. The panelists often use descriptor terms that are not familiar to consumers (although these terms could become familiar with some training).

Expert panels find their biggest use in quality control, wherein the manufacturer strives to maintain a product with sensory integrity, so that the sensory characteristics of the product remain the same from batch to batch. Expert panels also find a use in new product development. The manufacturer may be interested in creating a product with a specific sensory profile (e.g., of specific fragrance, with a specific texture). Rather than resorting to expensive consumer panels time after time to evaluate each trial prototype, the manufacturer may call upon the expert panelist to assess the product and to guide the product developer. (For instance, in the opinion of the expert the product may have too little of a specific fragrance character.) In these situations, the expert panelist acts as a bridge between consumers and the manufacturer, not totally representing consumers but at least identifying where the product developer may be "off target."

Relating Consumers and Experts

Expert panelists are not consumers. Most manufacturers now recognize that fact. It may seem somewhat obvious to state that experts are not consumers, but in the early days of product testing the expert often substituted for the consumer. It was the express wish of many manufacturers to use the expert data in place of the consumer data. By doing so the manufacturer would be able to speed up the product development process, and save significant amounts of money.

Researchers soon found that the judgments of experts were not necessarily the same as the judgments of consumers. The experts would often describe characteristics that were not easily perceptible or understandable to the consumer. When an expert would point out a product defect, often the consumer evaluating the same product might not perceive the defect at all, or not necessarily perceive the defect as a defect. This discrepancy between the response of the expert panelist and the consumer's response often led to the introduction of products that were later withdrawn because they did not meet with sufficient consumer acceptance. By shortcutting the consumer evaluations, the manufacturer was only left with the opinions of the experts, which may or may not have reflected consumer opinions.

The Rationale for Relating Experts and Consumers

Expert panels represent a significant corporate investment that many manufacturers are loath to eliminate. Yet, in order for the expert panels to be useful as more than a simple quality control panel, their ratings must be relatable to the ratings that would be assigned by a consumer. Unfortunately, for the cosmetics and health and beauty aids fields, little has been done to relate these two sources of information about a product.

If expert panel ratings can be validly related to consumer ratings, then the manufacturer stands to reap a bonanza in terms of shortened development and testing times, as well as reductions in testing costs. Expert panelists are generally corporate staff members whose salaries are already included in overhead. Consumer testing, in contrast, is an expense which recurs from test to test. It is no wonder that the astute business person is perennially interested in relating the less expensive expert panel data to the more expensive consumer data. (If truth be known, most manufacturers would prefer to eliminate both sources of data, expert and consumer, and rely on instrumental measures. Instruments are not considered overhead, can work 24 hours a day, and do not require vacations or medical benefits!)

The Reverse Engineering Approach

As we saw above for abrasive tooth gel, reverse engineering enables the researcher to relate instrumental to subjective measures. The same approach can be used to relate two types of subjective measures—consumer and expert. Furthermore, the two sets of data need not be quantified on the same scales. The experts can use their attributes and scales, whereas the consumers can use their attributes and scales, respectively. The results generate an estimate of consumer ratings, corresponding to a profile of ratings assigned by a consumer.

An Example—Oral Gel

Let us continue our example with oral gels, this time introducing data from expert panelists. In this study the expert panelists rated the appearance, flavor, and touch/usage characteristics of the same gels as had been rated above. Thus, the data set is complete. For the same set of in-market gels we have consumer data (liking, sensory attributes), instrumental measures, and now expert panel profiles. The starting data for this analysis comprises a rectangular matrix. Each row corresponds to a different oral gel. Each column corresponds to either a consumer attribute rating, an instrumental rating, or an expert panel rating.

The analysis proceeds in the same fashion as it did when we related the instrumental measures to the consumer ratings. We develop quadratic models relating the factor scores, their squares, and interactions to the expert panel ratings. The models look similar to the models created for the consumer and instrumental ratings. The database now comprises 10 more equations, for a total of 24 (eight sensor, six consumer, and 10 expert equations, respectively).

If we did not have the instrumental measures for gels, and had only the expert panel and consumer panel ratings, then we would go through the factor analysis of the expert panel data, as discussed above. We will skip those steps, and simply use the factor scores corresponding to the instrumental data.

Goal Fitting

We perform the same reverse engineering analyses for expert panel data that we did for consumer data. Again, the independent variables are the factor scores. The dependent variables are, in turn, the consumer ratings, the instrumental measures, and now the expert panel data.

It does not matter which set of data will act as the goal to be matched. As far as the reverse engineering algorithm is concerned, all dependent variables are equivalent.

For this exercise we show how the expert panel profiles can be used to estimate consumer ratings (liking, sensory, even perhaps image characteristics if they are asked in the questionnaire). Table 11 shows the results of the exercise. Table 11 shows three different profiles generated by expert panels (e.g., in their evaluation of competitor products, or in their evaluation of different batches). The reverse engineering algorithm identifies the likely values of the factor scores (within the range tested) and, having found those values, estimates the likely profile of all three sets of data (consumers, experts, instruments).

5. Interrelating Consumers and Experts for Fragrance Development

Introduction

Expert panel data is often used in fragrance research to identify the characteristics of a fragrance for product development. Experts possess their own descriptive language for fragrances. In contrast, consumers often have an entirely separate lexicon, based upon what they are exposed to in advertising. By relating

Table 11 Example of Reverse Engineering to Interrelate Expert Panel Data (Goals) to Consumer Ratings

Factors	Expert goal 1	Expert goal 2	Expert goal 2 (weighted)
F1	−0.04	−0.14	0.77
F2	−0.79	−0.05	−0.67
F3	−1.00	−0.96	−0.13
Instruments			
Sensor 1	178	168	191
Sensor 2	251	264	300
Sensor 3	154	164	175
Sensor 4	39	46	60
Sensor 5	580	510	606
Sensor 6	188	184	196
Sensor 7	220	220	222
Sensor 8	133	134	141
Consumer sensory			
Dark	54	51	58
Shiny	47	48	50
Easy to apply	56	55	70
Thick	42	54	56
Sticky	62	63	64
Firm	58	57	51

	Est	Goal	Est	Goal	Est	Goal	Wt
Expert sensory							
Wet	6.0	6.0	6.1	6.0	6.0	6.0	10
Spreadable	7.3	8.0	8.1	6.0	7.4	6.0	10
Compact	2.4	5.0	1.0	1.0	1.0	1.0	10
Firm	13.4	9.0	14.2	9.0	13.8	9.0	1
Amount of particles	3.5	6.0	2.5	3.0	2.4	3.0	1
Size of particles	3.7	4.0	5.7	6.0	1.9	6.0	1
Density	8.9		9.9		8.5		
Slipperiness	0.6		0.7		1.0		
Adhesiveness	10.4		10.8		9.6		
Manipulable	7.4		8.4		7.2		
Cohesiveness	12.1		12.1		12.8		

Goal = Goal to be matched.
Est = Estimated rating.
Wt = Weight (relative importance).

experts and consumers, the product developer can traverse from the domain of the emotional to the domain of the sensory. Given the very limited and very general fragrance language possessed by consumers, and the substantial qualitative nuances that fragrances may possess, it is a major contribution to relate these two domains in a cost-effective fashion.

A Case History—Fine Fragrance

The case history presented here deals with a fine fragrance created by Wood, Inc. The objective is to develop a fragrance that possesses certain notes which support a marketing position of "youthful," "carefree," almost "recklessly abandoned." These are the consumer words, used by marketing to define the position, and to guide the creatives in communicating the product by advertising and packaging. But what about the liquid inside, the actual fragrance itself? What are the keys or accords to be used in the fragrance? Can the perfumers, marketers, evaluators, and researchers at Wood, Inc. come to a mutual understanding? Is there a link between the fragrance as smelled by a consumer and the language used?

Preliminary research began with 10 local consumers in the Miami area (home of Wood, Inc.) using six fragrances profiled on a list of image attributes. There was differentiation between the fragrances. Consumers clearly perceived some fragrances to be more "exuberant," more "carefree," etc. When asked to describe the "personality" of the fragrance, consumers were consistent with each other in terms of how they described each fragrance. This finding, not new to Wood, Inc., but of increasing interest, spurred the hope that researchers might be able to link personality or image characteristics to different fragrance components. Marketing and the fragrance group abandoned hope of linking specific chemicals in the fragrance with image characteristics, but did hope to link more complex building blocks, or keys to these image characteristics.

Stimuli—"In-Market" Fragrances or Experimentally Designed Blends?

The previous section on oral gel dealt with an assessment of in-market products. In the oral gel research, the products were all fairly similar to each other and could be characterized by a limited number of common attributes. The issue with fragrances is much more complex. Fragrances have a myriad of nuances, and the language for one type of fragrance is often far different from the language for another. Although a researcher might use the same set of attributes for many fragrances, in actuality the profile would be so different from fragrance to fragrance that the chances of developing usable data from in-market fragrances could be nil. The "in-market" fragrances were thought to be too perceptually dissimilar to each other. Therefore, the marketing and technical groups decided to develop fragrances which varied on a set of similar "themes."

Interrelating Data Sets

Experimental Design for Fragrances

The study used 20 fragrances, comprising combinations of five basic keys (here labeled A-E). Each key was present in every blend, ranging from 10% to 30% of the blend. All blends had to add to 100%. Table 12 shows part of the de-

Table 12 The Experimental Design for Four Fragrances, as Well as Attributes and Ratings

	Fragrance blend #			
Key	1	2	3	4
A	0.30	0.10	0.10	0.30
B	0.20	0.20	0.20	0.30
C	0.30	0.30	0.30	0.10
D	0.10	0.30	0.10	0.20
E	0.10	0.10	0.30	0.10
Cost of goods	115	131	143	118
Consumer liking	50	62	69	51
Fragrance "image"				
Older	43	45	52	44
Brash	47	45	43	52
Evening	46	52	51	57
Sophisticated	57	66	51	63
Uninhibited	47	48	47	50
Reckless	40	39	32	68
Extrovert	44	43	58	55
Erotic	41	30	70	35
Consumer "feeling"				
Uplifted	53	40	59	39
Confident	33	63	36	49
Aroused	53	68	37	63
Youthful	31	45	44	36
Energized	66	62	57	39
Alluring	58	61	57	32
Awake	70	39	42	59
Comfortable	55	55	74	68
Expert sensory				
Overall intensity	46	46	46	56
Earthy	36	37	17	38
Citrus	48	51	30	42
Sweet	33	36	22	26
Woody	68	57	74	53
Lingering	65	36	75	44

sign. The fragrance keys all blended together. Although there was some initial trepidation that this approach would be analogous to "painting by numbers," it was obvious that a success in blending using this technology would put Wood way ahead of the competition. Consequently, management approved the project and encouraged the participants.

Questionnaire and Test Procedure

The fragrance evaluation acquired two sets of data—consumer ratings of "image/feeling" and expert panel data for sensory attributes (using the lexicon of terms developed by the experts). The study was run with five experts and 20 consumers. Since this was a feasibility study to interrelate experts and consumers, both management and the technical team wanted to be financially conservative.

Table 12 shows part of the database as well, for four of the 20 fragrances tested.

Developing the Fragrance Mixture Model

One of the key benefits of a systematic variation of physical factors is that they are statistically independent of each other. That independence and the fact that there are a large number of fragrance blends relative to the number of components (20 blends vs. five components) enables the researcher to create a model. The model relates the fragrance keys and their interactions to the expert-rated attributes, and to the consumer-rated attributes. The fragrance model comprises a set of equations which looks like the equation below:

Attribute = $k_1(A) + k_2(B) \ldots k_5(E)$ + pairwise terms as needed (e.g., $k_n(A*B)$, or $k_m(A*A)$)

We have seen these equations before for product testing and optimization. They are now used for fragrances. The equation above does not have an additive constant (k_0) because the five independent variables (or keys A–E) must add to 1.

Relating Experts and Consumers

The task now is to relate the expert and consumer data by means of the "goal-fitting" approach. The consumer's language comprises advertising-based phrases and emotional phrases. The expert's language comprises descriptive characteristics. Both depend upon the specific combination of five different keys (A–E).

Here are two sets of problems and some suggested answers that can be obtained from experimental design and reverse engineering.

Problem 1—Using Consumer "Image" Ratings as Targets to Match. We have a marketing brief to create a fragrance with a given image. If we profile that image

Interrelating Data Sets

on a set of attributes, can we discover how the experts would describe the fragrance, and what we must mix in order to create a fragrance that will match this image? Table 13 shows four such "image prescriptions" (#1–#4) and the resulting estimates of expert profiles, and levels of the fragrance keys.

Table 13 Reverse Engineering—Formulations and Expert Panel Profiles Corresponding to Consumer Image Profiles

Key	Consumer profile as goal							
	#1		#2		#3		#4	
A	0.24		0.10		0.10		0.12	
B	0.30		0.26		0.15		0.12	
C	0.26		0.27		0.28		0.30	
D	0.10		0.18		0.27		0.30	
E	0.10		0.19		0.20		0.17	
Cost of goods	115		134		141		139	
Consumer liking	42		63		65		61	
	Est	Goal	Est	Goal	Est	Goal	Est	Goal
Fragrance "image"								
Older	40	40	50	50	52	60	48	70
Brash	55	60	50	50	44	40	41	30
Evening	56	50	53	60	52	65	52	75
Sophisticated	56	40	56	50	65	65	70	70
Uninhibited	52	60	45	40	40	30	40	25
Reckless	48		47		54		50	
Extrovert	47		62		59		58	
Erotic	51		44		41		40	
Consumer "feeling"								
Uplifted	55		51		47		51	
Confident	53		44		52		57	
Aroused	63		60		61		65	
Youthful	34		48		53		54	
Energized	56		56		78		84	
Alluring	49		50		49		44	
Awake	71		54		43		38	
Comfortable	38		56		44		27	
Expert sensory								
Overall intensity	47		45		47		47	
Earthy	40		29		32		32	
Citrus	48		36		45		49	
Spicy	29		25		25		27	
Woody	66		62		53		46	
Lingering	59		70		54		45	

Est = Expected Rating.
Goal = Goal to be matched.

Problem 2—Using Expert Panel Profiles to Evaluate Fragrances, and Estimating Likely Consumer Acceptance. We have a trained expert panel that is profiling new fragrance creations. Given the data set (Table 12) and the models created from that data set, can we estimate likely consumer acceptance (and thus avoid having to test each fragrance)? If we can estimate consumer acceptance based upon expert ratings, we can save a great deal of money. Table 14 shows four such "expert panel profiles" (#5–#8) and the resulting estimates of the consumer's ratings and levels of the fragrance keys.

6. An Overview

The attempts by researchers to relate different types of data for the same product (e.g., consumer, expert, instrument) have not generally proven successful using traditional methods. These methods have looked for simplistic relations between variables. When there are many predictor variables from which to choose (e.g., instruments) and when the relation between variables is not linear (e.g., liking as rated by consumers), the task becomes even harder. It is no longer an issue of finding correlations between two variables, but rather fitting non-linear functions of several predictor variables to a consumer-rated attribute. In many cases there is no hint even where to start (e.g., if the attribute to be predicted is consumer ratings of "masculine-feminine," then what are the appropriate instrumental measures or expert panel ratings to use?).

The logic and technology of "reverse engineering" and equation development may provide a feasible way to overcome some of the problems involved in relating different types of data. This approach differs from conventional methods. It assumes simply the existence of a set of ad hoc equations relating an independent, parsimonious set of predictors to every attribute. These predictors are either ingredients that are independently varied or, more normally, factor scores derived from a set of instrumental, consumer, or expert panel data. In any case, the predictors have the requisite properties of statistical independence and parsimony.

Potential Applications of the Approaches

Reverse engineering and interrelating data bases have potential applications in a variety of areas. Here are five:

1. Substitute expert panelists for consumers, in an ongoing quality control program.
2. Substitute machines for consumers, in the same ongoing quality control program. (As noted in the chapter, it makes little or no difference what type of data is used in reverse engineering and goal fitting.)

Table 14 Reverse Engineering—Formulations and Consumer Image Profiles Corresponding to Expert Panel Profiles

Key	Expert Panel Profile As Goal							
	#5		#6		#7		#8	
A	0.12		0.15		0.27		0.11	
B	0.24		0.30		0.22		0.30	
C	0.15		0.16		0.26		0.15	
D	0.30		0.11		0.15		0.30	
E	0.19		0.29		0.10		0.14	
Cost of goods	139		140		118		133	
Consumer liking	71		64		52		71	
Fragrance "image"								
Older	57		54		46		53	
Brash	49		55		52		46	
Evening	52		55		52		51	
Sophisticated	61		55		59		63	
Uninhibited	46		55		49		46	
Reckless	53		59		53		45	
Extrovert	60		63		54		60	
Erotic	37		66		55		34	
Consumer "feeling"								
Uplifted	50		54		53		49	
Confident	54		58		37		51	
Aroused	46		57		57		43	
Youthful	52		48		42		48	
Energized	69		51		56		63	
Alluring	51		41		51		52	
Awake	31		50		63		34	
Comfortable	62		55		63		66	
Expert sensory	Est	Goal	Est	Goal	Est	Goal	Est	Goal
Overall intensity	50	50	53	55	49	60	50	65
Earthy	35	35	30	30	35	35	37	40
Citrus	41	45	38	40	43	45	44	45
Spicy	30	35	30	40	27	50	34	40
Woody	68	70	70	70	60	60	60	65
Lingering	60	60	54	55	60	60	49	50

Est = Estimated rating.
Goal = Goal to be matched.

3. Reproduce one's own product in a new formulation. The objective here is to create a formula with the same sensory characteristics as one's current formula, but with a new ingredient system. In this example the product developer would actually systematically vary the new ingredients according to an experimental design, create the model, and then use the sensory profile of the current product as the goal or target to be matched. Reverse engineering then determines the specific ingredient combination within the new ingredient system that generates the requisite match.
4. Copy a competitor in one's own system. The sensory or image profile of the competitor becomes the target profile to be matched.
5. Create a combination of ingredients which provides a desired "image" profile. Historically, researchers have been interested in the relation between sensory attributes or liking and physical variables. Reverse engineering enables the researcher and product developer to use the more ethereal, less sensory-based image attributes as goals to be matched and "concretized" by an actual physical formulation.

References

Andersson, Y., Drake, B., Granquist, A., Halldin, L., Johannson, B., Pangborn, R.M. and Akesson, C. 1973. Fracture force, hardness and brittleness in crisp bread with a generalized regression analysis approach to instrumental–sensory comparisons. Journal of Texture Studies, *4*, 119.

Bruvold, W.H. 1970. Laboratory panel estimation of consumer assessments of taste and flavor. Journal of Applied Psychology, *54*, 326–330.

Burton, J. 1989. Towards the digital cheese grader. Dairy Institute International, *5*(4), 17–21.

Cardello, A.V., Maller, O., Kapsalis, J.G., Segars, R.A., Sawyer, F.M., Murphy, C. and Moskowitz, H.R. 1982. Perception of texture by trained and consumer panelists. Journal of Food Science, *47*, 1186–1197.

Congers, S.S. and Zook, K. 1968. Laboratory preference and acceptance panels: a case in point. Food Technology, *22*, 189–192.

Cunningham, D.G., Acree, T.E., Barnard, J., Butts, R.M. and Barell, P.A. 1986. Charm analysis of apple volatiles. Food Chemistry, *39*, 137–148.

Hare, L.B. 1974. Mixture designs applied to food formulation. Food Technology, *28*(3), 50.

McBride, R.L. 1979. Cheese grading versus consumer acceptability: an inevitable discrepancy. Australian Journal of Dairy Technology, *34*, 66–68.

Murphy, E.F., Clark, B.S. and Berglund, R.M. 1958. A consumer survey versus panel testing for acceptance evaluation of Maine sardines. Food Technology, *12*, 222–226.

Naes, T. and Kowalski, B. 1989. Predicting sensory profiles from external instrumental measures. Food Quality and Preference, *4/5*, 135–147.

Palmer, D.H. 1974. Multivariate analysis of flavor terms used by experts and nonexperts for describing tea. Journal of the Science of Food and Agriculture, *25*, 153–160.

Panasuik, O., Tally, F.B. and Sapers, G.M. 1980. Correlation between aroma and volatile composition of McIntosh apples. Journal of Food Science, *45*, 989-991.

Piggott, J.R., 1990. Relating sensory and chemical data to understand flavor. Journal Of Sensory Studies, *4*, 261-272.

Powers, J.J. 1976. Experiences with subjective/objective correlations. *In:* Correlating Sensory and Objective Measurements—New Methods for Answering Old Problems, J.J. Powers and H.R. Moskowitz (eds.), STP 594, pp. 111-129. American Society for Testing and Materials, Philadelphia.

Schutz, H.G. 1987. Predicting preferences from sensory and analytical data. *In:* Flavour Science and Technology, M. Martens, G.A. Dalen and H. Russwurm, Jr. (eds.), pp. 399-406. John Wiley & Sons, Chichester, U.K.

Szczesniak, A.S. 1973. Instrumental methods of texture assessment. *In:* Texture Measurement of Foods, A. Kramer and A.S. Szczesniak (eds.), pp. 71-108. D. Reidel, Dordrecht, The Netherlands.

Szczesniak, A.S. 1987. Correlating sensory with instrumental texture measurements—an overview of recent developments. Journal Of Texture Studies, *18*, 1-15.

Szczesniak, A.S., Brandt, M.A. and Friedman, H.H. 1963. Development of standard rating scales for mechanical parameters of texture and correlation between the objective and sensory methods of texture evaluation. Journal of Food Science, *28*, 397-403.

Taguchi, G. and Wu, Y. 1980. Introduction to off-line quality control. Central Japan Quality Control Association.

Vickers, Z.M. 1988. Instrumental measures of crispness and their correlation with sensory assessment. Journal of Texture Studies, *19*, 1-14.

Weisberg, S. 1985. Applied Linear Regression. John Wiley & Sons, New York.

Wieseman, C.K. 1971. Identifying and controlling product quality attributes using preference taste panels. Food Product Development, *5*, 21-22.

Williams, A.A., Rogers, C.A. and Collins, A.J. 1988. Relating chemical/physical and sensory data in food acceptance studies. Food Quality and Preference, *1*, 25-31.

8
Streamlining the Product Development Process

1. Introduction

With increasing competition among manufacturers for a share in the market, and with increasing costs associated with the "human" part of product development, manufacturers continually look for methods that streamline the time and costs involved. Manufacturers no longer want to or even can support the great overheads of five, ten and twenty years ago. Today the operative phrase is "lean and mean." Given this new competitive spirit which calls for doing "more with less," how can the manufacturer compete in a business that requires new products on a regular basis?

The Easy Way Out—Line Extensions

A simple way to get off the competitive treadmill is to merely modify one's products in a disciplined manner to create line extensions. The lipstick manufacturer creates and markets three or four new colors, and feels virtuous because this effort constitutes (at a superficial level) a "new products effort" with a significant outcome.

If manufacturers merely had to modify their current products slightly (e.g., by adding a new color, a new flavor, etc.), then there would be no problem. Of course consumer demands would not be fulfilled. The first manufacturer to introduce significantly new products would have an advantage.

Streamlining the Development Process 425

Line extensions can be developed by simply changing one or two aspects of a product. What happens, however, when the manufacturer wishes to go beyond the appearance of simply another "fragrance" or another "bar of soap," and produce a product to satisfy some new needs? How does the manufacturer go about the task? Are there any methods which enable the manufacturer to do so quickly? Inexpensively? With some assurance of market success? And if possible, allow the developer to go "out of the box" (the conventional products)?

Focus on Process, Not on Product

Let us step back for a minute and look at the way we have stated the problem. We do not know what product the manufacturer will invent. Therefore, there are no tried and true test methods to indicate the potential degree of success. We are lost at first because we are operating outside the conventional boundaries of product testing. Yet, we want to incorporate our knowledge of how to test with consumers, in order to give the product its best chance of market success.

We must now focus on the *process* of product development, not on a particular product. We are no longer devising a test to confirm or deny a hypothesis, or to measure an entire category. Our thinking is now more strategic.

Traditional Approaches vs. the Accelerated Program

Many product testers have been imbued with the "scientific method" as a deified approach to be revered and obeyed at all costs. Many tests are often over-engineered by newly minted product researchers who, in their zeal, want to follow the scientific method to its ultimate conclusions.

In some new product development work, rigorous adherence to the principles of testing may work well and produce valid data. However, ask a product researcher heavily involved in an accelerated new products program about the scientific method. The answer will typically involve "compromise"—compromise between the needs of the project (speed, efficiency, market correctness) and the discipline of testing (product design, panel composition, questionnaire, statistical robustness etc.). Rarely does the practitioner in an accelerated program follow the canons of disciplined product evaluation that are taught in schools or presented in books. Usually the practitioner devises or adapts methods that work, tries to be reasonably (but not unduly) rigorous, and tries not to violate the canons of product testing—at least not too strongly!

2. A Rapid, Two-Step Program for Accelerated Development

Introduction

The two-step program described here in this chapter works best when the manufacturer is faced with the following two requirements and two constraints:

1. *Requirement 1*: develop a new product.
2. *Requirement 2*: insure that the concept is reasonably acceptable and fits the product (or vice versa—that the product fits the concept).
3. *Constraint 1*: limited budget, requiring shortcuts in the development and testing process.
4. *Constraint 2*: multiple paths (concept options, product options) available, but the appropriate direction is not clear.

The two step program can be summarized as follows:

1. *Concept*: explore and optimize different concept "nuggets" or core ideas, each with its own set of elements or features. Explore the different concept nuggets in one large study comprising several parts.
2. *Product*: explore different product designs, again in the same study. If at all possible, test the concepts along with the products, to insure concept-product fit.

3. A Case History—Oral Protective Compound

Background

The Franck Company was founded by Marilyn Franck, an entrepreneurial product developer with a successful history of developing new products for companies. In 1988, Ms. Franck founded her own product development group with the express aim of working with other product developers in order to commercialize new product ideas.

In 1991, Marilyn came across a patent for an oral protective compound which appeared to reduce plaque quite significantly. At the time, she had been engaged in applying some of her knowledge on a consulting basis with a small oral care company. When she learned of this new patented product, she obtained the rights to use the product for a 3-year period on an exploratory, product development basis. She then brought the compound to her client, and in a management meeting the client decided to move forward with a product development effort. The compound appeared to promise significant benefits to consumers in terms of oral protection. Furthermore, the nature of the products that could be developed with this compound were consistent with the client's other products.

Deciding What to Create

The immediate questions that arose concerned the nature of the product that would incorporate the new compound. What would it be? Some of the marketers suggested that the compound be incorporated into a mouth rinse. Other marketers opted to use the product as part of a toothpaste. Still others thought that the product might find a niche as part of a dental floss.

Streamlining the Development Process 427

Questions of this nature perennially plague manufacturers who want to create and market a new product. It is generally not enough to have the rights to use a new compound, no matter how spectacular its properties may be. The true test of product development success comes from the successful use of the compound in a consumer acceptable product. The most promising compound in the world could fail miserably if it is incorporated into a product which fails to meet with consumer acceptance.

Faced with the problem of what carrier should be used to incorporate the new compound, the marketing group recognized that it would have to commission primary research on consumers' desires, as well as some secondary research on sales of different plaque reducers. Marketing was hoping that the secondary research would guide the company towards selecting one of the three vehicles. What emerged was the standard result—the largest potential lay in toothpastes, the second largest potential lay in mouthwashes, and the smallest potential (at least in total volume) lay in dental floss. However, it also become obvious that the cost to enter the toothpaste market was substantially greater than the cost to enter the dental floss market. Furthermore, there was so much competition in the toothpaste market that it appeared that the advantage in total potential of the toothpaste market was probably a mirage. Although the toothpaste market appeared to enjoy a greater total potential, there were so many competitors in the market that the likelihood of the compound making a real commercial difference appeared nil. Marketing and management were wary of the lure of a large potential that carried significant competition with it.

At this point, marketing and the Franck Company decided that the prudent thing to do would be to optimize concepts for the three different products that could incorporate the compound: toothpaste, mouth rinse, and dental floss.

Early Stage Concept Development—IdeaMap

Given the rapid timetable for development, marketing suggested that they execute three simultaneous concept optimization studies for the compound—one each for toothpaste, mouthwash and dental floss, respectively. The objective of each of these three studies was to identify the consumer relevant "hot buttons" for each product type that could be used should the manufacturer wish to incorporate the compound into that particular product.

As discussed in Chapter 1 on concept testing, development and optimization, the conventional concept optimization test comprises an in-depth analysis of one category, rather than three categories. Furthermore, the three product types are significantly different from each other. Features that might be attractive for a dental floss have little or no relevance to a mouthwash.

Recognizing that there was little overlap among concept statements belonging to the three categories (other than perhaps ultimate oral health benefits), the manufacturer decided to progress along three fronts simultaneously. Recognizing that money was limited, the manufacturer suggested doing three IdeaMap projects simultaneously with the same consumer—one each for toothpaste, mouthwash and dental floss, respectively. Each of the three products would have its own set of statements that could appear in the concept. There was some overlap as stated above, primarily in terms of oral health benefits. For the most part, however, the elements across the three product categories were quite different, primarily because many of the concept elements dealt with the particular characteristics of the product itself, rather than with global health issues.

Creating Elements for Each Product Type

The initial task for the "3-cell" IdeaMap study was to create a representative set of elements for each of the three product types, respectively. For this exercise, Franck engaged the services of a marketing "boutique" which specialized in creativity and concept development.

To facilitate the task, the creative group ran three different ideation sessions with four types of participants in each session. The participants were chosen from marketing, product development, marketing research and outside consumers who had been screened to be especially "creative" and verbal. Each creative group comprised eight participants and two facilitators from the creative boutique company. One group was devoted to mouthwash, another to dental floss, and a third group to toothpaste. Different participants (but the same facilitators) participated in the three sessions.

Each of the sessions lasted 2 hours, and followed the procedure listed in Table 1. The procedure had been developed by the creative boutique, and was designed to stimulate the participant's free flowing ideas and "creative juices."

Incorporating Specific or Tailored Knowledge About the Compound

The procedure outlined in Table 1 is sufficiently general to spark and capture creativity. One of the unique issues in this project was that the manufacturer had the rights to a patented compound. That compound, in one way or another, through one statement or another, had to be part of the concept, or at least have a chance to be tested. That is, unlike many other concept development and optimization projects in this study, there was an additional "agenda"—namely to exploit proprietary technology. Most participants were unfamiliar with the compound. The facilitator from the creative boutique gave the participants a brief description of the compound, its history, composition, and benefits. The participants read this information at the start of the session, and then discussed it for a few minutes prior to the formal part of the creative session. Afterwards, the participants put down the briefing document and proceeded to the formal part of the creative session. The short discussion of the new compound and its benefits imprinted on the participant's mind the features of the compound. The

Streamlining the Development Process

Table 1 Procedure for Developing Dental Product Ideas: "Creative Session"

Setup—Prior to the Session
1. Select the participants for the sessions.
2. Give the participants a brief or description of the product benefits (in general), and a short statement about the product form (toothpaste, dental floss, rinse, etc.).
3. Ask the participants to return 1 week later, after reading the brief and an accompanying set of five short articles on the general area of tooth/plaque protection.
4. Ask the participants to write down 15 benefits, phrases, statements to bring with them.

At the Session
5. Introduce the problem once again.
6. Go around the room, asking each participant to give one idea (short benefit, product description, etc.). Do this without having the panelist open the envelope in which they placed their 15 statements.
7. Go around the room again three times, asking each participant in turn to give an idea, element, phrase, etc. These should not be full concepts, but rather snippets of ideas.
8. Open up the discussion, recording every idea or element.

At the End of the Session
9. Identify categories of elements (e.g., length of protection, etc.).
10. Categorize every element as belonging to a category.
11. Polish the elements, and incorporate them in a database.

---

briefing insured that during the course of the discussion there would be mention of the new compound and its benefits. However, the short discussion did not overly focus the participant's attention on this compound to the exclusion of other features or benefits.

Results of the Creative Development Session

Each session generated several hundred elements. During the course of each session the participants eagerly threw out different ideas, and even small portions of ideas. The session was fairly free-flowing so that in the end the result was a list of words and phrases, some of good grammatical construction, but others appearing to be free-floating ideas in need of a little polish (and in some cases some "filling out" to bring them to a core idea). Table 2 shows a list of representative elements for each of the three product categories.

During the course of the creative ideation sessions the panelists did not categorize the elements. (A recording secretary present at the session picked up these elements, typing them into a computer.) After the session, the project

Table 2 Examples of Elements for Three Different Dental Protection (Anti-Plaque) Products

Toothpaste

Cleans your teeth
Whitens your teeth
Contains ProGuardin - anti-plaque
Cleans and protects teeth
A full day's protection every morning
Unique compounds for better oral health
Washes away plaque
Benefits you can see when you smile
Twenty-four hour protection
Plaque-Guard

Oral Rinse

A clear, green, smooth-tasting rinse
Rinses off plaque
Flavored to make each rinse experience pleasurable
Right-Rinse
Non-alcoholic rinse
A rinse without the astringent taste
Astringent, so you know it's working
Your teeth deserve the best!

Dental Floss

Waxed, for comfort
Thin and thick floss on two bands
Minty flavor to make flossing more fun
Floss for the health of it
Your teeth deserve the anti-plaque guard on our floss
Unique floss with anti-plaque
Now your floss works both ways—mechanical and protective
Two kinds of floss available—flavored and unflavored
New floss from the company that knows tooth protection
With an anti-plaque compound recommended most by dentists
A product from Plaque-Guard, the trusted name in tooth protection

leader at the creative boutique met with the manufacturer, bringing with her the typed list of elements. The marketing research manager and the project leader sat together, classified the elements into different categories, and eliminated both the infeasible and the redundant elements. The objective was to reduce each list for a product to a maximum of 80 elements. In full-scale quantitative studies with concept optimization the list of elements is often significantly larger—approaching or exceeding 300. In this particular project, where time and resources were limited and where it was vital to assess three products simultaneously, the marketing research director thought it prudent to limit the three sets of elements so that the total was 240. Had there been more time and more money, the editing process would not have been done with such a heavy hand, and each category might have comprised 150-250 elements.

Creating Restrictions

Since the project used concept optimization via IdeaMap, it was necessary to define those combinations of elements that could not go together. The project director from the creative boutique identified many illogical combinations of elements. The manufacturer identified many additional infeasible pairwise combinations of elements. Often it takes a person experienced with a product category to spot combinations that do not work together. An outsider can bring perspective and technique to a project, but generally does not bring the depth of understanding. That depth of understanding lies in the purview of the client manufacturer, who is more intimately involved with the category, the product and the project.

Implementing the Three-Category IdeaMap Project

Chapter 1 on concept development and optimization presented the approach for rapid, interactive development of concepts. The example involved one product category only, albeit with many alternative elements. The current project is far more complicated. There are actually three categories—mouthwash, dental floss, and toothpaste. When testing one category, matters are fairly straightforward. The only question involves the number of elements to test and the number of "iterations" per panelist. (Recall that each iteration comprises 20 elements, combined into 25 combinations according to a Plackett Burman screening design.) With one category a panelist can go through 1, 2, 3, 4 or even 5 or 6 iterations, and by so doing assess many, if not all of the elements. With three different product categories, how can the researcher effectively cover these adequately?

The marketing research group suggested two strategies to answer the problem:

Strategy 1—Run Parallel Studies, One Per Category. This strategy called for three different sets of consumers, so that each category could be evaluated extensively by one set of consumers. Each panelist would evaluate only a single category, but in depth. This strategy involved having the panelist go through six IdeaMap iterations, evaluating 150 combinations of elements pertaining to that category. With 50 panelists per category, this design required a total of 150 panelists.

Benefit Each panelist would rate many elements within a single category, providing a detailed measurement of that panelist's reaction to elements in the category. Thus the data would be "solid" on an individual-by-individual basis.

Drawback No ability for the researcher to compare the responses of a single panelist to the three different categories. A panelist would evaluate only elements from one category. There was discussion at this point about having the panelist evaluate two sets of concepts—one set of experimentally varied concepts for a single category, and a second set of reference concepts covering the three categories. This option was rejected when the group realized that, despite the evaluation of multiple concepts from the three categories, the most solid data would be obtained from a single category.

Strategy 2—A Panelist Evaluates All Three Categories. Each panelist evaluates concepts from all three categories, by going through two IdeaMap iterations for each category. This strategy has panelists evaluate the same total number of concepts, but do 1/3 of the evaluations for each category.

Benefit The researcher obtains information from the same panelist on all three categories. This enables a comparison of all three categories without need for "sample" matching.

Drawback Less solid data for each category on a panelist-by-panelist basis (although the same total number of elements per category are evaluated).

After extensive discussions the marketing group opted for Strategy 2 for the following three reasons:

1. Instead of 50 consumers rating concepts for each category, there would be 150 consumers rating concepts for each category. That increased base size felt "comfortable," even though the marketing group recognized that the data would be less robust for any one individual.
2. The data would allow the researchers to compare the acceptance scores from category to category without worrying about sample balancing. This design is called a "within-subjects design," because the same subjects (or panelists) participate in all conditions. Within-subjects designs are more immune to panel-to-panel differences which pervade studies using two or more "matched panels" to evaluate different stimuli.
3. The data would allow the researchers to segment consumers on the basis of their responses to the elements. Although the researchers and marketers did not plan to do any segmentation at this early stage, but rather planned simply to move forward and select a category and concept for the compound, there was the potential for additional analysis with the larger base size. To researchers and to marketers alike, large base sizes are more attractive because they better represent the ultimate marketplace. (There is always a worry, justified or not, that too small a base size in a product or a concept test will fail to represent the population. By choos-

ing Strategy 2 and having the greater base size, the group felt that they would avoid any possible under- and incorrect sampling of the target consumer population.)

Results from the IdeaMap Evaluation

Table 3 shows the partial model for each of the three categories—specifically the top three and the bottom three elements for each category in terms of persuasive elements. Table 4 shows an optimized concept for each product type, based upon the initial optimization of "interest in buying." Since the same panelists evaluated concepts from all three product categories, the highest possible score for each category, respectively, can be compared to one another. We see that the concept for mouthwash achieves the highest level of acceptance, whereas

Table 3 Additive Contributions of the Best Three and Worst Three Elements—Dental Care Products

	Product category = Toothpaste	
Additive constant		42
Benefit 4	Better brushing, and always working for you	7
Benefit 2	Removes plaque, even after you've finished brushing	7
Ingredient 3	Contains the ingredients dentists use most often	7
Problem 5	Plaque is serious business—try to avoid it at all costs	−3
Problem 2	Your teeth and gums deserve the best, not plaque	−5
Ingredient 6	Contains an alcohol-based surfactant for cleaning	−9
	Product category = Mouthwash	
Additive constant		36
Benefit 5	Rinses out, but continues to work nevertheless	9
Usage 2	Use in the morning, works all day for you	6
Benefit 4	Quick acting, just like you'd want it to be	6
Problem 2	Your teeth and gums deserve the best, not plaque	−4
Problem 6	Plaque can harm your teeth—why not prevent it?	−6
Problem 1	Dentists feel that plaque is today's #1 oral problem	−8
	Product category = Dental floss	
Additive constant		27
Benefit 1	Flossing is the number one activity for dental health	5
Benefit 6	Flossing gets out the plaque no other method can	5
Problem 1	Your teeth are critical to you—take care of them now	3
Problem 3	Dental problems are caused by plaque	−2
Problem 2	Want to avoid a "plaque attack?"	−3
Usage 4	Floss after every meal, and even in-between	−7

Table 4 Optimal Concepts for Dental Care Products (Concepts Based upon the Best Performing, Mutually Compatible Elements)*

A. Toothpaste concept

This new toothpaste removes plaque, even after you've finished brushing.
The toothpaste works because it contains the ingredients dentists use most often.
So, the toothpaste helps you to brush better.
And, what's more important, your toothpaste is working for you even after you've finished brushing your teeth.

B. Mouthwash concept

Our mouthwash is specially formulated to protect your teeth.
It's quick acting, just like you'd want it to be.
The mouthwash rinses out easily and pleasantly, with no harsh aftertaste.
Yet, the mouthwash keeps on working, with its residual power, to get rid of the plaque in your mouth, and keep your mouth healthy and fresh.

C. Dental floss concept

You know your teeth are critical to you. So we help you take care of them, with our new, pleasant-to-use dental floss.
You may be aware that dentists recommend flossing more than any other activity for maintaining the health of your teeth and gums.
Flossing gets out plaque that no other method can.
Use our dental floss—just as many dentists do.

* Concepts created by marketing based on pattern of winning concept elements, but with some "editorial" and "creative" license.

the concept for dental floss achieves the lowest level of acceptance. Keep in mind that these concept scores simply represent the percent of consumers who would be interested in purchasing the product, without pricing. Furthermore, these concepts have to be considered later on with an eye towards the difficulty of entering the market. It may well be that the dental floss concept, although substantially lower in maximum appeal, really holds the most promise because it is a relatively underdeveloped category.

An Overview to Early Stage Concept Optimization

When a manufacturer has to create a new product in a short period of time it is advantageous to travel down several paths simultaneously. The approach uses IdeaMap for rapid fire screening and optimization. The current approach then uses simultaneous experimental designs, one for each category.

Purists might decry the current approach because it appears to cut corners. The purist researcher wants to investigate a category in extreme depth in order to be sure not to miss anything that could lead to a market advantage. With the pressures of time and money perennially weighing on the marketing team, the perfect project could lead to disaster. A perfect project might take too long, or cost too much. It is up to the marketing team to decide how far away from "perfect" one can be and still feel comfortable with the results.

4. Rapid Product Development and Optimization

Introduction

By itself concept development only leads to promising statements and communications about the product. Can the product developer actually create the product?

The traditional methods for product development "under fire" consist of shoot, aim, and reload. The time and cost pressures in product development all too often force the product developer to create one or two prototypes. These are the manufacturer's best guesses. The test then selects the winning prototype from the limited set.

When the process of "shoot from the hip" is explored in more depth by interviewing the product developer, the perennial response to the question, "Why do it this way?" is that time and cost pressures dictate this strategy. When asked "Why not use experimental design?", the harried product developer often answers that there is neither sufficient time nor resources to do it right. All too often, however, the prototypes developed under extreme time and cost pressures are really not particularly acceptable. It is a source of continuing wonder that developers always have time to "do it again," but never time nor resources to "do it right the first time."

Fortunately, there are experimental designs which allow the product developer to learn a significant amount about the product, using a disciplined development protocol. In concept optimization we introduced the Plackett Burman screening design. Plackett Burman designs come in many types as Table 5 shows. Screening designs effectively quantify the effects of many different independent factors. For product development the Plackett Burman design is quite useful. The design enables the developer to assess a wide range of qualitatively different variables with a relatively small number of prototypes. Of course if the product developer still wishes to "shoot from the hip" then the simplest experimental design, requiring the fewest prototypes, is worse than one's own gut instinct. In this case history, however, we are talking about a disciplined development project, where product developers accepted the notion of experimental design and wanted to use it to get the product right.

Identifying the Relevant Variables in Each Category

The two-level Plackett Burman design is a good schematic to use for exploratory product development because it requires only relatively modest develop-

Table 5 Four Plackett Burman Screening Designs for Accelerated Product Development

Six Factors in Eight Prototypes

	Product					
	A	B	C	D	E	F
1	0	0	0	0	0	0
2	0	1	1	1	0	1
3	0	0	1	1	1	0
4	1	0	0	1	1	1
5	0	1	0	0	1	1
6	1	1	0	0	0	1
7	1	1	1	1	1	0
8	1	1	1	0	1	0

Ten Factors in 12 Prototypes

	Product									
	A	B	C	D	E	F	G	H	I	J
1	0	0	0	0	0	0	0	0	0	0
2	1	1	0	1	1	1	0	0	0	1
3	0	1	1	0	1	1	1	0	0	0
4	1	0	1	1	0	1	1	1	0	0
5	0	1	0	1	1	0	1	1	1	0
6	0	0	1	0	1	1	0	1	1	1
7	0	0	0	1	0	1	1	0	1	1
8	1	0	0	0	1	0	1	1	0	1
9	1	1	0	0	0	1	0	1	1	0
10	1	1	1	0	0	0	1	0	1	1
11	0	1	1	1	0	0	0	1	0	1
12	1	0	1	1	1	0	0	0	1	0

Streamlining the Development Process   437

14 Factors in 16 Prototypes

Product

	A	B	C	D	E	F	G	H	I	J	K	L	M	N
1	0	0	0	0	0	0	0	0	0	0	0	0	0	0
2	0	1	1	1	1	0	1	0	1	1	0	0	1	0
3	0	0	1	1	1	1	0	1	0	1	1	0	0	1
4	0	0	0	1	1	1	1	0	1	0	1	1	0	0
5	1	0	0	0	1	1	1	1	0	1	0	1	1	0
6	0	1	0	0	0	1	1	1	1	0	1	0	1	1
7	0	0	1	0	0	0	1	1	1	1	0	1	0	1
8	1	0	0	1	0	0	0	1	1	1	1	0	1	0
9	0	1	0	0	1	0	0	0	1	1	1	1	0	1
10	1	1	1	0	0	1	0	0	0	1	1	1	1	0
11	0	0	1	1	0	0	1	0	0	0	1	1	1	1
12	1	1	0	1	1	0	0	1	0	0	0	1	1	1
13	0	0	1	0	1	1	0	0	1	0	0	0	1	1
14	1	1	0	1	0	1	1	0	0	1	0	0	0	1
15	1	1	1	0	1	0	1	1	0	0	1	0	0	0
16	1	1	1	1	0	1	0	1	1	0	0	1	0	0

Table 5 Continued

16 Factors in 20 Prototypes

Product

	A	B	C	D	E	F	G	H	I	J	K	L	M	N	O	P
1	0	0	0	0	0	0	0	0	0	0	0	0	0	0	0	0
2	1	1	0	0	1	1	1	1	0	0	0	1	0	0	0	0
3	0	1	1	0	0	1	1	1	0	1	1	0	1	0	0	0
4	1	0	1	1	0	0	1	1	1	0	0	1	0	1	0	0
5	1	1	0	1	1	0	0	1	1	1	1	0	1	0	1	0
6	0	1	1	0	1	1	0	0	1	1	0	1	0	1	0	1
7	0	0	1	1	0	1	1	0	0	1	1	0	1	0	1	0
8	0	0	0	1	1	0	1	1	0	0	1	1	0	1	0	1
9	0	0	0	0	1	1	0	1	1	0	0	1	1	0	1	0
10	1	0	0	0	0	1	1	0	1	1	0	0	1	1	0	1
11	0	1	0	0	0	0	1	1	0	1	1	0	0	1	1	0
12	1	0	1	0	0	0	0	1	1	0	1	1	0	0	1	1
13	0	1	0	1	0	0	0	0	1	1	0	1	1	0	0	1
14	1	0	1	0	1	0	0	0	0	1	1	0	1	1	0	0
15	1	1	0	1	0	1	0	0	0	0	1	1	0	1	1	0
16	1	1	1	0	1	0	1	0	0	0	0	1	1	0	1	1
17	1	1	1	1	1	1	0	1	0	0	0	0	1	1	0	1
18	0	0	1	1	1	1	1	0	1	0	0	0	0	1	1	0
19	0	0	1	1	1	1	1	1	0	1	0	0	0	0	1	1
20	1	0	0	1	1	1	1	0	1	0	1	0	0	0	0	1

0 = Option 1.
1 = Option 2.

ment time and effort. The strength of the design is also its weakness, however. Since the screening design assesses many different factors rather than forcing the researcher to concentrate on just a few, often the hardest task is to figure out what variables to modify. Contrast this with the conventional state of affairs. When the developer has little control over the variables, or when the test investigates only one or two independent variables, matters are easy at the front end of development. The researcher knows what can be varied and merely has to identify the levels to vary. When the researcher has the luxury of varying many ingredients or process conditions matters become more difficult. The very plethora of factors to vary in an efficient manner opens up the floodgates of imagination. The exact opposite happens—the researcher begins to think of relevant factors that were only taken for granted before.

In this study the development group recognized that they had to vary the factors of products in different product categories. They recognized that many of the product prototypes they were to make would have minor cosmetic differences. The products would look, feel, and taste different, but might not perform differently. That is, the modifications that the developer would make would affect primarily surface characteristics, not the deeper functional characteristics.

For each category the product developer listed the different factors that were under control. Furthermore, the developer recognized that for each category she could only create a maximum of 12 products in order to remain within the development timeline prescribed by the marketing group. For three categories this required a total of 36 prototypes. Table 6 lists the physical dimensions that the R&D group felt to be most relevant for this product.

Creating a Design

The two-level rather than three-level design is efficient because it makes the product developer's job significantly easier. Rather than identifying three levels (e.g., for response surface optimization), the developer need only identify two levels of each ingredient/process condition. Furthermore the design works well if one or more of the variables is "discrete" (e.g., color of rinse). The procedure will work as long as the variable takes on one of two identified levels (continuous) or one of two identified options (discrete).

Testing the Alternatives

The development proceeded fairly rapidly. Using the screening design as the schematic, the product developer created the requisite prototypes. Furthermore, the marketing group suggested that they test two additional products in each category to represent the "in-market" competitors. These two products would

Table 6 Physical Variables Selected By R&D For The Tooth Protection Study—Each Product Form (Toothpaste, Mouthwash, Dental Floss) Varied On 6 Factors, Each Factor At 2 Levels (0,1, Option A, Option B)

Variable	Category = Toothpaste 0 Condition	1 Condition
A	Mint flavor	Cinnamon flavor
B	No active crystals	Active crystals
C	Opaque	Translucent
D	No stripes	Stripes
E	Low grit	High grit
F	Thinner consistency	Thicker consistency

Variable	Category = Mouthwash 0 Condition	1 Condition
A	Mint flavor	Cinnamon flavor
B	Low alcohol base	High alcohol base
C	Watery consistency	Slightly thickened
D	One cap dose	Two cap dose
E	Low active	High active
F	Colorless	Neutral color

Variable	Category = Dental Floss 0 Condition	1 Condition
A	Mint flavor	Cinnamon flavor
B	Thin thread	Thick thread
C	Weak thread	Strong thread
D	No colored sections	Colored sections
E	Matte appearance	Shiny appearance
F	No aromatization	Aromatization

act as benchmarks. The in-market products were disguised to seem like simply another prototype.

The marketing research director suggested testing each product among 20 panelists. In early stage, rapid development where time and costs are limited, the researcher strives to obtain an overall "snapshot" or picture of the prototypes. That snapshot will help the developer home in on the relevant levels. Low base sizes suffice to show the pattern of responses as a function of the physically measured variables, even if the base size of 20 ratings per product is considered "small."

Each person tested a randomized seven of the 14 products in one set. Each category required 40 panelists, which yielded 20 ratings per product. The total

Streamlining the Development Process

study required 120 panelists, each panelist chosen to be a representative category user of the product. There were 40 users each of toothpaste, mouthwash, and dental floss, respectively.

The Questionnaire

There were two major pieces of information to be gained from the product evaluation:

1. How well did the prototypes perform on an absolute basis? The objective was to determine whether the prototypes were adequately acceptable, and to compare the prototypes to the in-market competitors (the benchmarks).
2. How closely did the prototypes score in terms of "delivering what the concept promised?" The first part of the project optimized three concepts via IdeaMap. Each product category generated its own optimal concept (see Table 4). The optimal concept was then inserted into the product evaluation questionnaire as an "attribute." After the panelist tried each prototype, the panelist rated the prototype on attributes. Finally, the last question for each prototype was the following:
   "Read the description of the product. Now, thinking about the product you just tried, how *similar* is this product to the description that you just read? Use the following scale:
   0 = very big difference between what I just read and the product I just tried → 100 = the product I just tried is identical to the description of what I just read.

Test Sequence

In order to insure the quality of the fieldwork, the panelists reported to a central location in two groups of 20 individuals. A test session involved only one category. At the central location a "moderator" explained the purpose of the study, described the products, showed the panelists how to use the products, and then explained the questionnaire. Most consumers had no problem understanding the attributes. It took a little explanation to clarify the concept of "similarity" vs. "dissimilarity" of concept to product.

Once the moderator felt that the panelists understood what was expected of them, each panelist evaluated a practice product (outside the set), and completed a questionnaire. This final exercise assured that the panelists understood what was expected of them, and increased the quality of the results.

Finally, the panelists received the seven products, one product packed in each of seven envelopes. The envelopes were marked to show the sequence and day of evaluation. Each envelope also contained a questionnaire for that product. The panelists were instructed once again on how to use the product, and told to return 1 week later with the seven product questionnaires filled out. For their participation the panelists were paid upon completion of the seven evaluations. (This orientation and payment insures panelist cooperation and maintains the integrity of the data.)

Performance of the Products vs. Competition

The first question to ask is whether the product developer "got the product right." That is, when the developer created the mouthwash, the toothpaste or the dental floss, did the product prototypes score well or did they score poorly on an absolute basis? The norms for the category (blind tested, unpositioned product) are as follows, based upon previous studies and available data:

70+	Excellent, needs no further work
60–70	Very good, needs no work
50–60	Good, can use some modification (at low end)
40–50	Poor, can use significant modification
40 or below	Very poor product, may enjoy success in the marketplace from factors other than product

The ratings of liking appear in Table 7 for the 12 prototypes and two competitors. The experimental design clearly created a range of different products that vary considerably in the degree to which consumers will accept them (on a blind basis). The good news, at least for mouthwash (and to a small extent for toothpaste), is that some prototypes appear to do well on an absolute basis. This finding is promising. The greater the number of prototypes which perform reasonably well (e.g., 50+ on the 100-point scale), the greater the odds that the product developer has truly created a potentially successful product.

Table 7 also shows the performance of two competitors in each category. These two competitors are not part of the design. Rather, their ratings give the product developer and marketer a sense of how well the competition performs. In many studies the product developer might wish to incorporate many more than two benchmarks into the study, but here, with the tight schedule and funds, it was more prudent to limit the number of benchmarks to the two that seemed to be the most competitive with the new product. Table 7 shows that the competitor benchmarks differ from each other, and do not necessarily exhibit high acceptance.

Table 7 also shows the rating of "fit to concept." The higher the fit, the better the product will deliver what the concept promises.

Early Results—Setting up the Data Tables for Modeling Analysis

The basic data for each category comprises a rectangular matrix. Each row corresponds to one of the 12 products in the experimental design. There are three such matrices, one for each category. Table 7 shows part of the data table.

Table 7 Ratings of Overall Liking and Fit to the Optimal Concept: Dental Plaque Products

Physical variables (appropriate to product)							Ratings on attributes					
							Toothpaste		Mouthwash		Dental floss	
Product	A	B	C	D	E	F	Liking	Fit	Liking	Fit	Liking	Fit
1	0	0	0	0	0	0	28	42	48	57	32	43
2	1	1	0	1	1	1	35	37	54	50	36	40
3	0	1	1	0	1	1	42	51	56	57	24	47
4	1	0	1	1	0	1	47	46	58	53	36	43
5	0	1	0	1	1	0	46	39	52	55	30	50
6	0	0	1	0	1	1	38	43	48	52	42	52
7	0	0	0	1	0	1	56	46	44	48	42	47
8	1	0	0	0	1	0	32	38	47	46	36	43
9	1	1	0	0	0	1	30	35	57	58	39	41
10	1	1	1	0	0	0	36	42	52	51	36	45
11	0	1	1	1	0	0	31	45	61	63	45	46
12	1	0	1	1	1	0	44	40	59	58	40	42
Competitor 1							56	60	53	57	46	48
Competitor 2							58	52	54	50	41	45

The left-half portion of the data matrix comprises the design. Within that section of the matrix there are 0's and 1's. A "0" under a column shows that the product does not contain the specific feature listed in the column. A "1" under a column shows that the product contains the specific feature. Each column (or feature) is statistically independent of every other column. Furthermore, each feature appears equally as often with every other feature. [This is one of the important properties of the Plackett Burman design].

The right-half portion of the matrix comprises the average ratings. Recall that each product was rated by 20 panelists so that each entry in the right of the matrix is the average from 20 observations. The ratings include overall liking and fit to the optimal concept. Only liking and fit (consumer data) are shown. However, the actual data comprises many more attributes.

Analyzing the Data by Regression Analysis

The data format in Table 7 leads naturally to a regression model. The independent variables are the features that R&D systematically varied. The dependent variables are the attributes rated by the consumer.

A key benefit of a Plackett Burman screening design is that the independent variables are clearly identified. Each independent variable can be coded as "1" (present, new level, higher level, depending upon the variable) or "0" (absent, current level, lower level, depending upon the variable). The data is set up for a dummy variable regression analysis, which uses 0's and 1's as the independent variable, and a continuous rating for the dependent variable.

The partial results from this analysis appear in Table 8. The table shows the contribution of each variable (in the "1" condition). The contribution of each variable in the "0" condition is automatically assigned a value of "0." The table shows two consumer-rated attributes—overall liking, and fit to concept. Thus, when flavor goes from "mint" to "cinnamon," liking decreases by –3. This means that for the toothpaste product, a change in the flavor decreases the average rating of "liking" by three points on the 100-point scale. Fit to concept decreases by five points.

Adding Objective Considerations to the Model

The first two models in Table 8 represent the relation between the factors under R&D control and consumer ratings. But what about other variables that are of interest, but which cannot be captured from consumer data? There are two distinct variables of interest:

1. Cost of Goods: This variable must be estimated by R&D separately for each prototype. Often R&D does not have exact numbers on the cost of goods, but either prior to or after the consumer evaluations, R&D may be able to provide the necessary numbers. These numbers have to be pro-

Table 8 Additive Model Showing the Part-Worth Contributions of Each of Six Formulation Variables to Consumer Ratings and R&D Estimates

Category = Toothpaste

	Constant	Mint A0	Cinnamon A1	No crystal B0	Crystal B1	Opaque C0	Translucent C1	No stripes D0	Stripes D1	Low grit E0	High grit E1	Thin F0	Thick F1
1. Like	34	0	-3	0	-4	0	2	0	9	0	2	0	5
2. Fit to concept	42	0	-5	0	-1	0	5	0	0	0	-1	0	2
3. Cost	52	0	4	0	5	0	6	0	8	0	12	0	9
4. Time	121	0	-8	0	-8	0	-25	0	-25	0	-8	0	-58

Category = Mouthwash

	Constant	Mint A0	Cinnamon A1	Low alcohol B0	High alcohol B1	Watery C0	Thickened C1	One cap D0	Two cap D1	Low active E0	High active E1	Colorless F0	Colored (neutral) F1
1. Like	45	0	3	0	5	0	5	0	3	0	-1	0	0
2. Fit to concept	54	0	-3	0	3	0	3	0	1	0	-2	0	-2
3. Cost	26	0	6	0	4	0	2	0	3	0	5	0	2
4. Time	58	0	0	0	-17	0	3	0	-17	0	-17	0	0

Category = Dental Floss

	Constant	Mint A0	Cinnamon A1	Thin thread B0	Thick thread B1	Weak thread C0	Strong thread C1	No colored section D0	Colored section D1	Matte E0	Shiny E1	No aroma F0	Aroma F1
1. Like	37	0	1	0	-3	0	1	0	3	0	-4	0	0
2. Fit to concept	46	0	-5	0	0	0	2	0	-1	0	2	0	0
3. Cost	32	0	6	0	15	0	4	0	7	0	3	0	6
4. Time	67	0	0	0	-17	0	-17	0	-33	0	-17	0	0

vided for each combination of factors. Once the numbers are available the analyst can estimate the part-worth contribution of each variable to the total cost. The parameters of the cost model are also shown in Table 8. Cost is very important as a factor in assessing product viability. If the optimum product costs too much to create, then it is no longer "optimum" from an overall point of view. The optimum may be the best from the consumer's viewpoint, but not from a business viewpoint.

2. Time to Create the Product: This variable is a measure of the delay that will be encountered in creating the product. Time is harder to estimate than cost. If it is difficult to accurately estimate the time, the researcher or product developer might simply wish to use three numbers—0 (short), 50 (medium) or 100 (long) as the time needed to create the product. These three numbers are qualitative levels. The models for time are shown in Table 8 as well.

Selecting the Optimum Combination of Features

The products presented in Table 9 show the best selections of features. The first goal (A) consists of selecting the features for each product category that maximize consumer acceptance, without any other considerations.

Just because the product is maximally acceptable (in comparison to prototypes and competitors) and scores high in absolute terms does not mean that the

Table 9 Optimal Combinations of Features for Three Different Plaque Treatments Based Upon Different Marketing / Development Objectives:

	Product category 1 = Toothpaste									
	Features present						Consumer rating			
Goal	A	B	C	D	E	F	Liking	Fit	Cost	Time
A. Maximum liking	0	0	1	1	1	1	52	48	87	5
B. Minimum cost	0	0	0	0	0	0	34	42	52	121
C. Minimum time	1	0	1	1	0	1	47	44	79	5
D. Like, cost, time	1	0	1	1	0	1	47	44	79	5

	0 Condition	1 Condition
A	Mint flavor	Cinnamon flavor
B	No active crystals	Active crystals
C	Opaque	Translucent
D	No stripes	Stripes
E	Low grit	High grit
F	Thinner consistency	Thicker consistency

Streamlining the Development Process

Table 9 Continued

	Product category 2 = Mouthwash									
	Features present						Consumer rating			
Goal	A	B	C	D	E	F	Liking	Fit	Cost	Time
A. Maximum liking	1	1	1	1	0	1	61	56	43	27
B. Minimum cost	0	0	0	0	0	0	45	54	26	58
C. Minimum time	0	1	0	1	1	0	57	59	40	7
D. Like, cost, time	0	0	1	1	0	0	53	58	31	44

	0 Condition	1 Condition
A	Mint flavor	Cinnamon flavor
B	Low alcohol base	High alcohol base
C	Watery consistency	Slightly thickened
D	One cap dose	Two cap dose
E	Low active	High active
F	Colorless	Neutral color

	Product category 3 = Dental floss									
	Features present						Consumer rating			
Goal	A	B	C	D	E	F	Liking	Fit	Cost	Time
A. Maximum liking	1	0	1	1	0	1	42	42	55	17
B. Minimum cost	0	0	0	0	0	0	37	46	32	67
C. Minimum time	0	1	0	1	1	1	33	47	63	0
D. Like, cost, time	0	0	1	1	1	0	37	49	46	0

	0 Condition	1 Condition
A	Mint flavor	Cinnamon flavor
B	Thin thread	Thick thread
C	Weak thread	Strong thread
D	No colored sections	Colored sections
E	Matte appearance	Shiny appearance
F	No aromatization	Aromatization

product is the best that can be developed. There may be at least two factors that militate against the selection of these products:

1. The products may be too expensive.
2. The products may not be sufficiently close to the concept.

The researcher can create a set of compromise prototypes, working with consumer acceptance, cost of goods, and closeness of the product and the concept. Goals B, C, and D show the combination of different features which satisfy other criteria. Goal B shows the cheapest product, Goal C shows the product that is quickest to develop, and Goal D shows the best compromise product incorporating acceptance, time and cost.

5. Creating a Truly New Product or New Category

Introduction

Throughout this book we have been dealing with products that are well-defined. There are, however, many cases where the researcher has to create a product or even a category that never existed before, rather than modifying an existing product or creating a product which represents a variation of an existing product. How can the developer accomplish this new product creation? Or, more poetically, how can the developer hold serendipity in the palm of his hand, and make the muse of creation a loyal companion rather than an infrequent visitor?

To continue our topic of oral care, consider the case history of the Bryson Oral Care Company, Ltd. The founder and owner, Dr. Lawrence Bryson, has the tough job of developing new products that can be sold to much larger companies outright, or licensed for use around the world for a generous licensing fee. Bryson Oral is thus in the business of new product development. Unlike conventional companies, Bryson Oral has to continually create new products to maintain its stream of revenues. These products must be new (not just simple "tweaks" on an existing product), must be patentable (to insure the licensing fee), and must be consumer-driven (because Bryson Oral's clients deal in over-the-counter health and beauty aid products).

Given the tough job of creating new products on an ongoing basis, Bryson recognized that he would have to "go out of the box" in order to increase the creativity of his marketing and development groups. The next section presents an approach which can be best summarized as, "Consumers may not be able to tell you what they want, but they can tell you when they see it!"

Creating New Products by Fulfilling Revealed Needs

We will continue the case history approach, remaining within the oral care category. With extensive secondary research, Bryson recognized that a growing opportunity area was "plaque protection." But—what was the specific product? Keep in mind that the objective is to create a truly new product. Although it may be argued that no one can really ever invent a "totally new" product, it may be possible to get closer to the ideal of "new" by combining elements or features from different product categories into a new combination that is acceptable to consumers for specific end uses. Bryson recognized that the combination of chemical and mechanical features could create a new product. The task was to identify what the new product should do (first objective), and then to create a combination of features that would deliver what was needed (second objective). As far as Bryson was concerned, the range of features to be included was open, and could accommodate elements from as many diverse product categories as was needed.

The approach for new product development follows these nine steps:

Step 1. Identify a Need or an Opportunity. The objective here is to identify a critical need that may or may not be known to the consumer. There are many ways to identify the need, but the ultimate objective is to find an opportunity. Table 10 shows one way—problem detection, in which the consumer panelist is presented with a large number of different "needs" or "benefits." The consumer rates each of these on three criteria: purchase intent, uniqueness, and relevance to the individual (i.e., is this need relevant to me). The consumers use a nine-point scale for each. After the data is aggregated, the researcher looks for the need or opportunity that is both relevant and unique (i.e., no one else in the category is currently fulfilling that need).

Step 2. Express the Need / Opportunity in a Sentence or a Succinct Paragraph (the Setup Concept). After Step 1 is completed the researcher has a set of needs or opportunities. These are typically short phrases. Step 2 requires that the researcher state this need clearly, to convey a "word picture" to the consumer, or even that he use a concept to convey this need or opportunity. This is the "setup concept." This concept, when read, will convey specific expectations to the consumer. Table 11 shows two different setup concepts, developed on the basis of results from Step 1. Each concept addresses a different need or end use that consumers felt to be relevant and unique. The concepts are short, to the point, and descriptive of end benefits, rather than flowery, and "sales oriented."

Table 10 Problem Detection Elements and Results (Partial List)

#	Opportunity/situation/problem	Interest	Uniqueness
1	On-the-go plaque protection—can't protect well when I'm traveling	64	34
2	Self-diagnosing tooth "health" kit	35	78
3	One rinse for 2 months so I don't have to worry every day	68	73
4	Device showing that it removes plaque so that I "know it's working"	56	54
5	Super oral hygiene kit	47	50
6	Device to make brushing easier all day, so that I can follow my dentist's advice	78	66
7	Selective plaque attacking system, so I can really get my teeth clean	67	51

Table 11 Two Different Setup (Opportunity) Concepts for Use as Attributes (Concepts Emerge from the Problem Detection Task)

Opportunity concept 1 (actual concept used in optimization plaque)

A treatment that will allow me to use it once every 3 days, and not worry about plaque.

Opportunity concept 2

A treatment that will give me ongoing feedback about how I'm doing and the state of health of my teeth.

Step 3. Create the Resource Bank Comprising Words, Pictures, Video, etc. These are elements to be used in concept optimization. In conventional concept testing and optimization the objective is usually well-defined, as is the product concept. In conventional concept testing the idea is to identify alternative features of the product. In new product development for the plaque protection, there is no single concept. There is a need. But the researcher has no idea what physical features will be relevant to that need. In actuality the researcher must work with many different elements, selected from a variety of different product categories. Table 12 shows a list of some of these elements. Keep in mind that by themselves some elements are quite relevant to plaque protection, whereas other elements are more tangential but could provide important clues about the product. In Step 3 the goal is to go "broad," looking at many different categories and many different, generally unrelated elements which *could possibly fit* a plaque protection product.

Table 12 Partial List of Wide-Ranging Elements

Equipment	Picture of toothbrush
Equipment	Picture of sandblaster
Equipment	Picture of foaming action
Equipment	Picture of a laser gun
Mode of action	Dissolves plaque
Mode of action	Combines with plaque to form inert compound
Mode of action	Combines with plaque to loosen it
Mode of action	Mechanical action against plaque
Applicator	Long handle with reservoir
Applicator	Liquid applied by brush
Applicator	Chewing gum
Ingredient	Soap-based surfactants
Ingredient	Gritty material enrobed in a gelatin base
Ingredient	Effervescent materials that work on contact

Streamlining the Development Process

Step 4. Categorize and Constrain the Elements. This is the standard procedure in concept optimization. The array of elements (from many different product categories) must be arranged in a way to allow these elements to be tested in concepts. Conventional concept optimization deals with simple categorization of elements (e.g., brand names, heritages). New product development with concept optimization can proceed in precisely the same way, even though the elements in a category might be quite disparate from each other. In Table 12, note that "active process" for assisting in plaque removal ranges from chemical to mechanical. Note also in Table 12 that the elements range from "close in" to the idea of oral health to "far out of the box" (viz., tangential ideas from other product categories that might have relevance, even if relevance is based on analogy, rather than actual substantiveness).

Pairwise restrictions are handled in the same way as they are with concept evaluation. The researcher identifies which combinations of elements cannot go together, and notes these as restrictions. The restrictions are usually developed on the basis of intuitive obviousness, rather than on the basis of technical correctness. Thus, the rules about what elements go together are fairly lax at this early stage. Many combinations that would not really make sense from a purely logical standpoint are allowed to occur, primarily because the objective is to arrive at a concept that fits an opportunity or a need. Although the particular combination may not be logically correct (e.g., it may combine elements which would never go together in "real life"), the combination itself may spark lateral thinking. Perhaps the combination itself doesn't work in a rational, technical sense, but it may have its own sense if we think of the combination as an analogy. And, since the research is very early stage, where all possibilities are open, the researcher definitely does not want to miss this potential combination because it may, by analogy, suggest a new combination of compatible features. Contrast this with the more conventional types of concept testing. In those conventional procedures, the final result must be a product or positioning concept that can be immediately implemented or converted into an advertisement or new business strategy. Consequently, it is vital in conventional work to weed out combinations that are illogical, technically infeasible, or are not consonant with current business strategy, because the conventional work will immediately lead to action.

Step 5. Combine These Elements into Small Concepts by Experimental Design.
This fifth step again parallels the standard concept optimization procedure. The concepts for this study are combined into small concepts (two to four elements) by means of the main effects experimental design shown in Table 13. This smaller design is preferable for new product development because of the nature of the concept. There is no conventional "architecture" for the new product concept. There is no orchestrated succession of categories to portray a coher-

Table 13 Experimental Design for New Product Concepts

Concept	Category			
	A	B	C	D
1	2	3	0	0
2	3	2	1	0
3	0	3	3	0
4	1	0	3	0
5	0	1	2	0
6	3	0	2	1
7	3	3	2	1
8	1	2	0	1
9	2	0	1	1
10	0	1	1	1
11	1	3	0	2
12	2	2	1	2
13	2	0	2	2
14	1	0	0	2
15	3	1	3	2
16	1	1	1	3
17	2	1	0	3
18	3	3	3	3
19	0	2	3	3
20	0	2	2	3

Note: 0 = element not present from the category.
1,2,3 = Three different elements from the category present in the concepts. A concept has either one or no elements from a category.

ent message, as there is for conventional research (where the concept has a name, heritage, use, composition, price). We don't know exactly what we are searching for, and we don't know the steps of the "dance" for this product concept. Therefore we do not want to run the risk of creating a silly, meaningless "hodge podge" of elements that are thrown together without thought, and which overwhelm the reader as a laundry list of features without a coherent organizing principle. Nor do we want to have coherence by parsimony—description of the concept by a single element. The new product concept must have substance, and must present a product, rather than a single benefit, or end use statement.

Step 6. Follow the IdeaMap Approach, but Have Panelists Rate Three Attributes for Each Test Combination. IdeaMap allows the researcher to test many different combinations of elements, with these elements arranged according to an experimental design (see Chapter 1). Each individual evaluates totally different combinations. (Thus with 100 consumers evaluating 100 combinations each, there will be a total of 10,000 different combinations of elements evaluated across the full set of consumers. No two will be the same, except by chance.) The IdeaMap procedure is perfect for this new product search, because it allows the exploration of many elements in many different combinations.

Rather than rating purchase intent or acceptance alone, however, the IdeaMap interview is extended to obtain more information. The panelist rates each test combination on the following three attributes, each on its respective 9-point scale:

1. Interest (1 = definitely not interested → 9 = definitely interested)
2. Uniqueness (1 = similar to many other products on the market →
   9 = very different from other products on the market)
3. Fit To Concept 1 (a need/opportunity) (1 = definitely does not fit →
   9 = definitely a perfect fit)

The rating attributes allow the consumer panelist to integrate all of the information and provide a single, unified response. The consumer need not "design" the product. Rather, the computer (via experimental design) throws up different potential combinations of features, and asks the panelist to rate whether these features are close to fitting a concept or far away from fitting a concept. Consumers only have to "intuit" their answers, not justify them. That is, a consumer need only feel that somehow one combination of elements is closer to the end use described by the attribute concept and another combination of elements is further from the end use concept. Although this seems like a strange way to develop a product, in actuality it is very simple and quite a bit of fun. The panelists never know what is coming up. Sometimes the combinations are simply amusing. Other times the combinations are quite close to describing a new set of features which truly fit an end need or fill an opportunity.

The actual evaluation for the oral care comprised four rounds of 20 test concepts each, with the panelist rating each test concept on the three attributes above. To insure ongoing interest, the panelists were pre-recruited to participate, paid at the end of the session (lasting 1 hour), and afforded four breaks during the course of the evaluations. Furthermore, to maintain interest, the 80 test concepts were interdigitated so that panelists were unaware that they were evaluating four sets of concepts that were experimentally designed. The concepts from the four different designs were so varied that panelists thought that there was no underlying design (although they reported that some elements did reappear from concept to concept).

In this project 100 consumers participated. The consumers approved of the idea of plaque protection.

Step 7. Create the Concept Model. The individual data from each panelist comprises ratings of interest, uniqueness and relevance to an end concept. Each individual's data can be summarized by three separate equations or "models." A model shows the contribution of every element tested by the panelist to an attribute. The model is therefore an equation (specifically a "dummy variable regression equation," where the predictors take on values of 0 or 1, and the response takes on values of 0 (would not buy) or 100 (would buy). Chapter 1 on IdeaMap goes through the details of the procedure). Suffice it to say here, each individual generates a model for each of the three rating attributes, with the model presenting the utilities of every one of the elements in the set. High numbers for an element mean that the element would lead to purchase (for the "purchase intent" scale), or that the element is perceived to "fit the concept" (for the "fit to concept" scale), or the element is perceived to add uniqueness to the product (for the "uniqueness" scale). Table 14 shows part of the model, summarizing the results from the entire panel of 100 consumers.

Step 8. Optimize the Concept Subject to Objectives. Here the researcher or the marketer creates new combinations of elements that satisfy the following three objectives:

a. The concept is highly acceptable.
b. The concept is unique.
c. The concept is perceived to answer the problem by "fitting the setup concept" (used as an attribute).

Table 15 shows two developed concepts which satisfy the foregoing objectives, along with the diagnostics. Unlike the conventional approach for concept development which comes up with actionable concepts but which explores a limited range, the concepts developed here are starting ideas or "seeds" for a new product. That is, the concepts in Table 15 need to be refined. They are not finished, polished concepts. Instead they are combinations of elements which present potential new products, or new ways of thinking about current products.

Step 9. Refine the Concept in Focus Groups. The results of this early development yield different combinations of elements. Some of these combinations may be "right on," whereas other combinations may have the spark of a new idea, but in reality may "miss the mark" on many but not all elements. Since, however, this is early stage research, the concept development task need not end with Step 6. There is always the ability to develop concepts, and to present these concepts to focus groups to dissect the idea, to find out what "works" and what does not. Furthermore, with the Concept Optimizer to act as a "mixing palette," the researcher can bring a portable laptop computer to the focus group, create new combinations "on the fly," expose the new concepts to consumers, and surgically

Table 14 Models Showing How Elements Drive Acceptance, Fit to Concept, and Uniqueness (Partial Data Only)

	Purchase	Fit end use	Unique
Additive constant	42	36	22
Image of the way it works			
Picture of toothbrush	4	4	−6
Picture of sandblaster	−3	8	8
Picture of foaming action	5	6	2
Picture of a laser gun	7	3	10
Cartoon picture—selective attack, by "little dots," of a tooth covered with plaque	6	2	3
Cartoon picture—ultrasonic device removing plaque deposit	4	9	4
Digital device showing the "amount of plaque"	9	−3	7
Mode of action			
Dissolves plaque	2	3	−3
Combines with plaque to form inert compound	3	4	−1
Combines with plaque to loosen it	3	2	−3
Mechanical action against plaque	7	4	1
Builds a protective layer against plaque	5	−4	2
Immunizes the body to prevent plaque	−5	2	8
Applicator device			
Long handle with reservoir	4	2	−3
Liquid applied by brush	−2	5	−1
Chewing gum	6	−1	3
Picture of a toothpick	−5	−4	−5
Picture of a squeeze tube	3	2	−3
Ingredient			
Toothpaste-type surfactants	4	3	−4
Gritty material enrobed in a gelatin base	−4	−2	2
Effervescent materials that work on contact	6	4	4
Natural ingredients found in seaweed	−2	−4	6
Protective chemicals that coat the teeth	−1	3	5

restructure the concept (e.g., by inserting elements or by removing elements). The result of Step 7 is an improved set of concepts which appeals to consumers, which are unique, and most importantly, which fit the setup concept.

A key benefit of this early stage development procedure is that it can cut across categories. Neither the researcher nor the consumer creates concepts. Rather, both the researcher and the consumer react to new combinations, by

Table 15 "Optimal Concepts" Which Fit the Objectives:

Concept # 1—Mechanical action (blasting/abrasive action)

	Purchase	Fit end use	Unique
Additive constant	46	32	22
Picture of sandblaster	–3	8	8
Mechanical action against plaque	7	4	1
Long handle with reservoir	4	2	–3
Toothpaste-type surfactants	4	3	–4
Sum of the concept elements	58	49	24

Concept # 2—Digital readout device coupled with a chemical protective action

	Purchase	Fit end use	Unique
Additive constant	46	32	22
Digital device showing the "amount of plaque"	9	–3	7
Combines with plaque to form inert compound	3	4	–1
Liquid applied by brush	–2	5	–1
Effervescent materials that work on contact	6	4	4
Sum of the concept elements	62	42	31

End use presented as the following attribute:
A treatment that will allow me to use it once every 3 days, and not worry about plaque.

simply assessing the degree to which the combinations of elements appearing on the computer screen move the concept closer to or further from the objectives. Creativity now comes in the form of being able to *react* to a "gestalt," or a "whole," rather than having to create a "gestalt" anew.

On the Monkey Typing Shakespeare

Quite often those whose business it is to create concepts complain that it takes a supremely creative individual endowed with insight, wisdom and articulateness to create new products. A "mechanistic" approach, such as that described above, seems to lack "soul." What the approach lacks is not soul, however, but the subjective ego of the individual who proclaims his/her unique "creativity." The approach democratizes creativity, empowering consumers to guide the course of development. Consumers need only be able to recognize that some combinations fit a setup concept, and others do not. In effect, the creative process has been liberated and empowered. The computer technology provides the means to expose the consumers to a continual stream of stimuli. The wide range of elements from different product categories insures some synergism or at least lateral thinking. The experimental design insures that the stimuli are presented in the appropriate fashion, and guarantees that the results can be analyzed to create a model (and thus insures that there will, in fact, be a result). The computer presentation allows for words, pictures, video, music and voice, so that the concepts possess some additional reality beyond a simple word description, or beyond a slightly better line drawing plus product description. Finally, the consumer acts as the integrator. Yet, the consumer is called upon to do relatively little—simply to act as an instrument to essentially provide feedback. The direction for new product development is assured by the setup and "homework" done ahead of time.

6. An Overview

Creating products under a rapid timeline and with minimal research efforts is a recurring challenge. How then does the manufacturer maintain success in the face of these demands? Simply guessing about the product and concept may work, but it depends upon the ongoing "genius" (or at least insight) of the individuals in charge of the project. This dependence requires luck on a consistent basis. Any gambler knows that if the odds of success are modest for one bet, then the odds of ongoing success in a sequence of bets rapidly declines to 0. With luck alone as one's guide, the developer and marketer are bound to fail—perhaps not at first, but eventually the law of odds catches up. No one can succeed by luck over the long haul. Yet, the monetary and time cost of research can be significant. Although many marketers pay lip service to the utility of research in enhancing luck and increasing the odds of success, in a stressful situation it is tempting to forego research and "shoot from the hip."

The best option in this harried scenario is to create products and concepts in a rapid fashion, exploring multiple scenarios, concept options and product options. Experimental designs recommend themselves. These designs force efficient exploration of many concept ideas and product options, even though they leave a lot of holes in the "design space." Within the range of design the odds are high that there will be combinations which meet with high consumer acceptance. In the eventuality that there are no winning combinations, the experimental design will quickly point out the problem. Rather than leading to frustrating, back and forth iterations, the experimental design provides immediate answers and direction. The researcher has substituted multiple paths for one path, significantly increasing the chances of success.

References

Baxter, N.E. 1989. Research guidance: not giving it your best shot. *In:* Product Testing with Consumers for Research Guidance, L. Wu (ed.), STP 1035, pp. 10–22. American Society for Testing and Materials, Philadelphia.

Best, D. 1991. Designing new products from a market perspective. *In:* Food Product Development, E. Graf and I.S. Saguy (eds.), pp. 1-27. Van Nostrand Reinhold/AVI, New York.

Box, G.E.P., Hunter, J. and Hunter, S. 1978. Statistics for Experimenters. John Wiley & Sons, New York.

Carr, B.T. 1989. An integrated system for consumer-guided product development. *In:* Product Testing with Consumers for Research Guidance, L. Wu (ed.), STP 1035, pp. 41-53. American Society for Testing and Materials, Philadelphia.

Claycomb, W.W. and Sullivan, W.G. 1976. Use of response surface methodology to select a cutting tool to maximize profit. Journal of Industrial Engineering, *98*, 64-65.

Draper, N.R. and Smith, H. 1981. Applied Regression Analysis. John Wiley & Sons, New York.

Hlavacek D. and Finn, J.P. 1989. Hitting the target more frequently: a systematical approach to research guidance tests. *In:* Product Testing with Consumers for Research Guidance, L. Wu (ed.), STP 1035, pp. 5-11. American Society for Testing and Materials, Philadelphia.

Joglekar, A.M. and May, A.T. 1991. Product excellence through experimental design. *In:* Food Product Development, E. Graf and I.S. Saguy (eds.), pp. 211-230. Van Nostrand Reinhold/AVI, New York.

Moskowitz, H.R. 1989. Sensory segmentation and the simultaneous optimization products and concepts for development and marketing of new foods. *In:* Food Acceptability, D.M.H. Thomson (ed.), pp. 311-326. Elsevier Applied Science, Barking, Essex, U.K.

Plackett, R.L. and Burman, J.D. 1946. The design of optimum multifactorial experiments. Biometrika *33*, 305-325.

Wittink, D.R. and Cattin, P. 1989. Commercial use of conjoint analysis: an update. Journal of Marketing, *53*, 91-96.

9
Advice to a Young Researcher

1. The Importance of Historical Perspective and Future Vision

Too many researchers work in a vacuum, without a historical perspective or a vision of the future. Many researchers are content to spend their working days carrying out the details of an experiment in accordance with well-established scientific practices. The projects are carried out by rote, with a modicum of technical expertise, sometimes approaching executional perfection yet without "soul."

Probe for a moment beneath the research execution to the design of the study, to the reasons behind the design, the historical perspective in which the research has been designed, and the implications of the results for ongoing decisions. Many of those who execute perfectly do so in a rigid, unthinking manner. Much of their effort goes into process, not product. The execution of the task at hand becomes their main focus. They lose sight of the fact that research is commissioned to answer a specific problem, that there is a context of the research critical for the particulars of the study, and that there exists a broad historical framework which dictates what research should be done, why it should be done, and how it should be analyzed.

How then can the researcher avoid becoming a technician without vision? Or even better, how can the technician become a true researcher with a broader vision, rather than just an executor or manager of the project? Here are five suggestions:

1. Ask questions about every facet of the research execution. Question not just "how" but also "why." There are reasons for each aspect of a research design. Some are good reasons, rationally reached. Others are mere conventions about how the research is done, with little or no scientific rationale.

2. Ask questions about the research design. Why is the design the way it is? Could there be a better design? What are the faults of the design? For instance, the question could be asked about the base size. Why does the research require the base size that it does? Is it because of the representation (a large base size insures that panelists of different types will be represented)? Or is the base size large because that's simply the way that "everyone does this type of research"? Another question might concern the attributes. Why are the particular attributes chosen? Is it because there is a laundry list of terms that is commonly used for this type of research? Or is it because these attributes were selected by a small group of consumers in a pre-test as being relevant for the project? Why the scale? What particular scale is being used? Will the scale yield the necessary discrimination between the products or the concepts?

3. Ask questions about the heritage of the research. Every research design has a history behind it, in terms of its scientific basis. On what scientific groundwork is the research based? Although much of the research is applied, very little of what we called "applied research" was developed specifically to answer applied questions. Most product testing research has its history in psychophysics, opinion polling, fragrance evaluation, etc. What is the intellectual history of the research? What antecedents of the research appear in the scientific literature? To what types of research does the current project link back? Is there an intellectual history of this type of research in the academic community?

4. Ask questions about the statistical analysis of the research, and the substantive analysis. Too many researchers confuse the substantive interpretation of the data with the statistical treatment of the numbers. The statistical treatment simply tells the researcher how the data will be combed, massaged and analyzed. It does not tell the researcher anything about the conclusions to be drawn, the actions to be taken upon finding the results, and the implications of the results for different paths. For the researcher to truly understand the research he must probe deeply using data analysis. What exactly does each step of the analysis show, and what does it not show? The answers to these questions should be phrased in terms of the product and the consumer, not in terms of the statistical results. For tests of difference between two products it is not sufficient to say that the analysis will show whether two products are significantly different from each other. That is statistical talk. Rather, it is better to talk about the statistical test as showing whether or not the two products come from the same distribution with a common mean. The statistical talk is shorthand, but all too often it tends to replace the deeper, more profound understanding of the project by a simpler, glib, professionally acceptable statement.

Advice to a Young Researcher

5. Ask questions about the specific uses of the research. Where does the research lead? Will it lead to a simple yes or no answer? Will it build a framework of understanding around the product or concept to answer questions that may be asked later? Is the research part of a world view of the product? (Very often the design of a research project, like the brush strokes of a painter, tells a lot about the way the researcher views the world.) Is the research simply a conventional test or does it have an art of its own which can become the basis for a new way of testing? It is a pity that so much effort is wasted on standardized "bread-and-butter" research, and less effort expended on new ways of answering problems. Those new approaches could revolutionize the testing business, as well as bring increased understanding and profit to the manufacturer.

On Doing Research Within the Framework of a World View

Much of this book deals with experiments and analyses within the framework of a world view. A world view organizes one's perceptions of reality into a coherent whole. When applied to research projects, a world view is a way to approach the problem, to set up the necessary experiment, and to analyze the data.

Researchers, guided by an implicit or explicit world view, conduct research in a manner that can be discerned by either the statements of the problem or their research designs. Most researchers operate and design their experiments according to a viewpoint. It is rare for an experienced researcher with many years of experiments and publications to approach research without a world view, no matter how well-hidden that view may be.

The world view may be direct and apparent, or it may be indirect and hidden. A substantial part of the researcher's world view can be gleaned from the literature sources in published research, or from discussions with practitioners regarding the "why's" of applied, commercial, non-published research. Most researchers will point to certain previous research or general types of problems that they want to answer. Research does not spring from the mind of the investigator without having an antecedent framework from which to emerge.

We don't often think about a researcher's world view, even though most researchers operate with one. Why not? For one, there is more of an emphasis on speed today. There is less time to reflect on the results of the study when the study must be presented quickly to corporate management. The luxury of thinking which leads to greater wisdom and insight disappears in an era of speed. Consequently, research results are reduced to a single page, and often to a single paragraph. The research user, typically a marketing person not a scientist, is neither interested in the why's and wherefore's of research, nor interested in the niceties. What is important are the results. The method for getting to the results is left in the hands of the trusted researcher who acquires and analyzes the data. The last thing the harried marketing or management person needs is to attend to aspects of the research other than the basic results and implications for the problem at hand.

There is another reason, however. Our scientific and applied work has become far less general in nature, and far more refined, tightened, and specific. Experts abound in all areas. Experts perform the same defined, small-scale tasks well and repeatedly. Experts are not often challenged. After their training has finished, it is time for their education to begin. Unfortunately, these highly trained (but not necessarily educated) experts become immediately absorbed by the process of collecting and analyzing specific, small-scale data sets from tightly defined studies. The expertise of the researcher leads inevitably to that individual being tightly slotted into a research area where he or she is most effective. The researcher is essentially "parachuted in," fully formed, and expected to begin producing. There is no time to become educated, to acquire a world view. Education and vision have been defeated by training and expertise.

There is a third reason, as well. This is the age of ease, of data wealth, of analytic power—all tools which promised great breakthroughs but which led to myopic analyses. Whereas 30, 40 or 50 years ago data collection was difficult and analysis was tedious, today the researcher can acquire data quickly, analyze it just as quickly, and report the results. Too much data may be worse than too little. Data points are no longer precious. Years ago each data point had to be obtained with one's lifeblood (or so it seemed). Each point was so painfully acquired that there was a lot of up-front thinking in each study. The effort to get the data made the results ever more valuable. Thought took over in an attempt to reduce the effort in data collection and analysis. Today, data is so easy to obtain that the focus has shifted to managing the data acquisition process and the data itself, rather than to understanding it. Talk to any researcher and all too often the conversation is about the newest methods for automating and analyzing, not about thinking and interpreting.

A world view can compensate for a lot of the problems caused by the automation and the simplification of data acquisition. To some extent the world view prescribes the appropriate types of research. A world view eliminates much of the random acquisition of meaningless data. A world view provides a framework in which to design experiments, to acquire data, and to analyze it. A world view does not help creativity, but it does harness the random, aimless scientific wanderings.

2. Escaping the Illusions of the Past

Too many practitioners today bemoan the fact that opportunities for good research are simply not as available today as they were in the past. Many researchers carry around with them an idealized version of what the situation was 10,

20, 30, even 40 years ago. All too often these disillusioned people use the changed environment as an excuse not to proceed.

In applied as well as in basic science, it is vital to look towards the future. The past is the past. The good work that was done remains, but in truth much of the newer types of research are significantly better than the research conducted a decade or more ago. The practitioners are better educated, and there are now theoretical foundations and validations for research procedures that were lacking in the "good old days." Knowledge of statistical methods is more widespread. The advent of the personal computer has put advanced statistics into the lap of any practitioner willing to invest just a few hundred dollars in "off the shelf" software.

If things really are better today in terms of education and tools, why then do so many practitioners cry and bewail their fates, repeatedly claiming that the field of product testing has gotten worse? There are at least three different reasons:

1. There is greater competition of approaches. Researchers speak in cacophony. Many qualified individuals are now "out there" in the research community as consultants or "suppliers," each one proclaiming his wares more loudly than the next. What was a relaxed, almost gentlemanly approach to product testing in the years gone by seems to now be a free-for-all, inhabited by tough businessmen rather than by the ivory tower oriented academics who somehow (unbeknownst to themselves) found that they had landed in the world of practical application.
2. The audience of research users is more sophisticated than it was previously, but not so sophisticated that these research users can do things themselves. Many practitioners are trained, some are educated. Those who experienced the past know that when they were less sophisticated they had less to worry about. Now they know more. Many practitioners have an inkling that they do not know all that they should. This is anxiety provoking. A little knowledge is not only dangerous, it leads to a conscious retreat back into slumber, simply to reduce the anxiety.
3. People have an innate propensity to hearken back to an idyllic past. One need only talk to people and the phrase "good old days" keeps coming up. Whether the past was good, bad, filled with excitement or just plain boring, it's the past, finished, over with and thus safe. In product testing there was just as much trouble in research 10, 20, and 30 years ago as there is now. However, most researchers who are around to recount the tales of what has gone by feel secure with the trials and tribulations of that period. The demons that haunted research in those days are long since gone, dead and buried. They don't have the impact of the present.

3. Distinguishing Between Actionable and Inactionable Research

Introduction

Researchers conduct studies for a variety of end uses. Most researchers, however, believe that their research is fundamentally "actionable." By actionable we mean that the research leads directly to specific actions. In contrast, inactionable research is information that does not lead to action.

Actionability of research is not inherent in the research itself, but in the use of the research. What makes some research "actionable" and other research "inactionable?" Why does some research lead to specific direction and is used again and again, while other research languishes on the shelf and is never consulted? What is it about the research itself (rather than the ingenuity of the researcher)? Note that inactionability condemns the research. It means that the data from the research cannot be immediately acted upon. The data leads nowhere in terms of immediate utility.

This author believes strongly that actionable research has an obvious connection with factors under the researcher's control, whereas inactionable research has no such obvious connection. Researchers can control stimuli. Thus, actionable research tends to deal with products and concepts. Researchers cannot control attitudes, at least not directly. Attitude research, therefore, is basically inactionable research. Attitude research may tell the researcher a lot about the inner psychological workings of the consumer, but the research does not directly lead to a course of action in terms of modifying a concept or a product. Ultimately, attitude research may lead to changed concepts or products, or may lead to a changed program of marketing and communication. However, the specific actions used to change marketing and communication do not immediately emerge from the research. They may, however, emerge from one's intuitive analysis and integration of the research data.

Three Levels of Actionability

At the most fundamental level, actionable research deals with the effect of specific, controlled changes in the stimulus. For example, experimental variations of the stimulus in a systematic fashion generate patterns relating the stimulus variations and the consumer response. When the researcher uncovers these patterns, he has developed an actionable "road map" which tells him what to do to the product in order to achieve a desired consumer response (assuming, of course, that the stimulus has been sufficiently varied along the proper stimulus dimensions). This level of actionability requires relatively little intuition or understanding. The structure of the data set itself dictates the changes that must be made to the product. Actionability is great because the data is formally structured to map consumer responses (as dependent variables) to independent variables directly under the researcher's control. It is no wonder, then, that experimental designs of product variations lead to experiments and databases that are consulted for months and years after the study has been completed.

One step removed from this immediately actionable data are studies which impose a pattern on the stimuli. The stimuli themselves just remain different executions or variations of a product, not necessarily connected to each other on an *a priori* basis. However, the researcher can locate these stimuli on a variety of imposed dimensions (e.g., perceived color, perceived fragrance intensity). The underlying pattern can then be uncovered, but changes in the level of a stimulus on the imposed scale do not immediately translate to defined and specific changes in the physical stimulus. At this level of actionability, more emphasis must be placed on hypotheses about relations or patterns extant in the data. However, there still is a quantitative connection between the physical aspects of the stimulus and the dependent variable.

A further step removed is inactionable data, such as attitudes about products. The data do not connect factors under the researcher's control with consumer responses. Attitudes may not translate to stimulus levels. Any patterns which emerge are based almost entirely on intuition and insight, rather than on formal quantitative relations between stimulus properties (attitudes) and responses (consumer acceptance).

A similar distinction between actionable and inactionable data can be made for concept research as well. The most actionable data for concept work consists of experimentally designed variations in the concept. Less actionable data consists of evaluations of fixed concepts, where the effort is made to relate diagnostics of the concepts (strengths and weaknesses) to consumer reactions (e.g., purchase intent). The least actionable data comprises general attitudes toward a category. Attitudes themselves do not tell the researcher exactly how to create concepts. They suggest directions, but only in a general way.

Why Bother with the Distinction of Actionable and Inactionable?

In today's quickening business environment, research is critical. However, research must "pay out" in terms of what it delivers. Typically the pay out of inactionable research is long term, contingent on insight and interpretation. Inactionable research may eventually pay out, but all too often it does not lead to relevant action. In contrast, actionable research typically pays out, can be tested for validity, and enjoys immediate use. This distinction does not mean that inherently inactionable research should be abandoned. Rather, it means that the researcher should think through the research objectives and the degree to which the research can be directly applied to answer those objectives. The time is past when research for the sake of research is accepted in a business environment. Research must be linked with payout. An analysis of actionability can indicate the degree to which the payout can be realized within a reasonable period of time.

4. Distinguishing Between First and Second Order Problems

More than 2300 years ago, Aristotle stated that the virtuous approach to life consisted of aiming towards the "middle," where moderation replaces stinginess on the one hand and extravagance on the other. This dictum holds for research as well. Not every question needs to be researched, no matter how powerful the tools available to answer the question are. Some questions can be "armchaired," other questions are not even worth researching, some questions definitely require research, and still others are not researchable at all.

How does the researcher know what is worth being researched? There are several criteria to apply to a question. These criteria are not necessarily precise nor universal. They simply provide a way to distinguish that which is important from that which is trivial.

The key question is the following: Is this a first order problem, or a second order problem? By "first order problem" we mean a major problem upon which many issues hang that need to be resolved, which will impact the state of the business and, potentially, will guide decisions involving a great deal of money. First order problems worth investigating include the following three:

1. How well does my product perform? (This is always a major question. Poor performance can lead to major reformation or repositioning efforts.)
2. What are the key drivers of product acceptance in my category, and do my products have the right levels of these key drivers? [This is also a major question because poor performance of one's products on the key drivers means that there are competitors which do better and are potential threats to one's product.]
3. Are there clusters of product acceptors and rejecters in my category? Have I identified the appropriate groups of consumers in the category? (This is also a major question because there may be ways to divide consumers in a product category in order to identify new, unsatisfied niches where one has a chance to gain a competitive advantage.)

Let's now look at the same problems, phrased in a way to reduce them to second order problems. That is, the researcher can rephrase the question to make the answer significantly less powerful than it is as phrased above.

1. How well does my product do against the market leader (or some other defined product used as the standard of comparison)? Is my product better, the same, or worse? This is the typical question. The question is not

first order because poor performance against a market leader does not tell the researcher the absolute level of performance of the product. The market leader may only be the market leader because of high expenditures within the trade, and not because of any intrinsic merit of the product itself.
2. Do I have a "win" vs. competitors on any key attributes? If so, then perhaps that "win" is what makes my product better overall.
3. How well does my product perform in different markets? Although this is a standard question, the performance in different markets is difficult to integrate into a single overview. Most researchers, in fact, simply look at market-to-market performance to discover "exceptions" (products that perform well in most markets, and perhaps aberrantly poor in a single market). Even when confronted with these exceptions the researcher has a difficult time understanding precisely why the aberrant score was obtained. (Usually the aberrant data point is chalked up to statistical variability and the inherent variability between consumers.) By looking at market-to-market variations without an organizing principle for individual differences, the researcher converts a potentially valuable question and problem into a simple statistical breakout with relatively little value for insight.

It is hard to define exactly what will be a first order problem versus a second order problem. The definition or classification is primarily intuitive rather than fixed. However, if the researcher cannot do anything with the data except remark on a pattern or lack thereof, then the odds are that the question is second order. If, in contrast, the researcher can fit the results into an understanding of the product category, and if the results can be interpreted within this framework and lead to definitive action (e.g., reformulation, new product development, new niche marketing), then the odds are high that the question being asked is a first order one.

5. Too Little, Adequate, and Too Much Research: When Is There Enough?

Research is an intensely personal thing to many people. Some product developers and marketers abhor research, preferring decisions made on the basis of "intuition." In many cases the success of intuition is absolutely astounding, and leads to questioning of the value of research for making decisions. Unfortunately, we never hear about the marketer or product developer who used intuition and failed miserably. Somehow that failure doesn't seem to be as noteworthy as the success attributed to wisdom, insight and intuition! Other product developers and marketers go overboard in the opposite direction. They absolutely refuse to make any decision whatsoever without having a mountain of research to back them up. Part of this mountain comprises good research, necessary to make the decision. Other research appears to be completely superfluous.

How then can the researcher avoid spending too much money for unnecessary research, or not spending enough money to do the right research? Is there a proper amount of research that should be done, or is the research appropriate for any particular project to be left to the whims of the practitioner and biases of the product developer?

Research is not a commodity—we cannot measure its utility by the pounds of interviews or the heaviness of the report that results from the research project. There is no right or wrong amount of research to do. There is, however, a point in any research project which divides "not enough research" from "an adequate amount," and another point which divides "an adequate amount" from "too much." Each project involving research should be considered separately. There are some guidelines.

Key Guidelines for Amount and Type of Research

Objective 1—Understanding. If the objectives are to understand the dynamics of a product category before the manufacturer enters the category, then the more (proper) research done, the better. The more profoundly a manufacturer understands the dynamics of the category, the better off the subsequent development and marketing efforts will be when the company allocates resources to its entry.

The objective "to learn as much as possible" does not mean, however, that all research is equally valid. What is necessary here is research which leads to a broad view of the category, not research which focuses on a narrow view. It is far less instructive to test only the market leader with many consumers than to test the full array of products in the category, but with fewer consumers. (This topic is dealt with extensively in the chapter on category appraisal and benchmarking.) Unfortunately, most manufacturers fail to grasp the fact that in the early stages of development, an investment in research can pay back tremendously later on.

This early stage research can be summarized by the following nine points:

1. Oriented towards learning
2. Scope more important than precision of any one measure
3. Many products more instructive than a few products
4. Many attributes more instructive than a few attributes
5. Relations between products or between attributes as important as the score achieved by a given product
6. Research done in the early stages of a project, before many decisions are made

7. The information obtained from the research will be used to guide additional work. The data becomes a "bible" about the product.
8. The data is analyzed in various exploratory fashions by researchers searching for patterns in the data which they hope exist. The research is thus characterized by uncertainty, rather than certainty.
9. Typically the data is sufficiently comprehensive to maintain its value for years to come, after the research has been completed. Often the researcher (or even his successor in the company) uses the data to assess the validity of new hypotheses not conceived when the research was originally commissioned.

Objective 2—Confirmation of a Decision or Point of View. This research is usually conducted when the project is far along and many decisions have already been made. (Some of these decisions may, in fact, be based upon previous research conducted by the manufacturer.) The manufacturer already knows a lot about the product, and conducts the research with another objective in mind. The objectives may be to confirm one's initial decision (e.g., select prototype A instead of prototype B for further research; test the current product versus the competitor for advertising claims, etc.). This research is not conducted for broad learning.

This research may be summarized by the following five points:

1. Precision is absolutely vital. The validity of the research is a function both of what is measured and the precision with which it is measured.
2. The research is undertaken with an ulterior motive in mind, and not for the sake of knowledge or guidance per se. New knowledge may emerge, but this new knowledge is usually more an unwanted "gift" (or embarrassment) than a welcome finding.
3. The focus of the research is tightly limited, usually dictated by the specific end use of the data.
4. The analysis of the data is typically dictated by standard, well-respected and non-creative statistical methods. The data will be used by third parties, who will judge the research by means of conventional and strict measures of "correctness" or "validity."
5. The data is typically "time-limited." The research is not done with a global point of view. The research is conducted to address a specific point, and typically only that point. Often, as soon as the point has been answered, the research usually loses its further utility. (The utility of the research lies in the use of the answer it provides or the position or point of view that it supports, rather than in the merits of the work itself.) There is rarely a subsequent consultation of the research to answer other questions. Most parties involved in this type of research know that the project was commissioned for a limited problem or to answer a specific, focused question.

6. For the Novice Researcher: How to Follow the Vision, Despite Reality

Most researchers in industry are forced to conduct their research under less than ideal circumstances, especially with regard to new ideas and approaches. As the pressure of business increases and the funding for research decreases, the researcher must make compromises. Although the novice researcher, newly minted from graduate school, may enter the field fired up with ideas and desires to do a better job and to open up new vistas, all too often the reality of the situation intrudes and forces the novice to subdue his vision and excitement. This loss of vision is applauded by more seasoned colleagues who label it "maturation" and "recognition of reality."

In product and concept testing there is a desperate need for new ideas, new approaches, and the sustained push which expands the "envelope" of current technology. In both academic and corporate research there is far too little risk taking. The novice researcher coming into an organization, filled with the excitement of approaches developed and refined in the university, is often punished for any research other than maintenance research (e.g., paired comparison testing of current versus new products). Maintenance research always comes first because it is the "bread and butter" of the testing group. Through this service research, the group keeps its funding.

For the science of product and concept testing to survive and grow, there must be a different approach to testing. Time is past when the bread-and-butter research alone sufficed, and low-cost data acquisition was the ultimate goal. Although paired comparison (and other similar) research addresses many commercial problems, it cannot move the company ahead. The researcher must try new procedures, even if they are risky and are not hallowed by 30 years of tried and proven history.

All of the above plaints underscore the need for the researcher to have a vision about the research, and not to embrace techniques blindly and obediently, just because the current political climate in the company dictates the use of "tried and true" methods. For the researcher to make a true contribution he must go beyond the traditional methods, stretch himself intellectually, and think laterally. It is vital to try new methods, to experiment, and even to make mistakes, all the time growing and maturing until one's research really begins to make a difference, rather than just being the well-executed work of a technician.

Putting the Vision into Practice—A Strategy

It is easy to talk about a vision and just as easy to prescribe doing the right thing. But how does the novice in a company put this vision into action, especially when there is no funding and when all the novice's efforts are carefully scrutinized by superiors, many of whom are "burnt out" and bitter from years of intellectual decay? The novice cannot simply ask his superior to fund what may appear to be "blue sky" research, especially during a time when research is being cut back. How can we prevent "burnout" and "rust-out"?

The best way to proceed is to start very small. By starting small we mean that the researcher might put into a test some additional attributes or products beyond those that are currently being evaluated. The objective here is to obtain the necessary (albeit less than complete) data with which to do an analysis. For instance, in an evaluation of two products the adventurous researcher may test four products instead of two (perhaps with a lower base size for each product to keep the cost down), and have consumers evaluate the products on a large series of attributes. The data that emerges comprises ratings of four products on 8–15 attributes each. (Ideally the same panelists would evaluate all four products.)

From this supremely muted experiment can emerge a superb analysis. For example, the researcher could then do the following two analyses which add value to the data, and which demonstrate the utility of new ways of thinking:

1. *What sensory inputs are important in decision making?* Here the researcher relates overall liking to attribute liking by a linear function, and compares the slopes of the linear function across different attribute liking (e.g., liking of appearance, liking of fragrance, liking of feel). The slope identifies which sensory input gives the "greatest bang for the buck."
2. *What sensory attributes drive overall liking?* Here the researcher relates overall liking to sensory attributes by a quadratic function. The results show how to change products to improve acceptance.

Both of these analyses generate insights that lead to the acceptance of new research procedures. Once the researcher demonstrates what can be accomplished by evaluating more than just one or two products, and can present the results succinctly and cogently, management will sit up and take notice. Management rarely will accept hypothetical cases with the same openness that it accepts actual data.

7. Presenting Proposals, Data, Results, and Interpretations

In any researcher's career he will be called upon numerous times to present proposals for research projects, to share data from experiments with colleagues and clients, to present final results, and to offer interpretations of the results. It seems so simple to lay out these responsibilities that we forget that the young researcher often enters the world of applied research without the skills or experience necessary to make the public presentation persuasive and successful. Even the most compelling data, no matter how well-supported statistically, will fail to convince the client of its validity unless the researcher can present the results in the appropriate format.

The Range of Written Presentation Approaches

Presentations of data on paper range from the elegant, simplistic, and compelling to a document that resembles a computer printout of statistics. Researchers have many individual styles of presenting their results. Some researchers prefer to skim the data, present an overview, and leave the details for an appendix. In the appendix the presentation can go into excruciating detail, with tables pulled to present the data in a variety of formats, with different comparisons, different statistical tests, etc. Other researchers prefer to present the data in logical, well-thought out but very detailed discussions, each discussion supported by several substantiating tables.

Which method is more correct, or, more realistically, which method is more effective? There really is no answer, although simplicity tends to win out over completeness and complexity. Whenever a researcher writes a report, it is first vital to anticipate the audience. How much information does the reader want to take on the journey of understanding? Are all the tables truly necessary? Most readers have a limited span of attention. They need to focus on the contents and implications of the results. An overly detailed presentation of the data may confuse, and may detour the reader away from the major point that the research seeks to show.

Knowledge of the audience is important from another standpoint—namely the total amount of detail that is needed. Management often prefers to look at the trends—the so-called "big picture." Management needs to be convinced of the validity of the data, but this convincing is more to demonstrate the overall competence of the researcher (and thus the research), rather than to demonstrate the validity of each point. Product development, however, may need the fine details—specifically, the performance of various products, the specific trends and other detailed information that will guide them to make a better product. Finally, the statistician who is interested in the *process* of data analysis may need all of the tables, including raw data, detailed computer printouts of the various statistical tests, and data summaries.

Product testers must be very cognizant of these different needs. A report designed for top management may be perceived by a statistician as sparse and superficial. In contrast, a report designed for the statistician and for others who want to dig deeply into the data may intimidate or just plain bore management, who will shrug off the report in quick order simply because it looks too formidable.

8. Introducing a New Test Strategy

Introduction

Marketers live in a different world from developers and researchers. The marketer's job is to "sell" the product to the consumer. The developer's job is to create the best product. The researcher's job is to discover and report the competitive pros and cons of the product.

Most developers and researchers have been educated with the sense that their mission is to search for the truth, and to develop the best possible products. This is the "gold standard," inculcated in the aspiring mind of the young student and maintained by professional organizations the world over. Most developers and researchers want to maintain the knowledge and skills that they have developed, and often supplement their day-to-day jobs with courses, additional reading, and informed "experiments" performed in their free time.

The Business Environment—A World Apart

Business differs from science. Although many business professionals want to produce the best possible products, and publicly champion doing so by scientific procedures, the business community recognizes one primary measure of success. That measure is profit. It is an unfortunate outcome, however, that sales and profit may not necessarily correlate with good scientific research and development, at least in the short run (when profitability is usually measured).

The Role of Product Testing in a Business Environment

This book stresses business-related product and concept development. The major emphasis is on procedures that move the research project along from inception to definite outcomes. All of the procedures detailed in this book are actionable, and lead to testable products and concepts.

Yet for many companies, even the procedures outlined here are too elaborate, despite their track records and straightforward ease of implementation. For many companies, the cost of any testing is perceived to be too expensive. Many companies make decisions simply on the basis of the director's intuitive or "gut" feel. Many corporate managements give lip service to the idea of scientific product development and testing, as long as those scientific procedures do not interfere with corporate day-to-day activities, and do not in the least affect the scheduling of product shipment to the market.

Faced with the reality of management that says it wants to test, but in fact throws up roadblocks, what should one do? How can the novice product developer or researcher ever hope to use the more sophisticated, perhaps more expensive techniques to create and test products? There are so many obstacles to overcome.

The Concept of "Adequate" Technology

One of the best strategies for introducing new approaches to a time- and cost-pressured management consists of implementing only one part of the strategy at any one time. The developer need not execute all of the relevant tests at the start. It pays to begin small, with a project or demonstration that is adequate for the task, but not too advanced. Three benefits of this "adequacy" strategy are:

1. The strategy of testing solves an immediate problem.
2. The strategy costs relatively little for what it delivers.
3. The strategy yields a visible and demonstrable answer. This is "point-at-able." The researcher or product tester can point to the solution of the problem. The very concreteness of the result gives management faith to go forward on the next, and perhaps more expanded, project.

A Lesson from Science—Pilot Studies

The adequacy strategy advocated here is an example of a "pilot study." Pilot studies are commonly used by scientists to determine, in a low-cost, low-risk fashion, whether or not a larger-scale study will actually work. Few scientific researchers venture to spend large amounts of money unless they have conducted a pilot study, have shown that the method or equipment works, and have demonstrated that the results are promising.

In the business world the adequacy study is the counterpart of the pilot study. An adequacy study demonstrates that the test works. It gives management confidence to move forward.

The business sequence differs, however, from the scientific application. The scientist running a pilot study is truly interested in the outcome of the test, and uses the pilot study to determine whether or not the study is worth pursuing. The pilot study is done at the very beginning of a research project to determine whether or not there is "pay dirt" out there. In contrast, in business the adequacy study comes later. The procedure has already been validated. The approach is known to work. What is necessary is to *convince* other people that the method will work. In a sense the adequacy study consists of putting one's best foot forward, underpromising what the approach can deliver, delivering (or even overdelivering) on results, and then hoping for a more receptive and convinced audience later on, based upon the success.

9. Statistics and Statisticians—The Old and New Roles

Product testers rely upon statistical analyses for insight and guidance. In the "good old days" (1960's and earlier), many researchers who used statistics did so to answer a simple question— "Is Product A significantly preferred to Product B?" The level of statistical knowledge was modest at best, and most of the use of statistics involved hypothesis testing.

Advice to a Young Researcher

The Traditional Role of the Statistician

In many companies the statistician is the corporate guardian of truth. In the more enlightened companies the statistician acts as a catalyst for advanced thinking. Companies fortunate enough to hire these forward-looking people make signal advances in their research. The statistician becomes the impetus for improved development of test methods, and is a valued member of the team. The open-minded statistician can deal with the ambiguities ever-present in research, and can make do with the compromises which forever plague any advance in technology.

There is another role that the statistician plays in less fortunate companies. Those companies use the statistician to suppress innovative thinking. All research designs must pass the statistician, who blesses the designs that pass "statistical muster." Novel designs that do things differently, that break rules, that explore new areas, are often rejected because the designs do not conform to the "statistical truth" that is being promoted by the statistician in charge. As a consequence, the statistician becomes the gatekeeper of knowledge, safely ensconced in the fortress of the traditional, virtually unassailable. Furthermore, this most conservative of individuals is safe in the fortress because the conservative statistician can hide behind a mass of formulas and statistical language which baffles all but the most intrepid scientists and product developers. Unfortunately for industry, these conservative-minded individuals do not change, because it is in the nature of the company to apply checks and balances to research and development, no matter how those checks and balances impede advancement.

The New Role of the Statistician

Today, statistics has taken on a far more significant and vital role in research. However, statistics has also changed. It is no longer the dry testing of hypotheses about products or concepts, which occupies many statisticians. Certainly researchers still test hypotheses with statistics, but there are substantially more applications, many of them involving exploratory data analysis and representation of data. These applications lead to insight, not just to verification of conclusions. As a result, statistics has been divided into at least two distinct branches:

1. Tests of hypotheses to confirm or deny decisions
2. Methods for representing data (e.g., multidimensional scaling, clustering, etc.)

The rapid advances in statistical (or better quantitative) thinking make it imperative that every researcher and statistician become familiar with new numerical methods and analysis procedures. The analytic techniques in statistics can lead to insights. Two examples are:

1. The representation of different products in a geometrical space (mapping). Although the statistician has the requisite tools to do the mapping, it is the product manager or developer who truly understands the implications shown by the map. Unfortunately, the product manager or researcher does not make full use of this information.
2. The clustering of different consumers into categories based on sensory preference patterns. The marketing manager can use this to plan products. However, most marketing managers are not familiar with clustering techniques. They may have encountered the techniques in graduate school, but do not have experience with them.

The new role of the statistician should be to promote the widespread use and understanding of these techniques among practitioners, rather than to hoard the techniques so that only he releases the information. Control is not productive. The statistician will not become any less valuable because of widespread knowledge of statistical procedures. Just the opposite. The statistician will be in the vanguard of the popularization of technology, not its sole keeper. Those statisticians who hold the application of quantitative technology close to their chest are not productive. The popularization of statistical techniques, especially those that lead to insight, will inevitably nullify attempts to maintain the facade of "purity" in the face of progress and application.

Using Statistics Properly

Statistics can be the researcher's friend but also the researcher's greatest foe. Properly used, statistics reveal to the researcher patterns in the data. Statistics can identify the existence and nature of clusters of similar-minded consumers, the pattern relating one sensory attribute to liking, or the pattern relating ingredients to sensory and liking ratings. But, when improperly used, statistics can hinder the thinking process. The precise quantitative nature of statistics, the speed with which it uncovers patterns, and the ease and availability of statistical packages can seduce the researcher in the process of data analysis. One consequence is that the researcher turns a blind eye to the ultimate product of data analysis, insight and conclusions. One begins to focus on the externals—data tables, statistical differences, regression coefficients.

What then should the researcher do? It is infeasible and downright silly to abandon statistics, such as fitting models, and go back to visual methods of fitting curves to data (the method of "ocular trauma"). The "knee jerk" Luddite response is to abandon tables of T tests in favor of visual inspection of differences and the pronouncement of differences simply on the basis of size of difference, without reference to an underlying distribution. We cannot recapture a statistics-free past, where beauty and truth somehow emerged from the intense cogitation of data, sans computer, sans data table, sans all methods of subsequent statistical analysis. Progress isn't made by reverting to a more primitive history.

To avoid getting fooled by statistics, the researcher might try the following 7-step exercise. It is known as a "thought experiment." Properly conducted, it will teach the researcher a great deal. The statistical input for this thought experiment will then be appropriate, and not a block to further thinking:

1. State the hypothesis clearly.
2. Set up the experiment on paper, and assume that you have run the experiment (but do not run the experiment yet).
3. Lay out a table of simulated raw data.
4. Draw the conclusions, and write up the results.
5. Return to step 3, and insert other data, and then draw new conclusions.
6. Continue Steps 3 and 4 until you feel that you understand the organic connection between what is done in the experiment and the types of data that can emerge, and the conclusions that would be drawn from the results. The conclusions should be stated in terms of what occurs with the product or concept, given the results of the "thought experiment." The conclusions should not be stated in terms of any statistical results whatsoever. (The statistics are only there to identify patterns, or to lead to stronger versus weaker conclusions.)
7. Once you profoundly "feel" (or intuit) that organic connection between the actions taken to vary the stimulus in the thought experiment and the results, and can write up scenarios that seem natural, then use the statistical packages to develop equations, test for significant differences, or put people into clusters. At this point the statistical package becomes a tool to use, rather than a device to suppress thought. You have done the requisite thinking in the "thought experiment." Like the collection of data, the statistical analysis flows automatically. It simply puts flesh onto a skeleton that has already been well-designed and clearly thought out.

10. Translating Numbers to Conclusions

Marketers, researchers and almost anyone who uses numbers feels comfortable when numbers are accompanied by some standards. These standards are themselves not typically numerical, but rather verbal. Indeed, if we were to look at the purpose of most research that is analyzed quantitatively, we will find that the numbers themselves are rarely used to support a numerical conclusion. Rather, the numbers are used to support a qualitative conclusion.

Numbers by themselves have no meaning. They are simply convenient ways to quantify some aspect of reality. Their real meaning, however, comes from comparing these numbers to norms. The norms are typically levels of numbers that correspond to some verbal classification (e.g., a score of 70 on a 0–100 point scale has far more meaning when the number corresponds to a "category," such as "this is a very good product").

Norms and verbal accompaniments are critical. Most young researchers who are quantitatively literate all too often lose themselves in the numbers that they generate. These young, avid researchers perform dozens of tests, analyzing the data in many different ways, trying to extract patterns from the data. The results of these analyses are not particularly valuable if presented in terms of a statistical chart or wall of numbers. They must be distilled down to verbal conclusions and implications.

This reductionist approach may disturb the purist novice who has spent months, and perhaps even years, learning about statistical procedures and "boning up" on all the modern methods. However, despite such knowledge and education, it is still vital to transcend the specifics of statistical analysis and present the interpretation and the conclusion. Statistics by themselves are dry. They are necessary to support a point. But they are not the answer. Decisions are not made on the basis of pure numbers alone. They are made on the basis of one's interpretation of the numbers. Numbers themselves are not the decision makers. It is the mind behind them.

11. Searching for Patterns vs. Searching for Statistical Significance

Many researchers are brought up with the idea of searching for statistically significant differences in their studies. Their goal is to show that two products differ from each other. The "knowledge" that emerges from this type of test is that two means differ. All too often, however, when questioned about the implication of this difference, the researcher does not quite understand the question. What is the more profound and fundamental implication of a statistically significant difference? Is it important to know that two products differ significantly from each other? Of what value is that particular piece of information?

The important thing to keep in mind is that the researcher is looking for patterns in the data that can be used to make decisions. Today, it is so easy to test for significant differences on a computer that all too often researchers lose sight of the meaning behind the questions they ask. If two products differ from each other, this implies that one product has an advantage not enjoyed by the other product. Therefore there may be a business problem.

1. If the product with the higher score is a competitor, and the attribute is "overall liking," then perhaps the competitor may claim superiority and erode one's own franchise by attracting one's users. In this case, "statis-

tical significance" means a threat in terms of market share or total number of cases of product to be sold.
2. If the product with the higher score is one's own product, then perhaps the product costs too much. If the product is truly very superior to competition (as assessed by ratings of liking), then perhaps the manufacturer can significantly cut the cost of goods, but diminish liking by only a small amount. In this case "statistical significance" means a threat in terms of profitability. Perhaps market advantage is being purchased at too high a price.

How to Search for Patterns When One's Education (or at Least Experience) Has Concentrated on Difference Testing

Given the nature of one's education, how can the young researcher expand his scope? A glib answer would be to take courses in experimental psychology or learn more about the foundations of statistics. With time pressures facing everyone today, however, those steps are more easily prescribed than followed. No one has time to read, and few people possess the technical background for an advanced study of statistics and statistical theory. Yet, unless the researcher expends the energy to develop and expand his view of the world, the odds are high that the researcher will eventually fall back into doing what is convenient—paired comparison testing.

But what of statistical significance testing? Should it be abandoned in favor of a more global viewpoint? Absolutely not. Statistical significance testing, in and of itself, is a perfectly valid tool in the arsenal of the researcher. To test for differences, and to assure oneself that the differences measured do not arise by chance alone, is a meaningful endeavor. What is dangerous, however, is the negation of thinking by statistical testing. If researchers stop thinking about and searching for patterns in nature, and simply look for one answer (e.g., is the difference "statistically significant"?), then they are in trouble. Statistical significance simply becomes a "buzzword," repeated without thinking or comprehension.

One way to get out of the rut is to plot the data to discover patterns. Plotting the data (either manually or on computer) lays out the points in a geometrical way. Intuitively, the researcher must seek patterns, or else the data only looks like a scattergram. It is inherent in human nature to search for an organizing principle. Plotting focuses attention on an organizing principle. Gestalt psychologists recognized this a half century ago or more. Give people a set of points on a piece of paper and they will attempt to fill in the missing space, to create a meaningful pattern. By laying out the data points on paper, the researcher forces himself to apply an organizing principle to otherwise random points. That exercise alone is worth all of the effort, for it brings to bear on the problem a level of concentration and insight not otherwise attainable.

Patterns in Time—The Long View

Researchers who look at market trends often use time series statistics which look for patterns over time. For instance, questions asked deal with a product's changing share of the market over time (which can be correlated with advertising), or the seasonal preference for a product (which can be used with the trade to identify promotional opportunities for a product). Most product and concept testers do not make use of the time series approach, however. That is, all too often the researcher schedules product or concept tests on an ad hoc basis, to answer specific, targeted problems without ever thinking of the potential learning that can be gained by testing the same product (or concept) again and again, year after year in a disciplined fashion.

One of the most interesting problems in consumer testing is the acceptance and rejection of sensory characteristics of products. In the fragrance category, for example, some fragrance qualities are "in" and others are "out." The market can change suddenly, however, so that the formerly less acceptable fragrance now becomes more liked. To find out what causes these changes is a major research project in and of itself. But to measure the pattern of changes over time for a limited number of different fragrance types is an altogether different and easier-to-answer proposition. One need only test the same fragrances season after season, year after year, in order to discern the pattern. A fragrance house could make up these basic fragrances to simulate the different types of fragrances generally available. (The same fragrance composition would be tested year after year, perhaps three times per year with a small group of panelists.) The result of this testing would reveal consumer acceptance of each of the basic fragrances over a 3-year period.

The results of this analysis can be very instructive—and indeed far more enlightening than conventional cross section analyses. For instance, one fragrance type might appear to be losing its popularity. In contrast, another fragrance type might appear to be holding its own, whereas a third fragrance type might appear to be gaining in acceptance. The results can also be "de-trended" by season to show whether there exists a seasonal preference.

It is truly unfortunate that management and researchers fail to recognize the value of long-term studies, albeit with smaller consumer base sizes. Too much of the research today is done reflexively, in a disorganized fashion, to answer momentary questions that come up because of management's reaction to category dynamics (e.g., a new competitor has entered the category; a leading competitor has reformulated, etc.). Longer term perspectives in product (and concept) research can identify emerging, relevant trends in the category. Discovering trends over time will make product development and marketing much more focused and proactive, instead of diffuse and reactive.

Advice to a Young Researcher

12. On the Cocooning of R&D

If we look around companies, focusing on the corporate function for consumer data and analysis, then we will discover two different structures. One structure comprises the R&D group which runs its own studies, and in general, attempts to collect consumer data at the lowest cost. In effect, many product developers and researchers become the "least cost supplier." The other structure comprises the marketing and market research groups. These groups contract out their studies, paying for an outside expert to analyze the data and report the results. In effect, the market researcher becomes a contracting agent for data acquisition and analysis, rather than retaining his identity as a true "researcher."

These two approaches to research are founded in quite different world views. They lead to radically different results. We will compare the two approaches and their consequences in the section below.

Technical Sophistication

In most cases the R&D product developer and sensory analyst (or consumer researcher) is technically quite sophisticated. Indeed, many product researchers working in R&D know a considerable amount about the product, what the product should accomplish, and how to translate consumer data into formulations. Consequently, the R&D group working with consumer data can make the data actionable for developers. Because the R&D facility is often small and their employees know each other, this technical sophistication quickly enhances the value of research.

In contrast, market researchers are usually less sophisticated. Many market researchers have never been in close proximity to product developers for long periods of time, and thus are often unfamiliar with the technical nuances of the products that they research. Consequently, the data acquired by many (albeit not all) market researchers is less immediately actionable than the data acquired by the product development group (e.g., in-house sensory analysts).

The Consequence of Being a Low Cost Supplier

Typically R&D and market researchers differ dramatically in this regard. R&D often attempts to be an "island unto itself," self-sufficient and able to do all of the product evaluations "in-house." R&D often runs its own studies, perhaps relying upon an outside field service to recruit the consumers who are then interviewed by the R&D researcher. Once the researcher has acquired the data, R&D analysts massage the numbers, search for patterns, and then report the results.

By the very nature of their studies the R&D product evaluations are typically much smaller than market research studies. R&D product evaluations typically occur in the early part of the research program when guidance data is needed. The budgets are smaller for these guidance studies. R&D researchers often point with pride to the fact that they are the "low cost suppliers" because they can turn around these evaluation tests far more cheaply than could be done by contracting the same job to outside suppliers. In the long run, therefore, when it comes to product evaluation, R&D keeps the project "in-house," preferring to field the studies virtually directly (without any middleman), using "cost" as a justification for doing so.

Let us contrast R&D's approach to the way market researchers do business. Technically unsophisticated, the market researcher, in charge of a medium to large budget, does not conduct the research "in-house." More often than not the market researcher contracts with an outside, full-service agency to design the project, perform the evaluations (among consumers), and then analyze and report the results. This means that market researchers really act as purchasing agents for data acquisition, statistical analysis and interpretation. In fairness to the market researcher, some of the data does not need technical sophistication. A great deal of market research data consists of attitudes towards a category, of tracking studies (to find out what people have purchased in the past few months), advertising testing, and even simplistic product tests (e.g., to determine the degree to which one product is better than, equal to, or worse than another in the consumer's opinion). There are many studies, however, which lie more properly in the purview of R&D research because they deal with consumer perception of products, or are designed for product development. It is simply an accident of corporate politics and corporate structure that the less technically sophisticated market researcher conducts these studies, rather than the more technically sophisticated product developer.

Market research studies are typically larger, more involved, and generally endowed with a greater budget than are R&D studies. Furthermore, market researchers use outside consultants or "suppliers" to analyze the data. Therefore, depending upon the ability of the outside consultant, the design, execution, and analysis of a product test by market research can vary from poor to superb. There are many cases where the higher budget of market research fails to yield any benefit in terms of improved data quality and better guidance. But, there are just as many cases where the budget allows the market researcher to hire the best brains, so that the project becomes far more valuable.

Consequences of Budgetary Constraints on Quality of Research

A generous budget becomes an investment in knowledge. A budget enables the researcher to convert dollars to results. We can see this relation in action by looking at the outcome of market research versus R&D studies. With its low budget, R&D often misses the chance to be fertilized with new ideas. There is the perennial constraint of trying to do too much with too little money. Consequently, creative energies focus on obtaining more data for fewer dollars. The effort concentrates on the production end of the data—how to get more for less. Simultaneously there is the recognition that with the limited budget the scope of the analysis is going to be shortened. Why? Simply because there is no one to do a detailed analysis. There is precious little incentive to delve into the data, find opportunities, and then spend time thinking about the implications. Since the budgets are low for R&D, and since there are so many projects to handle, R&D product evaluators have their hands full simply coping with the day-to-day requirements of the project. All too often the quality of the analysis slips, and degenerates into a quick glance at the numbers. R&D's low budget militates against using outside suppliers to design and analyze results. Consequently, over the years, the R&D studies often stagnate and become routine. The same individuals in R&D work on the projects, year after year, without a chance of being exposed to and implementing new points of view. Even the newly minted expert who joins the company full of ideas is soon relegated to working overtime to keep up with the job demands, but without budget, and therefore without support to pursue new ideas.

Let us turn to the case of marketing research. Those with no budget constraints are free to hire the best brains they can to do the job. Consequently, those researchers are continually challenged to improve what they do. Dollars become a means by which researchers convert corporate needs to opportunities. These fortunate individuals "leverage" their research dollars. Furthermore, because they deal with outside "suppliers" (vendors, consultants, whatever the term may be), it is to the advantage of the outside supplier to introduce new ideas, to create and validate new forms of research, and to "push the envelope" and try new methods. By those efforts the outside supplier assures himself of continued business, and of an ongoing advantage versus the competition. The client company often benefits handily from this state of affairs, because the need to convert knowledge to opportunity is served quite well. Certainly some of the new techniques will fail, but in the vast majority of cases the supplier's efforts will improve over time, as invalid methods are tried and discarded, and the truly valid and productive methods retained. We see this quite well in the new business attitudes of the 1990's, where there is great emphasis on "partnering" with suppliers, and allowing these suppliers the budget and latitude to help, rather than treating the suppliers as adversaries.

Index

Abrasive gel, case history, relating pairs of profiles to each other, 406
Actionability, different levels of, 464
Actionability of research, comparison to inactionability, 464
Additive model
   goodness-of-fit, 32
   micro-model, 123
   standard error of regression, 36
Adequate technology, 473
Advertising concepts, 50
Age segmentation, 322
Alexis Inc., case history, product optimization, 281
All-or-none responses, as attributes, 164
American Personals Company, 241
Amount of research, distinguishing between too much, just right, and too little, 467
Annoyance ratings, as measures of attitudinal importance, 366
Anti-perspirants:
   benchmarking study, 81

[Anti-perspirants]
   content analysis of, in-market products, 83
   importance of sensory inputs for, 328
Anti-perspirant fragrances:
   attribute list, 331
   factor analysis of attributes, 330
Arthur D. Little, 3
Artwork, use in concept testing, 25
ASTM, committee E-18, 4
Asymptotic relation, sensory attribute function, 258
Attribute importance, graphical display, for sensory attributes, 145
Attribute liking, in category benchmarking, 105
Attribute list, anti-perspirant fragrances, 331
Attribute order
   following temporal sequence of attribute, 163
   preference justification, 163
   recommended, 188

Attribute profile, corresponding to optimum product, 267
Attributes:
  appropriate number in a questionnaire, 162
  appropriate order in a questionnaire, 162
  for bath gels, 141
  developing a list, 177
  how many questions in a questionnaire, 161
  laundry list approach to using in questionnaire, 162
  pictures as, 181
  rotating order of, 187
  scaled using line scale, 164

Baby Wipes, case history, 74
Base size:
  appropriate, for a study, 190
  in product optimization studies, 249
  questions about and rationale for, 460
Bath gel, optimizing liking, 149
Bath gels, category benchmarking, 140
Baumann Inc., soap concepts, case history, 18
Benchmark products, in early stage product development, 439
Benchmarking:
  appropriate number of panelists, 87
  testing a product category, 80
Benchmarks, use in concept testing, 26
Benefit matrix, reversing and clustering, 138
Benefit space, discovery of holes, 135
Benefits, discovering unfulfilled benefits in a category, 134
Bias, preference justification, 189
Biases, in order of attribute, rotation to avoid, 187
Bipolar scale, 90
Blind vs. branded products, individual differences in, 339
Blind vs. identified testing, benchmarking study, 82
Blotters, use in fragrance testing, 350
Brand, vs. product, importance of, 337

Bryson Oral Care, case history, 449
Business vs. academia and science, 473

Case history:
  bath gels, 140
  Bryson Oral Care, 449
  fragrances, search for sensory preference segments, 350
  Franck Company, oral protective compound, 426
  lipsticks, TURF analysis, 383
  sanitary napkin, 241
  shampoos, product optimization, 280
  shaving cream, to create an optimal line, 377
  Siragusa Napkin Company, 298
  Wood Inc., fragrance, 416
Categories, as part of concepts, 22
Category benchmarking:
  comparing macro- and micro-models, 116
  8-step analysis, 92
  face and body soap, 108
  identifying drivers of acceptance, 104
  mapping, 100
  overview, 156
Category change:
  and norms, 203
  norms in the face of, 203
Category drivers, in category benchmarking, 104
Category labels, for attribute scale, 164
Category model:
  discrete features, 7-step procedure, 114
  using content analysis to develop inputs for, 114
  using discrete features, 113
Category norms, in category benchmarking, 106
Central composite design:
  example for sanitary napkin, 224, 226, 245
City block distance, in developing metrics of differences, 296
Classification, in psychophysics, 288

Index

Clustering:
  in benchmarking study, to reduce
    number of options, 84
  to find holes in a category, 133
  of products in a factor space, 103
  use in sensory preference segmentation,
    345
Clustering methods, use in concept
  response segmentation, 48
Cocooning, of R&D, 481
Cognitive worlds, individual differences,
  330
Communication attribute, in concept, 31
Companies, how they do concept testing,
  15
Compensation, for panelists, 173
Competitive frame, as part of product
  optimization study, 225
Concept, multi-national evaluation, 182
Concept development:
  in early stage, rapid product
    development, 427
  eight steps for design, optimization, 19
  incorporating specific knowledge, 428
  package design, 73
  stages, 19
  use of video, 73
Concept elements, examples of close-in
  vs. far-out, 450
Concept model:
  created by regression analysis, 31
  use in creating a new category, 454
Concept modeling:
  and concept response segmentation, 43
  creating high scoring concepts, 38
  development for key subgroups, 36
  estimation of part worth contribution,
    31
  synergy between elements, 37
  tonality, creating model with given, 37
Concept optimization:
  as a mechanistic process, 456
  assigning different importance to
    tonalities, 41
  hand lotion, 56
  for R&D, 56

[Concept optimization]
  in real time, 61
  overview, early stage development, 434
  subject to objectives, 454
  use of graphics in, 50
  using concept model, 38
  using concept response segments, 49
Concept Optimizer:
  use in concept creation, 39
  use in finding holes in a category, 138
Concept refinement, in focus groups, 454
Concept response segmentation: 43
  advertising concepts, 50
  clustering methods, 48
  derivation from sensory segmentation,
    45
  specifics for the procedure, 47
  use in concept optimization, 49
Concept size, 71
Concept testing:
  data base using tonalities, 29
  developing winners by design, 17
  example of questionnaire, 27
  experimental design, 18
  nonevaluative attributes, 26
  purchase intent scale, 26
  selecting winners, conventional way, 17
  use of benchmarks, 26
  use of diagnostic attribute ratings, 16
  use of in market products as norms, 14
  what happens in most companies, 15
Concepts:
  comparison of product and positioning,
    11
  creating them, 10
  creative boutiques, 13
  development by experimental design,
    24
  experimental design, using categories,
    elements, 22
  Gestalt, 13
  nail polish remover, 14
  promise testing, 11
  rated by both R&D and consumers, 11
  restrictions among pairs of elements, 22
  technical feasibility, 13

Conclusions, translating numbers into, 477
Conjoint measurement, in concept testing, 18
Constrained optimization, using factor scores, bath gel study, 149
Constraints:
  in experimental designs, 224
  mutually contradictory ones, in optimization, 275
  realistic limits in product optimization, 272
Consumer attribute, developing macro- and micro-models for, 118
Consumer interest, maintaining during an extended usage, 88
Consumer reaction, possibility of a pure consumer rating, 317
Consumer's language, merging qualitative and quantitative, 214
Consumers:
  changes as a function of over-participation, 316
  expertization of, 315
Consumers vs. experts, interrelations, 412
Content analysis:
  in benchmarking study, antiperspirant, 83
  regularizing the matrix, 115
  steps for, in category benchmarking, 86
  use in category model, 114
Contour plots, 238
Correlation, vs. functional relations, 300
Correlation analysis, between independent and dependent variables, 254
Correlations:
  between independent and dependent variables, 231
  between sensory and instrumental measures, 394
  limitations in sensory vs. instrumental data, 395
Cost constraints, in product optimization, 269
Cost vs. knowledge, in product optimization design, 244

Countries, problems of language, 183
Creating a model, art vs. science, in product optimization study, 234
Creating a product line, lipstick case history, 384
Creating a truly new category, 448
Creative session, schematic for conducting one, 429
Criterion biases, in discrimination testing, 290
Crossover experiment, to measure individual differences in importance, 341

Data display strategies, in product optimization studies, 251
Demographic breaks, 322
Demographic differences, used to create a line of products, 376
Dental floss, elements from early stage development, 430
Design issues, in product optimization, 222
Development guidance, and category benchmarking, 82
Diagnostic attributes, in category benchmarking, 89
Diagnostics, in concept tests, 17
Differences:
  vs. discrimination testing, 295
  meaningful vs. statistical, 195
Discrete features of a product, in category benchmarking, 113
Discriminant analysis, as a method for identifying key questions in a questionnaire for sensory segments, 370
Discrimination:
  in psychophysics, 288
  vs. product differences, 292
Discrimination testing, 289
  criterion or payoff related biases, 290
  and panelist experience, 294
  practical application, 295
  sorting task, 292
  triangle test, 291

Index

Dose response plots, 238
Drivers of liking, for bath gels, 143
Dropouts, occurrence in extended usage tests, 89
Dummy variable regression, in concept modeling, 31

Early stage data analysis, in product optimization study, 229
Elements:
　pairwise restrictions, in designed concepts, 22
　as part of categories, in concepts, 22
Empty space, in mapping, corresponding to category holes, 133
End use of data, as determinant of products to be tested, 161
Enhancement, in psychophysics, 288
Equation, model, in product optimization study, 232
Equations, sensory-instrumental relations, 397
Euclidean distance, in developing metrics of difference, 296
Experimental design:
　constraints in, 224
　in product optimization, 219, 243
　shampoos, 282
　use in concepts, 18
　use in developing concepts, 24
　use in IdeaMap, 66
Expert panelists, as evaluators of fine fragrances, 420
Experts vs. consumers:
　interrelations, 412
Extended tests, maintaining participation, 172
Eye shadow, 181

Face and body soap, category benchmarking, 108
Facial napkins, case history, 298
Facial soap, concepts, 10
Factor analysis:
　anti-perspirant fragrances, 330
　in category benchmarking, with sensory attributes, 100

[Factor analysis]
　as a heuristic approach, 102
　in category benchmarking, with sensory attributes, 100
　locating products in a factor space, 102
　on a panelist by panelist basis, 330
　use in sensory segmentation, 345
Factor models, examples, bath gel study, 150
Factor scores, as independent variables, in category benchmarking, 147
Fair share analysis, in new product development, 386
False positives, 299
Field execution, in product optimization studies, 249
Field setup, in product optimization studies, 248
Field work, sources of error, multiple product testings, 175
Fine fragrance:
　attributes used in describing, 416
　case history, Wood Inc., 416
　steps in evaluating, 352
First- vs. second-order problems, 466
Fit-to-concept, as attribute to optimize, 271
Fitting equations, a point of view, 405
Fitting equations to data, in product optimization studies, 255
Focus groups:
　for concept refinement, to create a new category, 454
　use in concept development, 64
Fragrance:
　database, for sensory segmentation, 355
　individual differences in ratings of, blind vs. branded, 339
Fragrance acceptance, pattern of liking, by sub-group, 358
Fragrance concepts, 50
Fragrance evaluation:
　questionnaire, 353
　steps in evaluating., 352
Fragrances:
　adding to an existing line, 381
　profiles, 180

[Fragrances]
  types of stimuli, designed vs. in market, 416
Franck Company, case history, oral protective compound, 426
Free choice profiling, 8
Future vision, importance of, 459

Gestalt concepts, 12
Ginny Cappello, case history, 350
Goal, out-of-bounds example, 277
Goal fitting:
  as an aspect of optimization, 240
  as an aspect of product optimization, 219, 414
  in product optimization, 276
  reverse engineering, 276
Goal profile, in sensory-instrumental relations, 407
Goodness of fit:
  additive model, concept, 32
  in product optimization studies, 255
Graphics, in concept optimization, 50
Guessing, in discrimination testing, 293

Habituation, 206
Hadden Soap Company, case history, 140
Half effect, and attribute use, example of shampoos, 333
Hand lotion, concept optimization, 56
Hand mousse, comparison of sensory and instrumental data, 394
Hedonic adaptation, role in boredom and switching, 205
Hedonic curve, as inverted U-shaped function, 307
Hedonics:
  individual differences, 325
  vs. Intensity, 306
Heuristic, factor analysis as, in category benchmarking, 102
Historical perspective, importance of, 459
Holes in a category:
  corresponding to product features, 132
  discovery, from category benchmarking, 132

[Holes in a category]
  fitting a hole, 153
Holes in a category benchmark map, 103
Home use testing, toothpaste, 361

IdeaMap:
  concept size, 71
  group model from, 73
  individual additive model, 68
  interactive stimulus presentation & data acquisition, 65
  multi-media technique, 77
  overview, 77
  pairwise restrictions, 71
  regression analysis, 68
  for sanitary napkin concepts, 65
  smoothing algorithm for estimating untested elements, 68
  technique for concept development and optimization, 64
  use in creating a new category, 452
IdeaMap project, implementation with multiple categories, 431
Identifying variables, product optimization, 242
Illusions of the past, 462
Image attributes, and orientation exercise, 167
Image profile:
  as dependent variable, in optimization, 152, 277
Image ratings, as targets to match with expert profiles, 416
In-house sensory panels, vs. consumers, 412
In-market fragrances, use in studies, 416
Independent variables:
  in product optimization study, 231
  in sensory-instrumental relations, 397
Individual additive model, use in IdeaMap, 68
Individual differences:
  assessing contributions of various factors, 340
  in attitudinal ratings, as measured by annoyance, 366

Index 491

[Individual differences]
  in cinnamon flavored toothpaste, 359
  in curve relating sensory intensity to liking, 342
  in importance of product vs. brand/package, 337
  liking of products vs. attitudes, 364
  in liking ratings, for toothpastes, 364
  overview, 389
  in scale usage, 323
  in sensory perception, 323
  in sensory ratings, for toothpastes, 363
Infeasible solutions, in product optimization, 275
Inferential statistics, 195
Ingredient limits:
  examples, toothpaste and lotion, 305
  use of power law to set limits, 304
Instrumental vs. sensory data, their interrelationship, 393
Integrated model, from category benchmarking, 143
Intensity functions, in psychophysics, 300
Interactions:
  between pairs of variables, in product optimization, 259
  in product modeling, 235
  in product optimization, 222
  in product optimization studies, 256
Interactive stimulus presentation, use in IdeaMap, 65
Interrelating data sets:
  experts vs. consumers, 412
  potential approaches, 393, 420
Interrelating experts and consumers:
  fragrance example, 414
  rationale, 413
  reverse engineering applied to, 413
Iso-hyphs:
  equal intensity contours, 238
  plots, 240

J&L Inc., case history, sanitary napkins, 368
Journals, relevant for product development, 4

K-means clustering, use in concept response segmentation, 48
Kathleen MacDonnell, case history, bath gels, 140

Leverage analysis, importance of individual liking attributes, 106
Lifestyle segmentation, 322
Likes and dislikes, in category benchmarking, by attribute, 105
Liking:
  as driven by sensory attribute, 111
  vs. sensory attribute, quadratic function, 111
Liking attribute, and leverage analysis, 106
Liking vs. disliking, 306
Liking of simple stimuli, individual differences, 325
Liking scale, norms for, 442
Lindberg Company, case history, sanitary napkin, 69
Line extensions, as example of quick product development, 424
Line of products, creating, 376
Line scale, 164
Linear equation, from product optimization study, 233
Lipsticks, case history, 383
Locations, multiple locations in product testing, 174
Lotion, experimental design, 227
Lotions, importance of sensory inputs for, 324
Low-cost supplier, consequences to R&D of becoming a, 481
Lynn Johnson, 241

Macro-model:
  in category benchmarking, 116
  used to design new products, 128
Magnitude estimation, as a scaling procedure, 166, 301
Mall recruitment, panelist screening, 167
Mapping:
  in category benchmarking, 100

[Mapping]
  reversing the procedure to find holes, 134
  using non evaluative sensory attributes, 100
Mapping products:
  in a factor space, in category benchmarking, 102
  incorporating liking in the map, 103
Market leader, and liking attributes, 107
Market research, vs. R&D, 482
Marketing boutique, use in early stage, rapid product development, 428
Marketing strategy, and fair share analysis, 387
Meaningful difference, 199
Memory, use in obtaining product rating, 88
Micro-model, in category benchmarking, 116
Mixture experimental design, for fine fragrances, 416
Mixture model, for fine fragrances, 416
Model, from product optimization study, 232
Modeling:
  accounting for interactions, 235
  bath gel study, 149
  development of data table to allow for, 442
  robustness vs. goodness of fit, 235
Moskowitz Jacobs Inc., 100-point scale, 202
Multidimensional scaling, 8
Multi-media concepts, in creating a new category, 451
Multi-national evaluation, of concepts and products, 182
Multiple groups, optimization for, 380
Multiple product testing, source of error, in field work, 175
Multiple rating attributes, use in creating a new category, 452

Nail polish remover, concepts, 14
Names, in concept optimization, 53

Need/opportunity, as a succinct paragraph or sentence, in concept study, 449
Nesbitt Corporation, case history, hand lotion, 56
New products, use of macro- and micro-models, to design, 128
New test strategy, approach recommended to introduce, 472
Niche, for sanitary napkin, 242
Nonevaluative attributes, use in concept testing, 26
Non-linearities:
  in product optimization, 222
  in product optimization studies, 256
Non-rejecter, screening for, in category benchmarking, 85
Non-rejecters, as panelists, 247
Norms:
  for concept testing, 14
  in concept testing, steps to follow, 15
  for liking, 100-point scale, 442
  in product testing, 202
  recalibration in the face of changing conditions, 203
Novice researcher, advice on following a vision, 470
Null hypothesis, 198
Number of products, advantages in increasing, for product testing, 160

Objective considerations, added to product model, 444
Objective constraints, in product optimization, 269
Olfactory research, and sensory mixtures, 312
Opportunities, corresponding to unfulfilled benefits, 134
Optimal concept, use as an attribute in product questionnaire, 441
Optimal line, use of TURF analysis, 377
Optimal line of products, 377
Optimal products, 238
Optimization:
  based upon factor scores, in category benchmarking, 148
  screening panelists, 246

Index

Optimizing:
  for overall acceptance, 263
  subject to constraints, bath gel study, 152
  using factor equations, bath gel study, 149
Optimizing a product, early stage product development, 446
Optimum product, estimating attribute profile of, 267
Oral care, case history, abrasive gel product, 406
Oral care product development:
  example of questionnaire, 441
  examples of formula variables, multiple categories, 440
Oral gel, example for interrelating experts and consumers, 413
Oral rinse, elements from early stage development, 430
Order of liking, as a display strategy, in product optimization study, 252
Order of trial, in fragrance testing, 353
Orientation:
  and image attributes, 167
  exercise to familiarize panelists, 166
  to product usage, 170
Orientation product, effect on stability of the means, 194
Overall rating, when should it be assigned, 189

Package design, by concept development methods, 73
Package features, in concept optimization, for R&D, 58
Package vs. product, importance of, 338
Paired comparison:
  benefit, 184
  transformation to scalar data, 186
Panelist cooperation, maintaining participation, 172
Panelist experience, and discrimination testing, 294
Panelist recruitment, screening concepts, 247
Panelist screening, by means of lists, 169

Panelists:
  appropriate number, in category benchmarking, 87
  screening in optimization studies, 246
  screening for participation, 167
Paradigm shift, 5
Part worth contribution, from regression analysis, in concept modeling, 31
Partial data, use of, 173
Partial payment, for panelists, when it is appropriate, 173
Patterns in data, vs. statistical significance, 478
Patterns in time, 480
Perceived intensity, as a psychophysical measure, 300
Performance and image, as attributes to optimize, 271
Personality descriptors, as attributes, 179
Physical features, as aid in experimental design, shampoo, 282
Pilot studies, 474
Plackett Burman design, use as a precursor for product modeling, 446
Plackett Burman screening design, use in early stage product development, 435
Plots, linear vs. quadratic, surface, 239
Power function exponents, table of 303
Power law, 7
Predictor variables, in category benchmarking, 147
Predictors, reducing the set, in category benchmarking model, 146
Preference justification, 163, 189
Principal components analysis, used to reduce the set of predictor variables, 147
Print ad, as a method to pull consumers in different sensory segments, 374
Problem definition, in product testing, 158
Product, multi-national evaluation, 182
Product concepts, 56
Product design, new products, use of macro- and micro-models, 128
Product development:
  example of rapid speed, 435

[Product development]
  identifying variables, oral care products, 435
  process vs. product, 425
  streamlining the process, 424
Product differences, 292
Product evaluation, blind vs. identified, antiperspirant study, 82
Product line, creating, with defined product differences, 381
Product model:
  equation, in product optimization, 254
  from product optimization study, 232
Product optimization:
  and constraints, 269
  identifying variables, 242
  and infeasible sensory constraints, 275
  optimizing fit to concept, 272
  questionnaire, for sanitary napkins, 248
  reverse engineering method, 6, 8, 219, 275
  steps involved, 220
Product rating:
  from memory, 88
  during usage, 88
Product selection, considerations in product testing, 160
Product setup, activities for, 248
Product shipping, in product optimization studies, 250
Product switching, correlated with boredom and hedonic adaptation, 205
Product testing:
  considerations, 158
  frequency of rating during the course of usage, 87
  norms, 202
Product usage:
  and explanation, 171
  as a method for identifying individuals in sensory segments, 369
  orientation to, 170
Product variables, in early stage product development, oral care, 440
Product × attribute matrix:
  display strategies, 251

[Product × attribute matrix]
  in product optimization studies, 250
Products, fitting personalities and pictures to products, 179
Products vs. brands/packages, individual difference in importance of, 336
Products tested, appropriate number per panelist, 87
Projective techniques, attributes, 182
Promise testing, as a concept-testing method, 11
Proposals for research, 471
Prototypes vs. competitors, performance, 442
Psychophysical approach, using dose response functions, 238
Psychophysical function, use of, 301
Psychophysics, 6, 288
Psychophysics of hedonics, 306
Purchase intent:
  conversion of to two box, 28
  driven by pictures, 53
Purchase Intent Scale, use in concept testing, 26
Purpose of test, consideration in product selection, 160

Quadratic functions, sensory-instrumental relations, soap data, 401
Quadratic model, from product optimization study, 233
Questionnaire:
  example for concept test, 27
  fine fragrances, 416
  in category benchmarking, 89
  language, 176
  for product optimization studies, 248
  qualitative research, 176
  used for product optimization study, 228
Questionnaires, as a method for identifying individuals in sensory segments, 369

R&D, rating concepts, 11
Rapid product development, two-step process, 425

Index

Reduced bias measure, for differences between products, 299
Regression analysis:
  to analyze experimentally designed concepts, 24
  dummy variable, use in concept modeling, 31
  in product optimization studies, 255
  used to create concept model, 31
  use in IdeaMap, 68
  use in modeling data from Plackett Burman design, 444
Relevant test population, rationale for, in benchmarking study, 84
Replicate samples, 191
Research, guidelines for, amount and type, 468
Research objectives, learning vs. confirmation, 469
Resource bank, elements, for new category creation, 450
Restrictions:
  among pairs of concept elements, 431
  in IdeaMap, 71
Revealed needs, fulfilling them to create a new product or category, 449
Reverse engineering:
  bath gel study, 152
  an overview, 409
  in product optimization, 275
  in sensory-instrumental relations, 409
Reversed matrix, in mapping, to fill holes in a category, 134
Richards Company, case history, antiperspirant, 81
Risk reduction, need in product development, 2
Robust models, vs. goodness of fit, 235

S. S. Stevens, 6
Sampling, as a method to pull people in different sensory segments, 374
Sanitary napkin:
  case history, 368, 241
  category benchmarking, 110
  concept development by IdeaMap, 65

Scalar data, transformation from paired comparisons, 186
Scale usage:
  in category benchmarking, 95
  in different cultures, 184
  and individual differences, 323
Scales:
  best vs. worst types, in questionnaire, 163
  in category benchmarking, 90
  unipolar vs. bipolar, 90
Scaling, 187
Scaling procedure, 184
Screening:
  of panelist, 167
  rescreening participants at the test site, 174
  use of non-rejecters, 86
Screening concepts, use in recruiting panelists, 247
Screening design, 221
Screening questionnaire, for sanitary napkins, used to identify individuals in sensory segments, 371
Segmentation methods, 322
Sensitivity analyses, examples, 263
Sensitivity analysis:
  based upon model, 235
  as method to bring life to equations, 259
  single variable vs. iso-hyphs, 238
Sensory adaptation, long term, with repeated use, 205
Sensory attribute:
  vs. liking, 145
  vs. liking, continuous function, 111
Sensory constraints, in product optimization, 269
Sensory Evaluation, ASTM Committee, 4
Sensory fatigue, 206
Sensory importance, and individual differences, 325
Sensory intensity, power law, 7
Sensory intensity vs. liking, individual differences, 342
Sensory liking curve, area subtended by, as measure of importance, 145

Sensory mixtures, laws of, 310
Sensory perception, and individual differences, 323
Sensory power law, 301
Sensory preference segmentation, 322
  example showing optimal levels, shampoos, 344
  as a global organizing principle, 342
  schematic of the approach, 343
Sensory satiety, 206
Sensory segmentation:
  as an approach for concept response segmentation, 45
  overview, 375
  vs. TURF analysis, to create a product line, 385
Sensory segments:
  case history, 350
  field execution of fragrance study, to find, 351
  finding new individuals who belong in, 367
  finding them by product usage, 369
  how to understand them, 345
  in preferences for sanitary napkins, 368
  panel considerations, 351
  plotting them in factor space, 345
  summary statistics, vs. other segmenting methods, 348
Sensory signature, in content analysis, 84
Sensory vs. instrumental data, their interrelationship, 393
Sensory-instrumental relations, relating pairs of profiles to each other, 406
Shampoo, concepts, 10
Shampoos:
  attribute list, 334
  case history, 280
  concepts, 11
  factor analysis of attributes, 332
  optimal attribute levels, for sensory segmentation, 344
Share, in concept testing, function of price, 15
Shaving cream, case history, to create an optimal line, 377

Siragusa Napkin Company, case history, 298
Skin lotion, concepts, 10
Slotting allowance, 282
Smoothing algorithm, use in IdeaMap, 68
Soap:
  category benchmarking, 108
  concept testing, experimental design, 18
  liking attributes and leverage analysis, 109
  sensory-instrumental relations, 398
Sorting task, in discrimination testing, 292
Speed, need in product development, 1
Stability of mean ratings, 190
Standard error of estimate, in product optimization studies, 256
Standard error of regression, use in additive concept model, 36
Standardization, of product usage, 171
Statistical parsimony, sensory-instrumental relations, 402
Statistical significance, vs. patterns, 478
Statisticians, old vs. new roles, 474
Statistics:
  as an aid to product development, 8
  inferential vs. mapping approaches, 475
Stepwise regression:
  in category benchmarking, macro- and micro-models, 122
  sensory-instrumental relations, 402
Stevens, S. S., and psychophysical intensity functions, 302
Subgroup analysis:
  in category benchmarking, 95
  vs. features of products, in category benchmarking, 98
Subgroups, additive model, concept, 36
Summary statistics, in category benchmarking, 93
Sunscreen, individual differences in rating, product vs. package, 341
Suppression, in psychophysics, 288
Suppression vs. enhancement, in taste mixtures, 311

Index 497

Synergy, in concept modeling, 37
Synthesized profile, corresponding to optimum, bath gel study, 151
Systematic product array, in product optimization, 221

T test, 198
Taste mixtures, suppression vs. enhancement, 311
Taste research, and sensory mixtures, 310
Technical attributes, explanation of, to panelists, 171
Terms in regression equation, choosing them, 259
Test execution, in product optimization study, 229
Test procedure, for fine fragrances, 416
Thresholds, individual differences, 323
Time preference:
 asking the proper question, 209
 in product design, 210
Time preference curves, use to estimate potential boredom, 208
Time series analysis, as a new area for consumer research, 480
Tonalities:
 assigning different levels of importance, 41
 as attributes, in a concept test, 29
Tonality, use in concept modeling, 37
Toothpaste:
 elements from early stage development, 430
 questionnaire for, home use test, 361
 study of individual differences in perceptions of, 359
Top two box, measure used in concept testing, 28
Trends, better revealed by quadratic than by linear equations, 236
Triangle test, in discrimination testing, 291

Tried first product:
 in analyzing results, 353
 ratings of intensity, to establish validity, 355
TURF analysis:
 case history, lipsticks, 383
 vs. sensory segmentation, to create a product line, 385
 theoretic nature of, 382
 total unduplicated reach, frequency, 377

Unipolar scales, 90

Validity, establishing it, in product optimization studies, 254
Variables:
 correlations between independent and dependent, 231
 setting levels, in experimental design, 244
Video, in concepts, 73
Videos vs. visuals, in concept development, 74
Visual research, and sensory mixtures, 310
Visualization, based upon model, 235

Wearout:
 short- vs. long-term satiety, 204
 with repeated product use, 204
Wheel and spoke design, 221
William Riha, case history, sanitary napkin, 242
Wood Inc., case history, fragrance, experts vs. consumers, 416
World view, as a structure in which to do research, 461

Young researcher, advice to, 459